Communications
in Computer and Information Science 123

Tai-hoon Kim   Sankar K. Pal
William I. Grosky   Niki Pissinou
Timothy K. Shih   Dominik Ślęzak (Eds.)

Signal Processing and Multimedia

International Conferences, SIP and MulGraB 2010
Held as Part of the Future Generation
Information Technology Conference, FGIT 2010
Jeju Island, Korea, December 13-15, 2010
Proceedings

 Springer

Volume Editors

Tai-hoon Kim
Hannam University, Daejeon, South Korea
E-mail: taihoonn@hnu.kr

Sankar K. Pal
Indian Statistical Institute, Kolkata, India
E-mail: sankar@isical.ac.in

William I. Grosky
University of Michigan-Dearborn, Dearborn, MI , USA
E-mail: wgrosky@umich.edu

Niki Pissinou
Florida International University, Miami, FL, USA
E-mail: pissinou@fiu.edu

Timothy K. Shih
National Taipei University of Education, Taipei City, Taiwan
E-mail: tshih@cs.tku.edu.tw

Dominik Ślęzak
University of Warsaw & Infobright, Warsaw, Poland
E-mail: dominik.slezak@infobright.com

Library of Congress Control Number: 2010940209

CR Subject Classification (1998): I.4, C.2, H.4, I.5, H.3, I.2

ISSN	1865-0929
ISBN-10	3-642-17640-2 Springer Berlin Heidelberg New York
ISBN-13	978-3-642-17640-1 Springer Berlin Heidelberg New York

springer.com

© Springer-Verlag Berlin Heidelberg 2010
Printed in Germany

Typesetting: Camera-ready by author, data conversion by Scientific Publishing Services, Chennai, India
Printed on acid-free paper   06/3180

Preface

Welcome to the proceedings of the 2010 International Conferences on Signal Processing, Image Processing and Pattern Recognition (SIP 2010), and Multimedia, Computer Graphics and Broadcasting (MulGraB 2010) – two of the partnering events of the Second International Mega-Conference on Future Generation Information Technology (FGIT 2010).

SIP and MulGraB bring together researchers from academia and industry as well as practitioners to share ideas, problems and solutions relating to the multifaceted aspects of image, signal, and multimedia processing, including their links to computational sciences, mathematics and information technology.

In total, 1,630 papers were submitted to FGIT 2010 from 30 countries, which includes 225 papers submitted to SIP/MulGraB 2010. The submitted papers went through a rigorous reviewing process: 395 of the 1,630 papers were accepted for FGIT 2010, while 53 papers were accepted for SIP/MulGraB 2010. Of the 53 papers 8 were selected for the special FGIT 2010 volume published by Springer in the LNCS series. 37 papers are published in this volume, and 8 papers were withdrawn due to technical reasons.

We would like to acknowledge the great effort of the SIP/MulGraB 2010 International Advisory Boards and members of the International Program Committees, as well as all the organizations and individuals who supported the idea of publishing this volume of proceedings, including SERSC and Springer. Also, the success of these two conferences would not have been possible without the huge support from our sponsors and the work of the Chairs and Organizing Committee.

We are grateful to the following keynote speakers who kindly accepted our invitation: Hojjat Adeli (Ohio State University), Ruay-Shiung Chang (National Dong Hwa University), and Andrzej Skowron (University of Warsaw). We would also like to thank all plenary and tutorial speakers for their valuable contributions.

We would like to express our greatest gratitude to authors and reviewers of all paper submissions, as well as to all attendees, for their input and participation.

Last but not least, we give special thanks to Rosslin John Robles and Maricel Balitanas. These graduate school students of Hannam University contributed to the editing process of this volume with great passion.

December 2010

Tai-hoon Kim
Wai-chi Fang
Muhammad Khurram Khan
Kirk P. Arnett
Heau-jo Kang
Dominik Ślęzak

SIP 2010 Organization

Organizing Committee

General Chair   Sankar K. Pal (Indian Statistical Institute, India)

Program Chair   Byeong-Ho Kang (University of Tasmania, Australia)

Publicity Co-chairs   Junzhong Gu (East China Normal University, China)
Hideo Kuroda (FPT University, Hanoi, Vietnam)
Muhammad Khurram Khan (King Saud University, Saudi Arabia)
Aboul Ella Hassanien (Cairo University, Egypt)

Publication Co-chair   Rosslin John Robles (Hannam University, Korea)
Maricel Balitanas (Hannam University, Korea)

Program Committee

Andrzej Dzieliński	Joonki Paik	Roman Neruda
Andrzej Kasiński	Joseph Ronsin	Rudolf Albrecht
Antonio Dourado	Junzhong Gu	Ryszard Tadeusiewicz
Chng Eng Siong	Kenneth Barner	Salah Bourennane
Dimitris Iakovidis	Mei-Ling Shyu	Selim Balcisoy
Debnath Bhattacharyya	Miroslaw Swiercz	Serhan Dagtas
Ernesto Exposito	Makoto Fujimura	Shu-Ching Chen
Francesco Masulli	Mototaka Suzuki	Tae-Sun Choi
Gérard Medioni	N. Jaisankar	William I. Grosky
Hideo Kuroda	N. Magnenat-Thalmann	Xavier Maldague
Hong Kook Kim	Nikos Komodakis	Yue Lu
Janusz Kacprzyk	Paolo Remagnino	
Jocelyn Chanussot	Peter L. Stanchev	

MulGrab 2010 Organization

Organizing Committee

General Co-chairs
>William I. Grosky (University of Michigan-Dearborn, USA)
>Niki Pissinou (Florida International University, USA)
>Timothy K. Shih (National Taipei University of Education, Taiwan)

Program Co-chairs
>Tai-hoon Kim (Hannam University, Korea)
>Byeong Ho Kang (University of Tasmania, Australia)

International Advisory Board
>Andrea Omicini (Università di Bologna, Italy)
>Bozena Kostek (Gdańsk University of Technology, Poland)
>Cao Jiannong (Hong Kong Polytechnic University, Hong Kong)
>Ching-Hsien Hsu (Chung Hua University, Taiwan)
>Claudia Linnhoff-Popien (Ludwig-Maximilians-Universität München, Germany)
>Daqing Zhang (Institute for Infocomm Research, Singapore)
>Diane J. Cook (University of Texas at Arlington, USA)
>Frode Eika Sandnes (Oslo University College, Norway)
>Jian-Nong Cao (Hong Kong Polytechnic University, Hong Kong)
>Lionel Ni (Hong Kong University of Science and Technology, Hong Kong)
>Mahmut Kandemir (Pennsylvania State University, USA)
>Matt Mutka (Michigan State University, USA)
>Mei-Ling Shyu (University of Miami, USA)
>Rajkumar Buyya (University of Melbourne, Australia)
>Sajal K. Das (University of Texas at Arlington, USA)
>Sajid Hussain (Acadia University, Canada)
>Sankar K. Pal (Indian Statistical Institute, India)
>Schahram Dustdar (Vienna University of Technology, Austria)
>Seng W. Loke (La Trobe University, Australia)
>Stefanos Gritzalis (University of the Aegean, Greece)
>Zbigniew W. Ras (University of North Carolina, USA)

Publicity Co-chairs
>Yang Xiao (University of Alabama, USA)
>Ing-Ray Chen (Virginia Polytechnic Institute and State University, USA)
>Krzysztof Pawlikowski (University of Canterbury, New Zealand)
>Guoyin Wang (CQUPT, Chongqing, China)

Han-Chieh Chao (National Ilan University, Taiwan)
Krzysztof Marasek (PJIIT, Warsaw, Poland)
Aboul Ella Hassanien (Cairo University, Egypt)

Publication Co-chairs
Jiyoung Lim (Korean Bible University, Korea)
Hae-Duck Joshua Jeong (Korean Bible University, Korea)

Program Committee

A. Hamou-Lhadj
Ahmet Koltuksuz
Alexander Loui
Alexei Sourin
Alicja Wieczorkowska
Andrew Kusiak
Andrzej Dzielinski
Anthony Lewis Brooks
Atsuko Miyaji
Biplab K. Sarker
Ch. Z. Patrikakis
Ch. Chantrapornchai
Chengcui Zhang
Chi Sung Laih
Ching-Hsien Hsu
C. Fernandez-Maloigne
Daniel Thalmann
Dieter Gollmann
Dimitris Iakovidis
Eung-Nam Ko
Fabrice Mériaudeau
Fangguo Zhang
Francesco Masulli
Gérard Medioni
Hai Jin Huazhong
Hiroaki Kikuchi
Hironori Washizaki
Hongji Yang
Hyun-Sung Kim
Jacques Blanc-Talon
Jalal Al-Muhtadi

Javier Garcia-Villalba
Jean-Luc Dugelay
Jemal H. Abawajy
Jin Kwak
Jocelyn Chanussot
Joonsang Baek
Junzhong Gu
Karl Leung
Kenneth Lam
Khaled El-Maleh
Khalil Drira
Kouichi Sakurai
Larbi Esmahi
Lejla Batina
Lukas Ruf
MalRey Lee
Marco Roccetti
Mark Manulis
Maytham Safar
Mei-Ling Shyu
Min Hong
Miroslaw Swiercz
Mohan S Kankanhalli
Mototaka Suzuki
N. Magnenat-Thalmann
Nicoletta Sala
N. Assimakopoulos
Nikos Komodakis
Olga Sourina
P. de Heras Ciechomski
Pao-Ann Hsiung

Paolo D'Arco
Paolo Remagnino
Rainer Malaka
Raphael C.-W. Phan
Robert G. Reynolds
Robert G. Rittenhouse
Rodrigo Mello
Roman Neruda
Rui Zhang
Ryszard Tadeusiewicz
Sagarmay Deb
Salah Bourennane
Seenith Siva
Serap Atay
Seung-Hyun Seo
Shin Jin Kang
Shingo Ichii
Shu-Ching Chen
Stefan Katzenbeisser
Stuart J. Barnes
Sun-Jeong Kim
Swapna Gokhale
Swee-Huay Heng
Taenam Cho
Tony Shan
Umberto Villano
Wasfi G. Al-Khatib
Yao-Chung Chang
Yi Mu
Yo-Sung Ho
Young Ik Eom

Table of Contents

Eye Tracking Technique for Product Information Provision

Seoksoo Kim

Department of Multimedia, Hannam University,
133 Ojeong-dong, Daedeok-gu,
Daejeon-city, Korea
sskim0123@naver.com

Abstract. This paper is about the study of the design of the product information provision system using eye tracking, which helps users in deciding over purchase of the product, with a system to provide product information that the user needs by tracking eye gaze of the user via Smart phone. The system provides the user with information of the product that attracts the user's eye, by means of users' eye tracking, user information confirmation using face recognition and user's preference for product. Therefore, once it is determined that the user requires a product, the server sends the product information stored in the product information database to the user's Smart phone to provide information the user requires. The customer, provided with product information in real time, can purchase the product that he/she wants efficiently, and avoid excessive consumption with accurate product information.

Keywords: Eye tracking, Product information Provision, Face recognition, Augmented reality.

1 Introduction

Recently, as information communication and ubiquitous system have been developed, information devices are widely used by users and various kinds of contents are released as a result. Especially, after Apple's iPhone was launched, the Smartphone market becomes hotter. It tends to gain tremendous popularity with convenient interface and many application programs available. The quality of contents loaded therein has been enhanced and the contents using augmented reality become increasingly popular lately.

The augmented reality augments the imaginary images that computer makes in real environment, to provide the user with new experiences. Ever since Ivan Sutherland had shown the first application of augmented reality using see-through HMD in 1960's, the researchers have developed many technologies related. Most of these technologies were included in tracking technologies or AR display technologies. The researchers have introduced the prototype augmented reality systems which were applicable to different application areas such as medicine, entertainment and training etc. The augmented reality systems in the initial stage simply covered a real object

T.-h. Kim et al. (Eds.): SIP/MulGraB 2010, CCIS 123, pp. 1–7, 2010.
© Springer-Verlag Berlin Heidelberg 2010

with 3D image made by computer, and was used for intelligence amplification which augmented information in the computer into a real object and facilitated works in real environment[1,2]. There was a researcher who interpreted augmented reality in terms of HCI (human computer interaction), awareness[3].

The expressive technology of augmented reality using Smartphone internally requires context awareness technology, and this kind of technology should provide accurate product information by tracking the product and the participants should understand the meaning of the shared information completely and correctly, based on smooth product information flows. Researches utilizing ontology are also on going to support efficient understanding of the meaning of information[4,5].

Context awareness computing, as a technology that analyzes context information of reality in imaginary space by connecting real world with imaginary space and provides user-centered intelligent service, provides how to describe the context of real world. It is characterized by a human-oriented autonomous service in application of feature extraction in context awareness and artificial intelligent techniques such as learning or reasoning[6,7]. Besides, the image-based searching technology finds feature points from a video, extracts features at the points and searches after providing the database with an index. Since the video data has more images than many texts or images, general index searching technique will not be useful. Therefore, important frames are only extracted out of the sequences of the total images[8].

As the purchasers of a product want to be provided with the product information promptly and conveniently to buy the product efficiently, the company selling the product needs to provide product information to the consumers.

However, most of the product selling companies is providing information to the consumers so that there is a limit for providing information to the purchasers promptly.

Therefore, this thesis is about to propose a research on product information provision system using eye tracking, which provides the user with the information of product that attracts the user's eye and is useful in selecting a product for purchase, by means of user's eye tracking, user information confirmation using face recognition and the user's preference for product. This is also a system to provide Smartphone with product information required by a user.

2 Structure of Eye Tracking Technique for Product Information Provision

This thesis was designed to solve problems in existing product information system. The purpose of this thesis is to study product information provision system using eye tracking that correctly finds a product where the user's eyes are focused and then provides the purchaser's Smartphone with accurate information of the product that attracts the user's eye in real time.

The product information provision system using eye tracking is featured by a customer information database that stores customer information such as infrared camera connected to the server of the product-selling company, eye tracking information sent through the camera, face recognition information, the server provided with product object information, and customer information including the customer's preference, visiting time and demanding articles, and a product information database that stores information of every products available in the store.

The figure below refers to the product information provision system using eye tracking based on this thesis. Figure 1 is an overall configuration for product information provision system using eye tracking. As illustrated in Figure 1, the product information provision system using eye tracking, tracking eye glaze of a customer who is about to purchase a product, receives information derived from eye-tracking with an infrared camera connected to the server of the company that sells the product. It also receives customer information database which contains the server calculating the favor degree, customer's face and detailed information and preference for the product, product information database of every products available, and the product information from the server and displays the information in customer's Smartphone.

- The infrared camera detects the pupil area by tracking the customer's face, tracks the position of customer's gaze and detects an object holding the customer's gaze to transmit information obtained by object detection to the server.
- The server measures the degree of the user's favor using the time the user's eye is caught and result values after comparing the received object information with the customer information contained in the customer information database, transmits information of the pertinent product to Smartphone if the user wants to purchase the product.
- The customer information database contains face recognition, purchase and the preference for products of the customers using the store.
- The product information database contains information of every products that are sold to the customers.
- Smartphone provides the product information received from the server to the user.

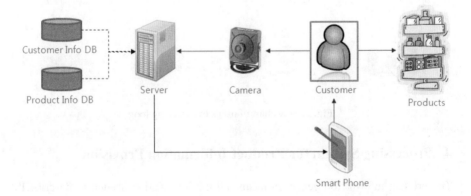

Fig. 1. Overall configuration for product information provision system using eye tracking

3 User's Eye Tracking Loop

Gaze detection is a method to recognize the position the user is looking by computer vision technique. There are many application areas for this gaze detection technology. A typical example is a computer interface for mentally disabled persons who cannot use both hands. Besides, it has the merit of improving processing speed, comparing to other input devices that use parts of the body including hands. Based on Yamato' results, most people's sight moving time from upper left to lower right on a 21-inch screen is within 150 ms, which is much faster than the movement by a mouse control[9].

The researches of the gaze detection technology, understood as a primary means of grasping the position the gazer concerns to the accompanying sensitivity, are actively in progress in USA, Europe, Japan and the like, and various researches are also conducted in the country. However, costly equipments including high-resolution camera, infrared camera or stereo visual system should be used to correctly grasp the position of eyes. In case a generally supplied WebCam is used, there are some limitations in real applications, such that the eye position needs to be set to the position already set up[10~12].

Fig 2 shows the flowchart of user's eye tracking loop applied in this thesis. A illustrated in Fig 2, the order of the user's eye tracking loop includes face recognition using Haar Wavelet, detection of pupil area within face region and gaze tracking using eye tracking algorithm.

Once the user's gaze was tracked, the time span the user's gaze was focused was measured, and then the product's object was detected by finding the points where the user's gaze was focused.

The product was identified after comparing information of the detected objects with the product information stored in the product information database.

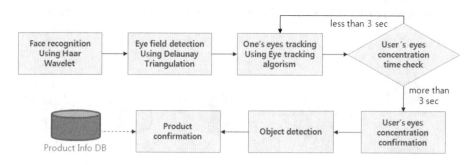

Fig. 2. Flowchart of user's eye tracking loop

4 Processing System for Product Information Provision

The product information generally means all the information needed to describe the product, such as structure information, technology information and process information etc., and it is comprehensive and vast. The uniqueness of each industry leads to different expression methods and rapidly increasing quantity of information that we have to manage. Besides, the information is characterized by its complicated interrelation.

The product information service using Smartphone, by perceiving information of the product received by Smartphone camera and providing related information with augmented reality technique, not only provides detailed product information and arouses user's purchasing needs, but helps the user purchase the product he/she really needs. For the provision of custom-made information for individual users, to have context information of both product and user and to provide matching information correctly and promptly are considered important.

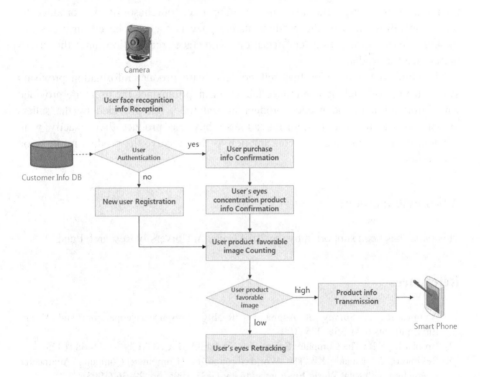

Fig. 3. System flowchart to provide product information

Fig 3 is a system flowchart to provide product information used in this thesis. After user authentication by comparing face recognition information received from infrared camera with the user's face recognition information stored in the customer information database, if the user is not registered, he/she is registered as a new user. On the other hand, if he/she is a registered user, the user's purchase information and information of the product his/her gaze is focused is identified, and his/her degree of favor with the product is calculated.

If the user's degree of favor is low, the gaze of the user is retracked and if the degree is high, the information is transmitted to the user's Smartphone.

5 Conclusion

The purpose of this thesis is to study product information provision system using eye tracking that correctly finds a product where the user's eyes are focused and then provides the purchaser's Smartphone with accurate information of the product that attracts the user's eye in real time.

This thesis has designed a product information provision system using eye tracking, which helps the user in deciding over purchase of the product by providing information of the product his/her gaze is focused, based on user's eye tracking, user information confirmation with face recognition and the user's preference for product.

Therefore, the companies that sell products with product information provision system using eye tracking can reduce labor consumption since the users are provided with information of the needed product in real time, not provided by the sellers directly as before. The customers can also buy the product they exactly want efficiently and avoid excessive consumption as they receive the product information in real time.

Acknowledgements

This paper has been supported by the 2010 Hannam University Research Fund.

References

1. Azuma, R.: A Survey of Augmented Reality. Presence:Teleoperators and Virtual Environments 6(4), 355–385 (1997)
2. Brooks Jr., F.P.: The Computer Scientist as Toolsmith II. CACM 39(3), 61–68 (1996)
3. Rekimoto, J., Nagao, K.: The World through the Computer: Computer Augmented Interaction with Real World Environments. In: UIST 1995, pp. 29–36 (1995)
4. Mizoguchi, R., Ikeda, M.: Towards Ontology Engineering, Technical Report AI-TR-96-1, I.S.I.R., Osaka University (1996)
5. Oh, J., Min, J., Hu, B., Hwang, B.: Development of Video-Detection Integration Algorithm on Vehicle Tracking. The KSCE Journal of Civil Engineering 29(5D), 635–644 (2009)
6. Dey, A.K., Abowed, G., Salber, D.: A Conceptual framwork and a tool kit for supporting the rapid prototype of context-aware applications,
   http://www.intel-research.net/Publications/Berkeley/101620031413_167.pdf
7. Yoon, S.: Design and Implementation of Internet Shopping Mall Based on Software Implemented Context Aware. The Korea Society Information Journal 14(1), 1 (2009)

8. Wei, X., Chung, M.L., Rui-Hua, M.: Automatic video data structuring through shot partitioning and key-frame computing. Machine Vision and Application 10(2), 55–65 (1997)
9. Yamoto, M., Monden, A., Matsumoto, K., Inoue, K., Torii, K.: Quick Button election with Eye Gazing for General GUI Environments. In: International Conference on Software: Theory and Practice (August 2000)
10. Fukuhara, T., Murakami, T.: 3D-motion estimation of human head for model-based image coding. Proc. IEE 140(1), 26–35 (1993)
11. Tomono, A., et al.: Eye Tracking Method Using an Image Pickup Apparatus. European Patent Specification 94101635 (1994)
12. Kim, K.N., Ramakrishna, R.S.: Vision-Based Eye-Gaze Tracking for Human Computer Interface. In: Porceedings of IEEE International Conference on Systems, Man, and Cybernetics, Tokyo, Japan, pp. 117–120 (October 1999)

A Study on Marker Overlapping Control for M2M-Based Augmented Reality Multiple Object Loading Using Bresenham Algorithm

Sungmo Jung, Jae-gu Song, and Seoksoo Kim[*]

Dept. of Multimedia, Hannam Univ., 133 Ojeong-dong,
Daedeok-gu, Daejeon-city, Korea
sungmoj@gmail.com, bhas9@paran.com, sskim0123@naver.com

Abstract. The problem of one marker one object loading only available in marker-based augmented reality technology can be solved by PPHT-based augmented reality multiple objects loading technology which detects the same marker in the image and copies at a desirable location. However, since the distance between markers will not be measured in the process of detecting and copying markers, markers can be overlapped and thus the objects would not be augmented. To solve this problem, a circle having the longest radius needs to be created from a focal point of a marker to be copied, so that no object is copied within the confines of the circle. Therefore, a marker overlapping control for M2M-based augmented reality multiple object loading has been studied using Bresenham algorithm.

Keywords: Marker Overlapping, Video Control, M2M, Augmented Reality, Multiple Objects Loading, Bresenham Algorithm.

1 Introduction

Augmented reality refers to a virtual environment where the sense of reality almost close to real situation can be provided through interaction [1]. That is, it is a technical field that provides object information in real time to improve understanding of the situation and cognitive power of 3-dimensional space [2].

At what point of the image that the camera captures the object should be augmented should be decided to realize augmented reality, and in this research, marker recognition [3] method was used. The marker recognition method is a method of augmenting an object on the marker while the computer recognizes the tag interface serving as a connection pointer. It is widely used with a high recognition rate and a easy-to-use function [4].

This marker-based augmented reality uses PPHT (Progressive Probabilistic Hough Transform)[5]–based on multiple objects loading technology to detect and copy the same marker, and in case only this method is used, markers can be overlapped and therefore the object will not be augmented.

For that reason, this research, utilizing Bresenham algorithm for M2M-based augmented reality multiple objects loading, an area was designated to prevent markers overlapping based on a focal point of the marker and markers overlapping control was made possible.

---

[*] Corresponding author.

T.-h. Kim et al. (Eds.): SIP/MulGraB 2010, CCIS 123, pp. 8–15, 2010.
© Springer-Verlag Berlin Heidelberg 2010

2   Related Study: Bresenham Algorithm

Most of the various line algorithms require floating point division operations having low performance. However, the rasterization which was proposed by Bresenham was a superhigh speed algorithm rasterizes straight line or circle or ellipse only with addition/subtraction operations of pure integers.

The fundamentals of the Bresenham algorithm [6] is accumulated 'error terms'. That is, the computer screen corresponds to a great 2-dimensional array of colored squares. An act of drawing a straight line (or another figure) at this point is a process of 'approximating' a continuous coordinate system of real numbers into a discontinuous coordinate system. Therefore, a dot usually provides an error between discontinuous coordinate systems printed in the original coordinate system of the line and the screen, and the key concept of Bresenham algorithm is to select a dot which best minimizes this error by drawing the straight lines (or another figure).

The process of drawing lines using Bresenham algorithm is as follows:

1) Array p1 and p2, the two dots representing the line in the order of coordinate axes. At this point, if the slope of the line is less than 1, array in increasing order of coordinate x and if greater than 1, array in increasing order of coordinate y. (Here assumed to be arrayed to increasing order of x)
2) Begin with the first dot out of the dots arrayed.
3) Make a dot at the present location.
4) Provide setting the next location. Then, increase the location of the present pixel by one to increasing order of coordinate x.
5) Calculate the error value at the next location. Here, the error term is the addition of the differences between y coordinate values of p1 and p2.
6) Compare the error terms and examine if the error portion is greater than one pixel. That is, after comparing the error terms up to now and the difference between x coordinate values of p1 and p2, increase the coordinate value by one to increasing order of coordinate y if the error term is greater than the difference.
7) Repeat 3) to 6) until the last coordinate is dotted.

For drawing of a quadrangle, the process of drawing four lines using Bresenham algorithm is repeated.

The process of drawing a circle represented by the equation, $x^2 + y^2 = r^2$, the fundamental of algorithm to be used in this research is as follows;

1) Begin with a fixed point on the top of the circle. Here, draw a quarter circle clockwise and repeat this circle four times.
2) Make a dot in the present coordinate.
3) Increase the coordinate by one to increasing order of coordinate x.
4) Then decide y coordinate. Decide one out of y or y-1 for y coordinate. If $x^2 + (y-1)^2 < x^2 + y^2 < r^2$ is valid, y becomes the next coordinate and if $r^2 < x^2 + (y-1)^2 < x^2 + y^2$ is valid, y-1 becomes the one. In other cases except for these, the error can be ignored with any dot between the two.
5) Repeat 2) to 4) until x==y.
6) Begin with the fixed point on the right side of the circle.
7) Make a dot in the present coordinate.

8) Increase the coordinate by one to increasing order of coordinate y.

9) Then decide x coordinate. Decide one out of x or x-1 for x coordinate. If $(x-1)^2 + y^2 < x^2 + y^2 < r^2$ is valid, x becomes the next coordinate and if $r^2 < (x-1)^2 + y^2 < x^2 + y^2$ is valid, x-1 becomes the one. In other cases except for these, the error can be ignored with any dot between the two.

10) Repeat 7) to 9) until x==y.

The algorithm of drawing an ellipse to be used in this research is similar to the case of a circle, and the algorithm was applied by changing the equation and calculating the point turning over the direction of axis.

3 Deduction of Markers Overlapping Problem

This research is based on the assumption that 8-bit input image is processed in binary image data, non-zero pixels are all treated as the same and N x 1 matrix is stored in a memory storage for the pointer of a space where copied markers will be stored. Fig. 1(a) shows the process of detecting the marker using PPHT and copying one

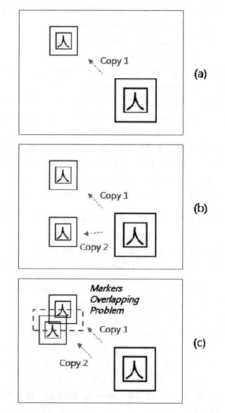

Fig. 1. Deduction of markers overlapping problem

marker. Two markers are copied through multi markers copying as shown in Fig. 1(b). In this way, n markers can be copied and therefore objects can be augmented, but if markers are overlapped as in Fig. 1(c), the object cannot be augmented.

Therefore in this research, to control markers overlapping problem, the marker area was set up with Bresenham algorithm applied and the problem was solved by designating non-overlapping area based on the marker's focal point.

4 How to Create Marker Area by Applying Bresenham Algorithm

4.1 Calculation of the Marker's Focal Point through Detection of Edge Points

For the calculation of the marker's focal point, the contour information of the marker area in the image was acquired by conducting the edge tracing algorithm and the edge points were extracted from the contour information inside the marker area.

By the Pythagorean Theorem, the first point is extracted as in Fig. 2(a), the farthest point as in Fig. 2(b) and then the farthest point is extracted as shown in Fig. 2(b). After three edge points is extracted, a point that maximizes the area of the quadrangle is extracted and becomes the last point, as in Fig. 2(d).

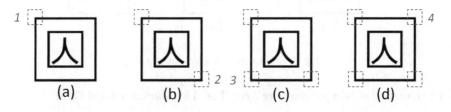

Fig. 2. Extraction of edge points

This research extracted the focal point using the template matching method which performs warping to the marker registered in the form of square from the edge points extracted within the image and then matches.

First, if the four edge points outside of the marker are defined as $(x_1, y_1), (x_2, y_2), (x_3, y_3), (x_4, y_4)$, $(x_1, y_1), (x_2, y_2), (x_3, y_3), (x_4, y_4)$, $(x_1, y_1), (x_2, y_2), (x_3, y_3), (x_4, y_4)$ and $(x_1, y_1), (x_2, y_2), (x_3, y_3), (x_4, y_4)$ respectively, the intersecting point of the line passing through (x_1, y_1) and $(x_1, y_1), (x_2, y_2), (x_3, y_3), (x_4, y_4) (x_2, y_2)$ and the line through (x_2, y_2) and (x_4, y_4) becomes the focal point of the marker. The marker's focal point is calculated using the linear equation, as followings;

$$C_y = \left(\frac{y_3 - y_1}{x_3 - x_1}\right)(C_x - x_1) + y_1 \tag{1}$$

$$C_y = \left(\frac{y_4 - y_2}{x_4 - x_2}\right)(C_x - x_2) + y_2 \tag{2}$$

C_x is obtained by calculating formula (1)- formula (2) and C_y is obtained by substituting this for formula (1) or (2).

If the height of a certain tag of the marker is h_r and the height of the marker's tag extracted is h_d, Sdiff (Scale difference ratio) depending on a changed plane during the extraction can be calculated as the formula (3).

$$S_{diff} = \frac{h_d}{h_r} \times 100 \tag{3}$$

The process of extracting the marker's focal point with the above formula is shown in the following figure.

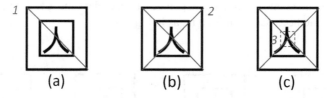

(a)   (b)   (c)

Fig. 3. Extraction of marker's focal point

4.2 Creation of Non-overlapping Area, Based on Marker's Focal Point

When calculating a circle's area from the marker's focal point in a square, the area should be divided every 45 degree. However, the marker extracted from the image is not recognized as a 100% circle, an ellipse's area needs to be considered. Since an ellipse cannot be calculated or divided into every 45 degree, when considering a line drawn at a point with a gentle slope, the slope $\frac{dy}{dx}$ should be regarded as 1 or -1.

In elliptic equation, $\frac{x^2}{a^2} + \frac{y^2}{b^2} = 1$, a point with the slope of -1 is $x = \frac{a^2}{sqrt(a^2 + b^2)}$, $y = \frac{b^2}{sqrt(a^2 + b^2)}$. Therefore, x should be an independent variable to this point and y should be one for the remained part. Other quadrants are drawn using its symmetry properties.

If the point is (x, y), the next point possible is (x+1, y) or (x+1, y-1), and it is determined by seeing whether the middle point of these points is fallen inside or outside of the ellipse.

The following figure describes the process of creating an ellipse with the discriminant, $d = F(x + 1, \frac{y-1}{2})$, for $F(x, y) = b^2 \times x^2 + a^2 \times y^2 - a^2 \times b^2$.

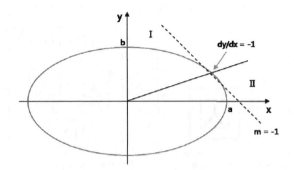

Fig. 4. Basic structure to create an ellipse

If d value is less than 0, (x+1, y) and greater than 0 (x+1, y-1) is selected. For the update of the discriminant c=to be calculated on the marker's focal point coordinate, if d_old is less than 0, it is d_new = $F(x+2, y) = d\_old + b^2 \times (2 \times (x + 1) + 1$ and if d_old is greater than 0, it is d_new = $F(x+2, \frac{y-3}{2}) = d\_old + l\,b^2 \times (2 \times (x + 1) + 1) - a^2(2 \times (y - 1)$.

As the independent variables of x are between 0 and -1 of the slope, for an interval of independent variables in y, the roles are just changed in the numerical formulas calculating independent variables in x. The following figure shows the process of creating an ellipse's area based on the marker's focal point.

Fig. 5. Creation of marker's focal point-based area

In case two markers are copied after the area created is designated as non-overlapping area, the applicable area is not interfered as in Figure 6 (a). Besides, if more than two markers are to be copied, as the number of markers printed in one screen is automatically limited, the markers overlapping problem can be solved and the number of copying operations can be regulated according to the scale of the marker.

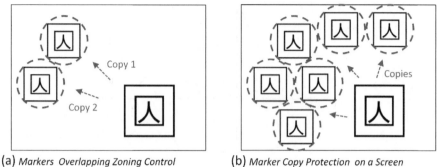

(a) *Markers Overlapping Zoning Control*   (b) *Marker Copy Protection on a Screen*

Fig. 6. Overlapping control during mark0er creation, by designating the created areas as non-overlapping areas

5 Conclusion

In this research, markers overlapping problem was deducted in PPHT based augmented reality multiple objects loading technology which realized one marker N objects loading. To solve the problem deducting, the method of setting marker area using Bresenham algorithm was researched, and the methods of calculating the marker's focal point through detection of edge points and of creating non-overlapping area based on the marker's focal point were suggested.

This research has the advantage that it can solve the problems that may occur during marker copying and limit the number of markers printed in one screen automatically.

Based on this research, a plan to increase or decrease the number of markers printed in one screen should be prepared in the further researches through the development of a technology of regulating the scale during the marker copying. Because of a real time, the nature of augmented reality, a research of very fast and correct algorithm is required.

Acknowledgement

This paper has been supported by the 2010 Preliminary Technical Founders Project fund in the Small and Medium Business Administration.

References

1. Azuma, R.T.: A Survey of Augmented Reality. Teleoperators and Virtual Environments 6(4), 355–385 (1997)
2. Caudell, T.P., Mizell, D.W.: Augmented reality:an application of heads-up display technology to manual manufacturing processes, vol. 2, pp. 659–669. IEEE Comput. Soc. Press, Los Alamitos (1992)

3. Kato, H., Billinghurst, M., Blanding, B., May, R.: ARToolKit, Technical Report, Hiroshima City University (December 1999)
4. Fiala, M.: ARTag, a Fiducial Marker System Using Digital Techniques. In: Proc. IEEE Intl. Computer Vision and Pattern Recognition Conference, pp. 590–596 (2005)
5. Fernandes, L.A.F., Oliveira, M.M.: Real-time line detection through an improved Hough transform voting scheme. Pattern Recognition 41(1), 299–314 (2008)
6. Bresenham, J.E.: A linear algorithm for incremental digital display of circular arcs. Commun. of the ACM 20(2), 100–106 (1977)

Single Image Super-Resolution via Edge Reconstruction and Image Fusion

Guangling Sun and Zhoubiao Shen

School of Communication and Information Engineering
Shanghai University, Shanghai, China
{sunguangling,shenzhoubiao}@shu.edu.cn

Abstract. For decades, image super-resolution reconstruction is one of the research hotspots in the field of image processing. This paper presents a novel approach to deal with single image super-resolution. It's proven that image patches can be represented as a sparse linear combination of elements from a well-chosen over-complete dictionary. Using a dictionary of image patches learned by K-SVD algorithm, we exploit the similarity of sparse representations to form an image with edge-preserving information as guidance. After optimizing the guide image, the joint bilateral filter is applied to transfer the edge and contour information to gain smooth edge details. Merged with texture-preserving images, experiments show that the reconstructed images have higher visual quality compared to other similar SR methods.

Keywords: super-resolution, edge reconstruction, dictionary learning, K-SVD, joint bilateral filter.

1 Introduction

Image super-resolution (SR) reconstruction, the process of combining multiple low-resolution images to form a higher resolution one, breaks through the limitation of traditional methods. With complementary information between multiple images, we can enhance image's spatial resolution under the confines of present imaging systems. Not only does the technology improves image's visual effects but has real meaning to subsequent image processings such as feature extraction and information recognition. Moreover, it now has a extensive range of application, including HDTV, biological information identification, satellite imaging, digital forensics, etc. In this paper, we focus on single image super-resolution.

Yang et al. [1] proposed a method for adaptively choosing the most relevant reconstruction neighbors based on sparse coding. Although it avoids over- or under-fitting so as to produce fine results, sparse coding learning over a massive sampled image patch database for dictionary is quite time-consuming. Similar to those aforementioned learning-based methods, they continued to rely on patches from the input image [2]. Instead of working directly with the image patch pairs sampled from high and low resolution images, they learned a compact representation for these patch pairs to capture the co-occurrence prior, significantly improving the speed of the

T.-h. Kim et al. (Eds.): SIP/MulGraB 2010, CCIS 123, pp. 16–23, 2010.

algorithm. However, their method based on dictionary of patch pairs turns out edge loss and aliasing when facing generic images.

The K-SVD algorithm [13], generalizing the K-means clustering process, is flexible enough to work with any pursuit method. Elad et al. [3] extended its deployment to arbitrary image sizes by defining a global image prior, making such Bayesian treatment a simple and effective denoising algorithm. Here, the K-SVD algorithm is applied to learn a dictionary of image patches, which is used to form an image with edge information. Joint bilateral filter (JBF) is then employed to transfer the edge and contour information to gain smooth edge details. As image's texture details are inevitably lost during any filtering, the zoomed image is merged with a texture-preserving image. The proposed method shows improved performance compared with other SR methods.

The rest of this paper is organized as follows. The principle of K-SVD and joint bilateral filter are briefly introduced in Section 3. The reconstruction framework is afterwards defined in detail. Experimental results and discussions are presented in Section 4 and conclusions are given in Section 5.

2   Related Work

Traditional approaches to generating a SR image requires multiple low-resolution images of the same scene. In a sense, the SR task is an inverse problem of recovering the original high-resolution image to the low-resolution images. The basic constraint for SR is that applying the image generation model to the recovered image should produce the same low-resolution images. However, SR reconstruction usually is a severely ill-posed problem due to insufficient low-resolution images, ill-conditioned registration and unknown blurring operators. Various regularization methods, such as [4] and [5], were proposed to further stabilize the inversion of such an ill-posed problem. Nevertheless, the results of these algorithms may be overly smooth, lacking important high-frequency details [6] when the desired zooming factor is too large or available input images are scarce.

Another category of SR approach is based on interpolation, such as [7], [8] and [9]. While simple interpolation methods like bilinear and bicubic interpolation tend to generate overly smooth images with ringing artifacts, interpolation by utilizing natural image priors usually presents favorable results. Such ways are effective in preserving the edges of the zoomed image but limited in modeling the visual complexity of the real images and tend to produce watercolor-like artifacts for fine-texture natural images.

The third category is based on machine learning techniques, which was first put forward by Freeman and Paztor [10], an example-based learning strategy that applied to generic images where the low-resolution to high-resolution prediction was learned via a Markov random field. Baker et al. [11] extended and put this learning-based SR reconstruction to face hallucination. Both attempted to capture the co-occurrence prior between high and low resolution image patches. However, they typically need millions of both resolution patch pairs to make the database over-complete. Motivated by manifold learning, Chang et al. [12] assumed similarity between the two manifolds in both high-resolution patch space and low-resolution patch space. Using this strategy, more patch patterns can be represented using a smaller training database. Anyway, using a fixed number K neighbors for reconstruction often results in blurring effects.

3 The Proposed Method

When working with color images, test images are first transformed from RGB to YCbCr. As the Cb and Cr (chrominance) channels are characterized by low frequency information, they are recovered just by bicubic interpolation. The edge reconstruction is then applied to the Y (luminance) channel. All three channels are finally combined together to form our SR result.

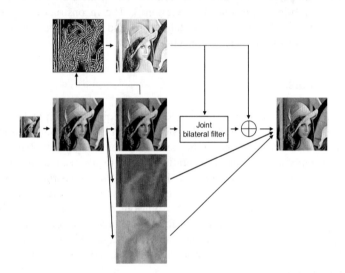

Fig. 1. The image 'Lena' is here used to illustrate the brief framework of our method. First, it is expanded by means of bicubic interpolation. The output is transformed into YCbCr, as is seen from top to bottom as Y, Cb, Cr. Second, after obtaining an edge image according to the Y image, we merge it back with the former Y to get a guide image. Third, joint bilateral filter is used to balance the edge and texture information. We merge its output with the guide image to form the Y image. The ultimate result is transformed from YCbCr to RGB.

3.1 Training Database

The assumption that signals such as images admit a sparse decomposition over a redundant dictionary leads to efficient algorithms for handling such sources of data. It's proven that image patches can be represented as a sparse linear combination of elements from a well-chosen over-complete dictionary. The theory of compressed sensing also suggests the sparse representation can be correctly recovered from the downsampled signals under certain conditions. As a result, the design of well adapted dictionaries for images has been a major challenge. The K-SVD algorithm [13], applied to obtain a dictionary that describes the image content effectively, performs well for various grayscale image processing tasks.

Using an over-complete dictionary matrix $D \in IR^{n \times K}$ that contains K prototype signal-atoms for columns, $\{d_j\}_{j=1}$, a signal $y \in IR^n$ can be represented as a sparse linear combination of these atoms. The representation of y may either be exact $y = Dx$ or approximate, $y \approx Dx$, satisfying $\|y-Dx\|_p \leq \varepsilon$. The vector $x \in IR^K$ contains the representation coefficients of the signal y.

The K-SVD algorithm, a way to learn a dictionary, instead of exploiting predefined ones as described above, leads to sparse representations on training signals drawn from a particular class. This algorithm uses either orthogonal matching pursuit (OMP) or basis pursuit (BP), as part of its iterative procedure for learning the dictionary. It includes the use of the K-SVD for learning the dictionary from the noisy image directly.

As our main challenge is focused on the luminance channel, we only use the K-SVD algorithm to deal with Y image, which ranges from 0 to 255. We find the best dictionary to represent the data samples $\{y_i\}_{i=1}^N$ as sparse compositions by solving (1).

$$\min_{D,X}\{\|Y - DX\|_F^2\} \; s.t. \; \forall i, \|x_i\|_0 \leq T_0. \tag{1}$$

In sparse coding stage, we use OMP to compute the representation vectors x_i for each example y_i, by approximating the solution of (2), $i = 1, 2, ..., N$.

$$\min_{x_i}\{\|y_i - Dx_i\|_2^2\} \; s.t. \; \|x_i\|_0 \leq T_0. \tag{2}$$

In codebook update stage, for each column $k = 1, 2, ..., K$ in D, we update it by computing the overall representation error matrix E_k by (3).

$$E_k = Y - \sum_{j \neq k} d_j X_T^j. \tag{3}$$

As the codebook length and the number of iteration play an important role in generating D, by numerous experiments, we find the relatively best their parameter combination. Usually, the learned dictionary D used for reconstruction is applicable to generic images. To better texture details, the low-resolution image is particularly learned, which makes the dictionary more practical.

3.2 Edge-Preserving Model

For most SR reconstruction, the luminance channel, containing both edge and texture information, plays a key role in image's quality. As a result, by using the learned dictionary D, we aim to improve the Y image with an edge-preserving image, which is called edge reconstruction.

Edge information is extracted through high frequency filter and convolution. After repeated and converged to find the sparsest representation, the edge image is obtained and displayed according to each block grid. Merged with the edge image, the original Y image is hence treated as a guide image.

3.3 Joint Bilateral Filtering

Bilateral filter is an edge-preserving filter, originally introduced by Tomasi and Manduchi [14] in 1998. It is related to broader class of non-linear filters such as anisotropic diffusion and robust estimation [15], [16]. The filter uses both a spatial (or domain) filter kernel and a range filter kernel evaluated on the data values themselves. More formally, for some position p, the filtered result is:

$$J_p = \frac{1}{k_p} \sum_{q \in \Omega} I_q f\left(\|p - q\|\right) g\left(\|I_p - I_q\|\right) \tag{4}$$

where f is the spatial filter kernel, such as a Gaussian centered over p, and g is the range filter kernel, centered at the image value at p. Ω is the spatial support of the kernel f, and k_p is a normalizing factor, the sum of the $f \cdot g$ filter weights. Edges are preserved since the bilateral filter $f \cdot g$ takes on smaller values as the range distance as well as the spatial distance increase.

Afterwards, joint bilateral filter (JBF) was introduced in which the range filter was applied to a second guidance image, I', for example, when trying to combine the high frequencies from one image and the low frequencies from another [17], [18]. Thus, the only difference between (2) and (1) is that the range filter uses I' instead of I.

$$J_p = \frac{1}{k_p} \sum_{q \in \Omega} I_q f\left(\|p - q\|\right) g\left(\|I_p' - I_q'\|\right) \tag{5}$$

Here, joint bilateral filter is used to transfer the edge information and balance it with texture details. After a number of comparisons, in the following experiments, the σ_s for spatial filter kernel is 8 and σ_r for range filter kernel is 4.

3.4 Optimization

The output image of a joint bilateral filter is denoised and edge-preserved but inevitably smoothed at the same time. Before obtaining the final result, we merge its output with the previously mentioned guide image to offset the loss of texture information. Truthfully, the image fusion here is linear and simply to seek the relatively visual quality while an adaptive fusion model is likely to be more favorable.

4 Experimental Results

In this section, based on twenty-four testing mosaic images, seen in Fig. 2, some experimental results demonstrate that our proposed method has better image quality performance compared with the previous zooming algorithms.

Fig. 2. Twenty-four testing images from Kodak PhotoCD[19] (referred to as image 1 to image 24, enumerated from left to right and top to bottom)

The first twelve testing images are chosen and modified to size 768×510, which then downsampled to size 256×170, for scale factor is 3 in the following experiments. As we work with color images, these images are first transformed from RGB to YCbCr. The comparison includes bicubic, Yang et al. [1] and our method, shown in Table 1. We also quantitatively compare the Structural Similarity [20] (SSIM) index of results with the original high resolution image.

Table 1. The comparison of PSNR and SSIM

	Bicubic		Yang et al.		Proposed	
	PSNR	SSIM	PSNR	SSIM	PSNR	SSIM
Image01	23.32	0.794	23.29	0.833	23.64	0.827
Image02	30.47	0.898	30.74	0.911	30.85	0.906
Image03	31.35	0.933	31.79	0.941	32.13	0.941
Image04	30.45	0.957	30.73	0.963	30.96	0.962
Image05	23.24	0.863	23.82	0.886	24.12	0.895
Image06	25.03	0.827	25.12	0.853	25.23	0.845
Image07	29.61	0.950	29.73	0.957	30.76	0.964
Image08	20.79	0.827	20.87	0.852	21.15	0.852
Image09	28.91	0.967	29.33	0.972	29.83	0.974
Image10	29.25	0.964	29.57	0.972	29.94	0.973
Image11	26.47	0.872	26.73	0.890	26.97	0.888
Image12	30.04	0.914	30.37	0.922	30.52	0.920
Average	27.41	0.897	27.67	0.913	28.01	0.912

Our results surpass Yang's method in PSNR but fall behind in SSIM. Actually, Yang's method has an advantage of texture preservation but is poor at edge reconstruction, seen in Fig.3 and Fig. 4. When it comes to edge, Yang's method creates unacceptable edge aliases. Our method, on the other hand, inevitably losing some texture details

Fig. 3. Test image is shown to demonstrate the edge reconstruction effects of the bicubic, Yang's and proposed method, enumerated from left to right

Fig. 4. Our method (right) makes the image smooth and sharpened and thus performs better than Yang's (left) visually in a sense

after joint bilateral filtering, keeps good edges and thus makes the whole image more agreeable.

5 Conclusion

This paper proposed an image SR reconstruction approach based on dictionary learning. While other similar SR methods show little advantage in preserving edges, SR method with bilateral filtering has a fine edge-preserving feature. Thus, edge and contour information are transferred, under the guidance of a joint bilateral filter, from a detailed edge image produced by an over-complete learned dictionary to the zoomed image. Considering that filtering may have influence on image's texture details, image fusion is enforced to avoid excessive texture loss. Compared with other SR methods, the proposed method, balancing the trade-off between edge information and texture details, has more favorable visual performance. Further research is to explore the proper parameters of the joint bilateral filter and seek an optimal learned dictionary for SR reconstruction.

References

1. Yang, J., Wright, J., et al.: Image super-resolution as sparse representation of raw image patches. In: IEEE Conf. Comput. Vision Pattern Recognit. (CVPR), pp. 1–8 (2008)
2. Yang, J., Wright, J., Huang, T., Ma, Y.: Image super-resolution via sparse representation. IEEE Trans. Image Process (to be published, 2010)
3. Elad, M., Aharon, M.: Image denoising via sparse and redundant representations over learned dictionaries. IEEE Trans. Image Process. 15(12), 3736–3745 (2006)
4. Hardie, R.C., Barnard, K.J., Armstrong, E.A.: Joint MAP registration and high-resolution image estimation using a sequence of undersampled images. IEEE Trans. Image Process. 6(12), 1621–1633 (1997)
5. Farsiu, S., Robinson, M.D., Elad, M., Milanfar, P.: Fast and robust multiframe super-resolution. IEEE Trans. Image Process. 13(10), 1327–1344 (2004)
6. Baker, S., Kanade, T.: Limits on super-resolution and how to break them. IEEE Trans. Pattern Anal. Mach. Intell. 24(9), 1167–1183 (2002)
7. Sun, J., Xu, Z., Shum, H.: Image super-resolution using gradient profile prior. In: Proc. IEEE Conf. Comput. Vision Pattern Recognit. (CVPR), pp. 1–8 (2008)
8. Hou, H.S., Andrews, H.C.: Cubic spline for image interpolation and digital filtering. IEEE Trans. Acoust. Speech Signal Process. 26(6), 508–517 (1978)
9. Dai, S., Han, M., et al.: Soft edge smoothness prior for alpha channel super resolution. In: Proc. IEEE Conf. Comput. Vision Pattern Recognit. (CVPR), pp. 1–8 (2007)
10. Freeman, W.T., Pazstor, E.C.: Learning low-level vision. Int. J. Comput. Vision 40(1), 25–47 (2000)
11. Baker, S., Kanade, T.: Hallucinating faces. In: Proc. IEEE Conf. Autom. Face Gest. Recogn., pp. 83–88 (2000)
12. Chang, H., Yeung, D., Xiong, Y.: Super-resolution through neighbor embedding. In: Proc. IEEE Conf. Comput. Vision Pattern Recognit. (CVPR), vol. 1, pp. 275–282 (2004)
13. Aharon, M., Elad, M., Bruckstein, A.M.: The K-SVD: An algorithm for designing of overcomplete dictionaries for sparse representations. IEEE Trans. Image Process. 54(11), 4311–4322 (2006)
14. Tomasi, C., Manduchi, R.: Bilateral filtering for gray and color images. In: Proc. IEEE Int. Conf. Comput. Vision (ICCV), pp. 839–846 (1998)
15. Durand, F., Dorsey, J.: Fast bilateral filtering for the display of high-dynamic-range images. ACM Trans. Graphics 21(3), 257–266 (2002)
16. Elad, M.: On the bilateral filter and ways to improve it. IEEE Trans. Image Process. 11(10), 1141–1151 (2002)
17. Eisemann, E., Durand, F.: Flash photography enhancement via intrinsic relighting. ACM Trans. Graphics 23(4), 673–678 (2004)
18. Petschnigg, G., Szeliski, R., et al.: Digital photography with flash and no-flash image pairs. ACM Trans. Graphics 23(3), 664–672 (2004)
19. Kodak Lossless True Color Image Suite, PhotoCD,
   http://r0k.us/graphics/kodak/index.html
20. Wang, Z., Bovik, A.C., et al.: Quality assessment: from error measurement to structural similarity. IEEE Trans. Image Process. 13(4), 600–612 (2004)

Scanning and Measuring Device for Diagnostic of Barrel Bore

Ales Marvan[1], Josef Hajek[1], Jan Vana[1], Radim Dvorak[1],
Martin Drahansky[1], Robert Jankovych[2], and Jozef Skvarek[2]

[1] Brno University of Technology, Faculty of Information Technology,
Bozetechova 2, CZ-612 66, Brno, Czech Republic
[2] University of Defense, Kounicova 65, CZ-602 00, Brno, Czech Republic
{imarvan,ihajek,ivanajan,idvorak,drahan}@fit.vutbr.cz
{robert.jankovych,jozef.skvarek}@unob.cz

Abstract. The article discusses the design, mechanical design, electronics and software for robot diagnosis of barrels with caliber of 120 mm to 155 mm. This diagnostic device is intended primarily for experimental research and verification of appropriate methods and technologies for the diagnosis of the main bore guns. Article also discusses the design of sensors and software, the issue of data processing and image reconstruction obtained by scanning of the surface of the bore.

Keywords: barrel guns, barrel diagnostic, image processing, sensors.

1 Introduction

Barrel firearm is exposed to complex mechanical stress and chemical effects by firing. This leads to intense wear of the bore and the main degeneration of its technical state.

The present market offers many modern methods and devices of technical diagnostics that can be used to refine the technical state of barrels, to detect faults, and to detect insufficient maintenance. The important demand of a modern barrel diagnostic system is to locate the exact place of damaged part of barrel bore with a suitable coordinate system. It leads to better analysis of geometry change of firing barrel.

2 Proposal of Diagnostic Equipment, Especially for Barrel

The entire mechanical design of the robotic diagnostic system for a barrel firearm is relatively complex and consists of two main parts:

- Sensoric head - in this part a sensoric system is located including a camera for capturing the inner surface of the barrel. The head is rotating so that it can scan the surface over the inner circumference of the barrel. This part of robot includes mechanics, which allows shifting the sensors in the direction perpendicular to the axis of the device.

T.-h. Kim et al. (Eds.): SIP/MulGraB 2010, CCIS 123, pp. 24–29, 2010.

- Robot's body - provides controlled movement of the diagnostic device in the barrel. Centering mechanism contains arms with wheel drive. Arms allow movement and centering of the axis of the barrel.

2.1 Universal Carriers

Carrying part of diagnostic equipment is versatile robot's body. Almost all electronics which is responsible for controlling the engine of the device and which communicates with the external environment are located on the body of the robot. The body has no sensors. It serves only as a movable part, which has the task of adapting to the caliber of the scanned barrel to ensure smooth movement within the barrel. It consists of several important parts:

- Versatile carrier (Fig. 1) is divided into two halves, which are identical in terms of engines.
- Driving system, which consists of six miniature geared engines and wheels.
- Centering mechanism (Fig. 1) which provides expanding of arms with driven wheels. The arms can be adjusted by the stepper engine from the caliber 120 mm to 155 mm.

On each arm of the body, there are two wheels and the first wheel is driven by a motor. The device has a total weight of about seven pounds, therefore it is important to have a sufficiently strong drive. Six driving wheels are powered by engines Maxon RE8 [1] with gearbox 4096:1. Expanding of arms is provided by two stepper motors NEMA 17 [2] located in both halves of the body of a universal carrier. Each engine NEMA expands three arms through mechanism levers.

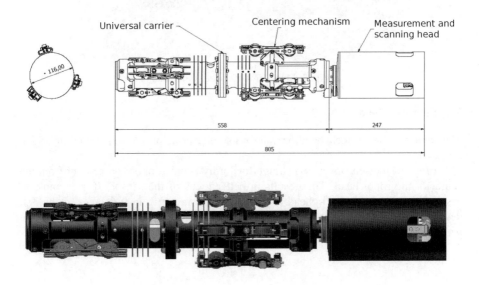

Fig. 1. Model of the barrel scanner

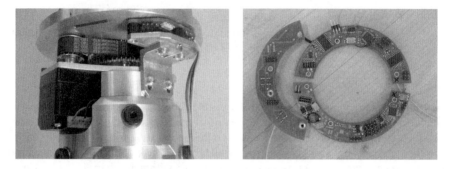

Fig. 2. Detail of the electronic semicircles and detail of the drive rotation of the head

Because of space saving, it was necessary to adjust the shape of the PCB. PCB is now like the ring which hugs the body of the robot. Head rotation provides also a NEMA 17 motor [2]. It is connected with the robot head by the drive belt pulley with the 1:3 transfer ratio. This allows the stepping movement of the head by 0.6 ° (Fig. 3).

Fig. 3. Mechanics of the sensoric head

2.2 Sensoric Head

Mechanics of the sensoric head operates on similar principles as expanding of driven arms.

Carriers of the sensors can be straightforwardly zoomed in or zoomed out from the center of the robot head. This results in better focus of the sensors (for example in case of a camera). Sensoric head is rotating in the range from 0 degrees to the 360 degrees. In order to set the head to the starting position, a magnet is placed on the head – that indicates the starting position of rotation.

3   Sensor System

Appropriate sensors have to be carefully chosen for such a type of diagnostic device. We propose three combinations of sensors which will be used in experimental research with the device:

- CCD camera with IR lighting, CCD camera with a bar laser lighting, triangulation laser to measure the inside diameter of barrel
- 3× induction sensor Bäumer IWFK20 for measuring the internal diameter [3]
- 3× potentiometric sensor for measuring the internal diameter [4]

The first group of sensors is used for mapping of a surface and a texture inside the barrel. First CCD camera with infrared lighting is used to obtain detailed texture while the second camera with bar laser lighting is used for precise surface reconstruction. It can be evaluated for damage detection and profile depth and width measuring. The last type of sensor in this group is a laser rangefinder for precise measurement of internal diameter (measured data are mainly used for calibration). The second and third group of sensors is equipped with single purpose sensors for measuring of the internal diameter. It uses resistance (contact) and induction (non-contact) sensors.

4   Draft Electronics and Data Processing Software for Diagnostic Equipment

The proposed electronic system can be divided into several parts:

- Modules for the movement of the robot in the barrel

- Mechanism to keep the robot in the center of barrel (expanding arms)

- Turning sensoric head

- Modules for collecting data from a sensoric system

All modules of the system communicate on CAN 2.0A bus using CANopen protocol.

4.1   Electronic Modules

Electronic modules for the movement of a robot in the barrel provide synchronous control of six drives (one module for three drives). Each drive has an engine, gearbox and encoder. Other modules are used for arms collapsing and expanding, controlling the sensoric head rotation engine and finally for sensor motion within the head part. There also is an electronic module used mainly for collecting data from sensors and controlling IR and laser lighting. Cameras have their own data channel.

4.2 Measurements and Data Processing

One complete measurement process of the barrel is almost fully automatic, whereas it requires manual intervention only at the beginning and at the end. The main and the most time-consuming part performs without human service, or supervision.

Precise measurement of the distance to the surface mainly through triangulation sensor is used to set the camera focus and detect whether the device is centered in the barrel.

Up to 600 images of barrel surface is obtained within one turn of sensoric head. These images are then folded into one image. All these folded images create detailed map of inner barrel. Two maps captured with a certain time period can be compared (automatically or manually) to check if there are some important changes on the barrel surface since last measurement.

Fig. 4. Reconstruction of elevation profile (profilometer)

CCD camera with bar laser lighting is used for mapping elevation profile (profilometry). Laser beam hits the barrel at certain angle. The camera picks up the barrel from the perpendicular direction. Laser beam is then detected in the obtained image and the height map is then computed from deviation of expected position. Calculated values can be corrected using data from triangulation distance sensor.

5 Conclusions

The proposed diagnostic device is a comprehensive solution from idea and design to implementation, mechanical design, electronics and software. Compared with existing diagnostic devices, this functional model is universal for different caliber with minimizing human assistance in measuring process. The aim was to facilitate experimental research, and gaining knowledge in this area by practical application.

Acknowledgement

This research has been done under support of the following grants: "Secured, reliable and adaptive computer systems", FIT-S-10-1, "Security-Oriented Research in

Information Technology", MSM0021630528 (CZ) and by the project called "dělo" from the Ministry of Defense, No. 0901 8 1020 R.

References

[1] Maxon motor, Sachseln, Program 06/07 (2009)
[2] Stepper motor NEMA 17,
    http://www.servo-drive.com/krokove_motory_
    rotacni_krokove_motory.php
[3] Inductive sensor Baumer IWFK 20,
    http://sensor.baumerelectric.com/productnavigator/downloads/
    Produkte/PDF/Datenblatt/Induktive_Sensoren/
    IWFK_20Z8704_S35A_web_DE.pdf
[4] Resistance sensor Megatron MM 10,
    http://www.megatron.de/export/Linear_Motion/MM10/
    DB_MM10_engl.pdf

Recognition of Hits in a Target

Vojtech Semerak and Martin Drahansky

Brno University of Technology,
Faculty of Information Technology, Bozetechova 2,
CZ-612 66, Brno, Czech Republic
xsemer03@stud.fit.vutbr.cz, drahan@fit.vutbr.cz

Abstract. This paper describes two possible ways of hit recognition in a target. First method is based on frame differencing with use of a stabilization algorithm to eliminate movements of a target. Second method uses flood fill with random seed point definition to find hits in the target scene.

Keywords: hit detection, OpenCV, digital image processing, computer vision.

1 Introduction

This paper describes algorithms, which can be used to recognize hits in a target in sport shooting. Camera was placed in front of the target at a distance of 25 meters. For recording camcorder Sony HDR-SR12E was used, which uses 12× optical zoom and can capture video in Full HD resolution. All videos were recorded in an indoor shooting range.

For the recognition algorithms, a very important factor to successfully run, are the light conditions. The task to recognize the hits in the black middle of the target is difficult, if the light environment is too dark. It is important to illuminate the target directly, like in the figure 1. In that case, the area of the hit is always darker than its adjacent pixels.

The next problem, which must be solved, is the movement of the target. Target is hanging on the target transportation device and it is waving and swinging due to an air circulation in the target area. If the movement of the target itself is greater than the diameter of the hit, the algorithm using only frame differencing [1] technique will fail to detect the hit.

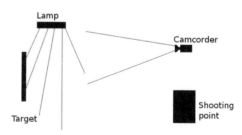

Fig. 1. Situation on the shooting range

T.-h. Kim et al. (Eds.): SIP/MulGraB 2010, CCIS 123, pp. 30–34, 2010.

2 Target Detection

The rough target detection was implemented via the template matching function in OpenCV. Adjusting to any other target would be easy – it can be done just by changing the template. To precise detection of the target position and size, Hough transform was used, which is implemented in OpenCV [5].

This solution is working for the circle shape targets only. For any other targets, like silhouettes, the center can be easily found by using the flood fill algorithm [2][1] and counting the center of gravity of the filled area.

Fig. 2. Target scene, 25 meters

3 Hit Recognition

The implementation using only the frame differencing algorithm has proved to be an insufficient one due to the target and camera movement. During the first recording the camera was placed too close to the shooter and the shock wave from the shot caused the camera to shake. After this shake however, most of the time camera (but not always) returned to the original position.

To eliminate this undesirable behavior, the camcorder with the massive tripod was used and placed next to the shooting point, therefore separated from the shooter by the barrier. Thanks to this approach the camcorder always returns to the original position and the shaking was reduced to a minimum.

3.1 Frame Differencing with SURF

To eliminate the target movement the frames before differencing are matched with the use of SURF algorithm [3].

There was a problem with insufficient amount of strong points in the target picture. As can be seen in the figure 3, strong points are found only in the upper left corner, where some text can be found. The rest of the target lacks the edges, corners etc., which could be used for matching. For that reason another target was used, which was marked with four crosses, one in each corner. These marks can be seen in the figure 2.

With this adjustment, the amount of strong points found on the target (figure 4) is sufficient and the matching can be done.

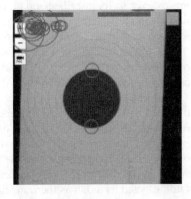

Fig. 3. Insufficient amount of strong *points*

To match the points the demonstration algorithm from OpenCV library was used, extended by the simple elimination of point mismatch. Its principle is based on Euclidean distance between two matched points. If the distance between them is too long, this match will be discarded. There is a lot of space for optimization, but this implementation is sufficient enough for running the recognition successfully.

After matching and subtracting frames it is necessary to clean up the resulting image. It is thresholded and the noise is reduced by using the morphological operations – erosion and dilatation [1][4]. There is applied the bit mask

Fig. 4. Extracted strong points

into the image at the moment, which restricts the recognition only on the selected area. This eliminates the false detections on the edges of the target.

This algorithm always compares the actual frame with the frame, which was saved during the initialization.

3.2 Segmentation Based on Flood Fill Algorithm

This algorithm has been designed for those cases, where it is impossible to mark up the target.

Seed points for the filling algorithm are chosen randomly for each frame. For the target from figure 3 two seed points are chosen. First seed point (P1 in the image 5) is located in the black middle area of the target, exactly in its inscribed square (random point in the square area can be generated much easier and faster than in the circle area). The second point has the same x coordinate as the first point (P1) and the value of P2 y coordinate is the sum of the P1 y value and a constant.

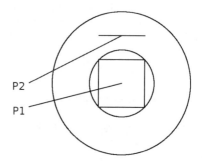

Fig. 5. Position of the seed points

It is assumed that the hit area is darker than its adjacent pixels. The filling criteria, which decides whether to fill the area or not, is more sensitive to lower and less sensitive to higher intensity values. Due to this, hits itself are not filled, but the white lines separating the hit zones do not stop the filling process. Intensity of the light is slightly changing across the target. The black middle of the target is a relatively small area, so it is not affected as much as the remaining white area. The intensity of every point in the black middle area is compared to the intensity of the seed point, whereas in the case of white area, the points are compared to the intensities of adjacent points.

The output of this process is the mask, which is saved into the temporary image. From all pixels of this image scalar value is subtracted in every cycle. Afterwards another scalar is added to the temporary image through the mask, which is greater than the first one and image is thresholded. This ensures that the algorithm will be stable, even if the seed is selected directly from the hit.

Results of this algorithm can be seen in the figures 6 and 7. It is clear, that the algorithm will have problems with detection, if the shot hits the digit marking the hit zone. This algorithm works well for the targets not containing digits or letters (the target from figure 3, result for this type of target can be seen in figure 7).

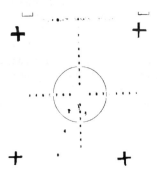

Fig. 6. Target without digit marking   **Fig. 7.** Target with strong digit marking

3.3 Scoring

These images are scanned for the connected components. Hits are represented by the bounding rectangle and they are saved in the linear list. Object, which is not overlapping with any other hit in the list, will be added to the end of it.

If the new object is stable, it will be printed on the screen. Last valid hit in the list will be marked as the last as the feedback for the shooter.

4 Results

First discussed method based on the frame differencing with usage of SURF is working properly. However, it was not possible to test it on larger number of videos, but this method has been successful with the used samples. From 38 hits only 1 hit has not been recognized. A weakness of this algorithm is its speed. It can process from three to five frames per second.

The second algorithm is faster, but its reliability is not very good. It can't recognize hits in the digits in the target and sometimes it fails to recognize the hit in the color boundary between the black and white areas. This algorithm will work as a backup one, if it is not possible to mark up the target for matching the frames.

Acknowledgement

This research has been done under support of the following grants: "Secured, reliable and adaptive computer systems", FIT-S-10-1 (CZ) and "Security-Oriented Research in Information Technology", MSM0021630528 (CZ).

References

1. Bradski, G., Kaehler, A.: Learning OpenCV. O'Reilly, Sebastopol (2008)
2. Krsek, P.: Zaklady pocitacove grafiky, Brno, Faculty of Information Technology, Brno University of Technology (2010)
3. Bay, H., Ess, A., Tuytelaars, T., Van Gool, L.: SURF: Speeded Up Robust Features. Computer Vision and Image Understanding (CVIU) 110(3), 346–359 (2008)
4. Parker, J.R.: Algorithms for image processing and computer vision. John Wiley & Sons, Inc., Chichester (1997)
5. OpenCV Wiki, http://opencv.willowgarage.com (on-line, 06/2010)

Disparity Estimation Method Based on Reliability Space Using Color Edge

Seungtae Lee, Youngjoon Han, and Hernsoo Hahn

Department of Electronic Engineering
SoongSil Univ, Sangdo-5Dong, Dongjak-Gu, Seoul, South Korea
newclee2@ssu.ac.kr, young@ssu.ac.kr, hahn@ssu.ac.kr

Abstract. In this paper, we propose a Disparity Estimation Method based on Reliability Space using Color edge. It first makes the disparity space image by using the difference between the Stereo images. It than calculates a section of continuous minimum value of the disparity space image, and the length of the section is calculated to produce reliability space. By comparing the parallax of the reliability space, the disparity data of each pixel selects the disparity data with the highest accuracy. Moreover, the parts that make regional boundary errors are corrected by classifying the boundary of each objects using color edge. The performance of the proposed stereo matching method is verified by various experiments.

Keywords: disparity space image, Reliability space, Disparity estimation method, stereo matching, correspondence.

1 Introduction

Recently, the three-dimensional restoration is one of the most interesting methods using cameras. It needs two-dimensional images of multi-view to realize three-dimensional restoration. Generally, it uses the binocular parallax image to make three-dimensional images. Binocular parallax image is the acquired image using stereo camera that has the human's eyes-like structure.

Selecting exactly corresponding points is the most important progress to acquire three-dimensional information in the multi-view images. This progress is the disparity estimation method. Disparity estimation method is divided the local method and the global method [1] according to used matching method.

The local matching method compare with reference image and standard image using matching window about the limited area. It guesses disparity value of each pixel using this value. It has fast progress speed and need small system resource. That is the advantage of this method. However, it has high probability of mismatching in area repeated same pattern or filled similar values. Add to that, it has an inaccurate value on the boundary of each object having different disparity. That is the disadvantage of this method. The most typical method of local matching method is block matching method [2] and feature based matching [3].

The global matching method is proposed to reduce mismatching error of local matching method caused by the limited search area. Then, it guesses disparity value of

T.-h. Kim et al. (Eds.): SIP/MulGraB 2010, CCIS 123, pp. 35–42, 2010.
© Springer-Verlag Berlin Heidelberg 2010

each pixel through searching width area without limiting search area. This method can guess exactly disparity value. That is the advantage of this method. However, it needs many system resources and it would take a long time. That is the disadvantage of this method. The typical method of global matching method is dynamic programming method [4] and graph cut method [5], belief propagation method [6], etc.

Dynamic programming method first makes the disparity space image by using the difference between the reference image and standard image. Then, it search the best disparity routine using HMM(Hidden Markov Model). This method has fast progress speed more than other methods. And, it can guess more exactly disparity information.

Graph cut method assigns the disparity value using the max flow / min cut algorithm. This method is one of the algorithms having high accuracy more than other methods.

Belief propagation method is the repetition algorithm based on message propagation method. This method almost has the best accuracy. But it needs so many repetition times to acquire the converged message. That is the disadvantage of this method.

In this paper, we propose a Disparity Estimation Method based on Reliability Space using Color edge. Its subject is what is extracting three-dimensional information faster and more exactly. It first makes the disparity space image by using the difference between the Stereo images. It than calculates a section of continuous minimum value of the disparity space image, and the length of the section is calculated to produce reliability space. By comparing the parallax of the reliability space, the disparity data of each pixel selects the disparity data with the highest accuracy. Moreover, the parts that make regional boundary errors are corrected by classifying the boundary of each objects using color edge.

2 Basic of Stereo Matching System

2.1 Stereo Image Calibration

The stereo image calibration is the method to reduce the processing time and make the accuracy higher when it search corresponding points as minimize the searching area through comparing the standard image with reference image. Generally, it is used the line matching progress that limit searching area as arrange in a line corresponded characteristic points between standard image and reference image for stereo image calibration. This searching line is the epipolar line through this progress in both images.

Fig. 1. Epipolar line

Figure 1 shows the epipolar line and corresponding points. Generally, it is used the physical stereo camera calibration to simplify this step. The physical stereo camera calibration is the method that solves the epipolar line matching problem by physical camera structure. This calibration reconciles epipolar line of standard image and epipolar line of reference image artificially.

2.2 Disparity Space Image

The disparity space image is made by using the comparison between the N line of standard image and N line of reference image.

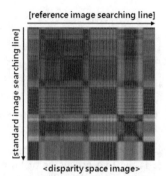

Fig. 2. Disparity Space Image

Upper figure 2 shows disparity space image. This image's each pixel value is the matching error about any N line of stereo images. In upper disparity space image, the dark area has small matching error. And, the right area has many matching error. This value is from equation (1).

$$D_{DSI}(I_S, I_R) = \left| I_{S-N\,line}(x_S) - I_{R-N\,line}(x_R) \right| \qquad (1)$$

This equation (1) is the formula to calculate the matching error. Equation (1)'s I_S is the standard image, I_R is the reference image. $I_{S-N\,line}$ is N-th searching line of standard image, $I_{R-N\,line}$ is N-th searching line of reference image. And, x_S, x_R is the position of pixel value classified by the parallax. Finally, this equation (1) is the formula to calculate the difference of standard image with reference image classified by the parallax.

3 Proposed Method

3.1 Algorithm Construction

Figure 3 shows the processing of proposed algorithm using flow chart. In Figure 3's first block, disparity space image has all of the matching error information about comparing the standard image with reference image. Therefore, it must choose the

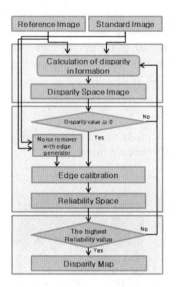

Fig. 3. Flow Chart of Proposed Algorithm

value that has high accuracy through classifying value of disparity space image. Because of this reason, it needs the two processes of how classifying. It chooses the effective value that is similar to zero value. Then, we can calculate a section of continuous minimum value of the disparity space image, and the length of the section is calculated to produce reliability space using the upper effective values. In the second part, by comparing the parallax of the reliability space, the disparity data of each pixel selects the disparity data with the highest accuracy. We can take final disparity map using the disparity data with the highest accuracy.

3.2 Color Edge

Sometimes, Stereo matching method using disparity space image has horizontal boundary violation errors. In any parallax, this problem is happened to same value of inner area and boundary and background area. Then, we cannot divide each area. Proposed method uses disparity space image. So, proposed method has horizontal boundary violation errors, too. To solve this problem, in this paper, it use color edge. Then it divides the regional boundary more exactly.

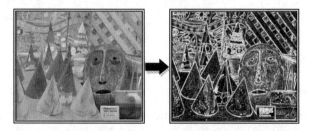

Fig. 4. Color Edge

Figure 4 is Color edge using sobel method. In this paper, generally, in many paper, it use gray edgy. But this paper doesn't use gray edge. Because color edge can make many boundary edge more than gray edge.

3.3 Noise Remover with Edge Generator

Color sobel edge has many noises. So we need noise remover for edge image.

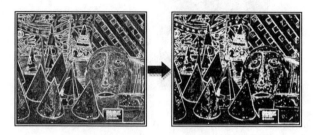

Fig. 5. Color Edge whit Noise Remover

The proposed noise remover is a modified mean filter. This method can effectively remove noise and leave the characteristics of the color Sobel edge.

$$I_{m-sobel} = \frac{1}{N} \times \left(\sum_{k=1}^{N} I_{sobel}(x_k) \right)$$

$$I_{m-sobel} = \begin{cases} I_{m-sobel} & ,if \ I_{m-sobel} \geq \theta_1 \\ 0 & ,otherwise \end{cases} \tag{2}$$

x_k is the edge value of a color Sobel operator on each pixel. N is 9 by using 3×3 mask. θ_1 is the threshold for removing noise from the results of the mean filter. This threshold is gotten through a determined value obtained from various experiments.

$I_{m-sobel}$ can divide the value into the true value or noise using the mean value. This method can remove small values of sporadically generated noise.

3.4 Reliability Space Image

In stereo image, each corresponding point of standard image and reference image is almost zero value in disparity space image. Case of some fixed area, this area has a section of continuous minimum value of the disparity space image at corresponding parallax. And, in the best corresponding parallax, it is made the longest section of continuous minimum value. Then, we can search exactly corresponding point using this value. In this paper, these values are arranged by reliability space.

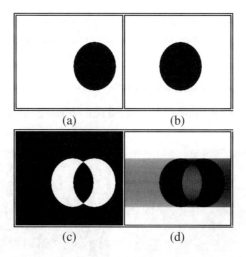

Fig. 6. Construction of Reliavility Space. (a) Reference Image, (b) Standard Image, (c) Disparity Space Image on any parallax, (d) Reliability Space Image.

Figure 6 shows construction of reliability space image using simple image. Figure 6 (c) is disparity space image about all of the line on and parallax. Figure 6 (d) is reliability space image. It calculates a section of continuous minimum value of the disparity space image, and the length of the section is calculated to produce reliability space.

4 Experiment

In this paper, we propose a Disparity Estimation Method based on Reliability Space using Color edge. This method uses the reliability space that is made through modifying exist method using HMM. Then, we try to upgrade that's speed and accuracy.

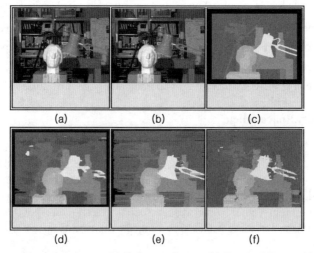

Fig. 7. Stereo matching result image. (a) Reference Image, (b) Standard Image, (c) Ground True, (d) realtime SAD, (e) Dynamic Programming.

Upper figure 7 shows the result of various stereo matching methods. Figure 7 (d) is the Real-time SAD[9]. And, figure 7 (e) is the matching method based on dynamic programming. And, figure 7 (f) is the Disparity Estimation Method based on Reliability Space using Color edge.

The proposed Disparity Estimation Method based on Reliability Space using Color edge speeds have improved compared to matching method based on dynamic programming by 27.9%. And, accuracy has also improved a few.

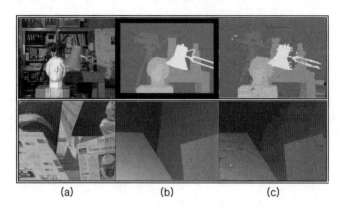
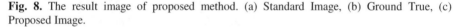

      (a)   (b)   (c)

Fig. 8. The result image of proposed method. (a) Standard Image, (b) Ground True, (c) Proposed Image.

Figure 8 shows the result images of proposed method. Used images are general reference images on the variety disparity estimation method. These images are each named tsukuda (384 × 284) and venus (434 × 284).

5 Result

In this paper, we propose the Disparity Estimation Method based on Reliability Space using Color edge. This method reinterprets disparity space image from reliability. It makes reliability space. Then, it can make disparity map by comparing reliability space.

The proposed method speeds have improved by 27.9%. And, accuracy has also improved a few. In other words, the proposed method has the advantage that is to keep that's speed and to improve that's speed.

Further study is needed to find better performance.

Acknowledgments. This work was supported by the Brain Korea 21 Project in 2010. This research was supported by the MKE(The Ministry of Knowledge Economy), Korea, under the ITRC(Information Technology Research Center) support program supervised by the NIPA(National IT Industry Promotion Agency)" (NIPA-2010-(C1090-1021-0010)).

References

1. Scharstein, D., Szeliski, R.: A taxonomy and evaluation of dense two-frame stereo correspondence algorithms. IEEE Trans. on Pattern Analysis and Machine Intelligences 25(8), 993–1008 (2003)
2. Young-Sheng, C., Yi-ping, H., Chiou-Shann, F.: Fast Block Matching Algorithm Based on the Winner-Update Strategy. IEEE Trans. on Image Process 10(8) (2001)
3. Czarnecki, K., Helsen, S.: Feature-based survey of model transformation approaches. IBM Systems Journal 45(3) (2006)
4. Torr, P.H.S., Criminisi, A.: Dense Stereo Using Pivoted Dynamic Programming. In: Microsoft Research, MSR-TR-2002-51 (2002)
5. Boykov, Y., Veksler, O., Zabih, R.: Fast Approximate energy minimization via graph cuts. IEEE TPAMI 23(11), 1222–1239 (2001)
6. Sun, J., Zheng, N.N., Shum, H.Y.: Stereo matching using belief propagation. IEEE Transaction on Pattern Analysis and Machine Intelligence 25(7) (2003)
7. Hirschmuller, N.H.: Improvements in real-time correlation based stereo vision. In: IEEE Workshop on Stereo and Multi baseline Vision (2001)

Web-Based IP Reuse CAD System, Flowrian

Ming Ma[1], Hyoung Kie Yun[2], Woo Kyung Lee[2], Jung Seop Seo[2],
Inhag Park[1], and Dai-Tchul Moon[2]

[1] System Centroid Inc., #414 Hanshin S-MECA, 1359 Gwanpyeong-dong,
Yuseong-gu Daejeon 305-509, Republic of Korea
[2] Hoseo University, 165, Sechul-ri, Baebang-myun, Asan, Chungnam,
S.Korea 336-795, Hoseo University

Abstract. IP reuse design in ready-to-use platform becomes very popular design methodology for SoC development. IP sharing (knowledge sharing) is very important to maximize design efficiency. So IP sharing and reuse environment has been requested for a long time. This paper demonstrates web-based CAD system, called Flowrian, originally developed for IP sharing and reuse design. The high speed internet infrastructure in South Korea is best in the world enough to introduce internet-based application even in specialized high tech area, such as IP-based SOC design. Flowrian consists of client personal computer and remote server computer which communicate each other through internet. The user on the side of client is able to design and verify SOC by sharing the IPs and tools installed on the side of remote server. Flowrian must be very feasible solution to validate the correctness of ready-to-use IPs installed in the IP DB center, such as Hoseo Engineering SaaS Center.

1 Introduction

ASIC design methodolgy has been changed over several decades - from the hiearchical logic design using schematic editor in 1980's, through the logic synthesis from RT-level HDL code in 1990's, to the platform based IP reuse SoC design in 2000's. Surely one of the key factor in SOC design is the reuse of IPs [1]. But it is very sensitive to get the IPs in open design environment because of their valuable contents condensing design experts' long-time experiences and knowledges. Until now several methods have been proposed to provide IP verification in try-before-buy way. None has accepted as the convenient way to verify IPs in protected environement.

Nowadays the infrastructure of high speed internet communication becomes very popular in worldwide enough to introduce new internet-based SaaS services even in specialized high technology area, such as on-line IP verification [2-3]. This paper proposed very new design environment, web-based CAD system, called Flowrian, which is invented by fusing EDA technologies and web communication.

Flowrian is the first web-based IP reuse ECAD system developped in Korea [4]. Flowrian consists of 3 basic elements - client PC, remote server, and internet communication. Users on client PC side can perform design capture, download application IPs from remote server, verify design by sending the design data to remote

T.-h. Kim et al. (Eds.): SIP/MulGraB 2010, CCIS 123, pp. 43–48, 2010.

server and debug the results. All the resources on server can be utilized as if it is built on user's client PC.

In this paper, the configuration of web-based CAD system will be explained in chapter 2. First of all, the overall configuration and operation flow will be introduced. Then the main mission of client and server will be explained in more detail. Chapter 3 shows results of audio encoding and decoding test design by reusing IPs on remote server. This can be only one example to prove web-based IP reuse design. Chapter 4 concludes by summarizing benefits of web-based CAD system.

2 Web-Based CAD System

2.1 The Overall Configuration

The web-based CAD system consists of several clients and a server communicating through the internet (Fig. 1).

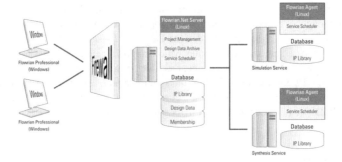

Fig. 1. Configuration of web-based CAD system

The client is an integrated design environment for designers to capture and debug Verilog/VHDL codes, analyze simulation results and monitor the execution in the server. The server acts like centralized service center where legacy and outsourcing IPs, expensive commercial CAD tools and high performance server computers are equipped as ready-to-use state. Two different softwares are running independently on the client and server computers. Although they carry out different parts of whole design process, their operations are unified into one continuous process thanks to the communication on the internet.

Fig. 2 shows the process to verify the design made by IPs installed in remote server. The jobs on client side are described on left side, and server works on right side. Once designers download IP interface data (IP Shell) from the server system, it is possible to capture HDL codes by reusing IPs on client. Some graphics tools running help designers capture and debug Verilog and VHDL codes in convenient way. The design data must be transferred to the server system because the real IPs are protected in server computer. A set of design files needed for simulation are collected, and transmitted to server through the internet. The server immediately queues the

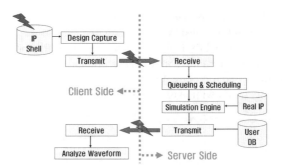

Fig. 2. IP verification flow

request and schedules the simulation. After performing the requested simulations, the results are sent back to the client. User can analyze the simulated results to determine whether the design satisfies user' specification.

2.2 Integrated Design Environment in Client

Even several methods have been proposed to protect core technology in IP from exposure, none is settled in the market. Our proposition can be a new IP trading model, on-line verification through the web. In this model there is a spatial gap between server and client computers. So the client program must operate differently from ordinary EDA tools to make designers feel free even the execution on remote server and try the function of IPs before purchasing.

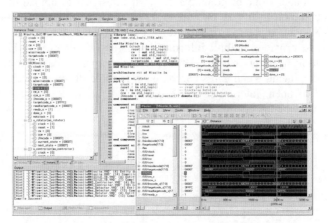

Fig. 3. Main window of the client software

The client is an integrated design environment that enables hardware circuit design – analysis of source codes, debugging of verification results, searching proper IPs, and simulation on the remote server. Fig. 3 shows the screen snap of client workspace. The main window serves designers to type and debug source code in Verilog and VHDL languages. The typed codes are analyzed to chase syntactic bugs.

The neighbor small windows provide graphical design information for analyzing and debugging, such as object tree, instance hierarchy tree, etc.

Some graphic tools are tightly integrated into the client, like building block editor, state diagram editor and stimulus editor. These graphic tools are used for automatic generation of Verilog/VHDL codes. Waveform analyzer is a must-have graphic tool to analyze logical operation with massive simulation vectors.

a. Architecture viewer & SPD   b. Binding data high-lighting

Fig. 4. Graphical debug aids

Some intelligent debugging aid functions are also provided. The HDL parser is embedded to generate several design aids, such as hierarchical logic architecture, signal propagation diagram, and same signal high-lighting in different views. Architecture viewer permits for designer to explore hierarchy so as to tracking connected signals as well as referencing the high-lighting lines in HDL codes (Fig. 4).

Library Uploader - one of graphic aids - helps designer to upload his own design modules as common IP library on remote server. This aids demand standard catalog format of Korean standard organization, TTA (Telecommunication Technology Association) to provides faithful information for outside engineers. Fig. 5.a shows library uploading window, which shows IP category, parsed data, symbol, exmaple, and HTML document. Vendor IPs can be open in three different levels of visibility. Namely, source open, internal structure open, and interface open are open policy options to determine the visibility of uploaded IPs.

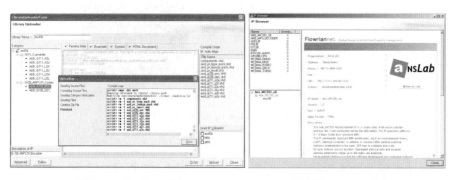

a. Library uploading window   b. IP browsing window

Fig. 5. Library uploading window

2.3   Centralized Services in Server

The server's role can be distinguished in three folds. The first is design analysis service. Upon request for simulation from the client, the server confirms the certificated user ID and accepts service request specification and bundle of design files. Many requests from different locations can be arrived at the same time, the server generates queue list and executes simulation one by one. At this time the real IP libraries are linked to user's design data. As the user's request, the result files can be backed up on user's account in server.

The second is IP library management. IP library are automatically installed on server by the request of client. IP libraries contain catalogs in HTML format, document files, various interface files and source or compiled design files. Each library keeps tag information to its owner. So only the owner has right to modify the library. Deletion right is given only to the administrator for security reason.

The third is server organization management - main or agent servers. Main server must exist as one in minimum and can provides all kinds of services. But the services can be distributed to several agent servers. In this case main server manages user's registration and login/logout. Each agent server can provide different IP libraries and services. If there are different design teams in one organization, different agent server can be allocated to each team.

3   Experiments

International standard audio codec G.771 and G.726 modules are design and installed as 'audio' library on server. Just reusing their interface data, test circuit is design like Fig. 6 – encoding 16 bit PCM data in 2, 3, 4, 5 bit ADPCM, then decoding to original 16 bit PCM. The below 3 different IPs are reused [5-6].

- AUD_G711_A2L : A-Law to Linear PCM Expander
- AUD_G726_ENC : G.726 Compliance ADPCM Encoder
- AUD_G726_DEC : G.726 Compliance ADPCM Decoder

The result fiels are simulation vectors and decoded PCM data. The former is analyzed in the waveform analyzer and the latter is converted to sound to check the quality with audio program.

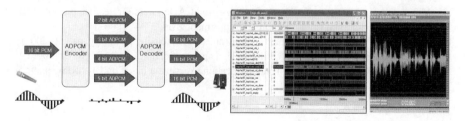

Fig. 6. Test design by reusing audio IPs

The test design and verification shows the possibility of web-based on-line IP reuse design. IP user does not need to make NDA contract to get compiled IP data neither setup CAD tools on user's site. About 20MB Flowrian client software installation is enough to evaluate function and performance of IPs through internet.

4 Conclusions

In this paper, we introduced web-based CAD system, Flowrian, invented for on-line IP verification. Flowrian enables designers on client PC Windows to utilize design resources such as IPs and CAD tools installed on remote servers as if they are on client with full protection of valuable IPs.

Flowrian provides several profits to customers: (i) PC windows is enough for IP reuse design and verification, (ii) no limitation on place and time because of 24 hour running server system, (iii) cheap payment according to pay-per-use rate, (iv) sharing licenses and server computers and consequently maximizing efficiency of design resource use, etc.

For future plan, we are working to develop C/C++ cross-compilation design service as well as platform for embedded system design. Engineering SaaS center is already setup in Hoseo university, which provides Verilog & VHDL design service for education purpose.

Acknowledgements

This paper is results of a study on the "Human Resource Development Center for Economic Region Leading Industry" Project, supported by the Ministry of Education, Science & Technology (MEST) and the National Research Foundation of Korea (NRF).

References

[1] Keating, M., Bricaud, P.: Reuse methodology Manual for System-On-a-Chip Design. In: KAP 2002, 3rd edn. (2002)
[2] Wirthlin, M.J., McMurtrey, B.: Web-based IP evaluation and distribution using applets. IEEE Trans. Computer Aided Design of Integrated Circuits and Systems 22(8), 985–994 (2003)
[3] Malahova, A., Butans, J., Tiwari, A.: A web-based CAD system for gear shaper cutters. In: 7th IEEE International Conf. on Industrial Informatics, pp. 37–42 (2009)
[4] Park, I.: On-line circuit design methodology and system reusing IP library. patent (2004)
[5] Park, I., Moon, D.C.: IP reused Verilog design and verification lab. Brain Korea (2009)
[6] Park, I.: ADPCM HW/SW design and verification experiment based on OpenRISC platform, Brain Korea (2009)

Medical Image Segmentation Using Modified Level-Set Model with Multi-Scale Gradient* Vector Flow

Rajalida Lipikorn[1,*], Krisorn Chunhapongpipat[1], Sirod Sirisup[2], Ratinan Boonklurb[1], and Nagul Cooharojananone[1]

[1] Department of Mathematics, Faculty of Science,
Chulalongkorn University, Bangkok, Thailand
Rajalida.L@chula.ac.th
[2] Large Scale Simulation Research Laboratory,
National Electronics and Computer Center
Pathumthani, Thailand

Abstract. This paper presents a novel method for medical image segmentation that can detect the edges or boundaries of all target objects(defined as high intensity regions) in an image by integrating multi-scale gradient* vector flow(MGVF) into a modified level-set model. The MGVF uses multi-scale images and the gradient of gradient magnitude of a scaled image to generate a vector flow field. This vector flow field is then substituted into a corresponding partial differential equation(PDE) of a modified level-set model that represents the active contour. The proposed method can effectively pull the active contour to attach to the boundary of each target object in an image, especially the boundary of an object that is very close to another object and the boundary of an object with low gradient magnitude. The experiments were tested on 1600 two dimensional CT scan images and the results have shown that the proposed method can accurately detect the boundaries of bones, colons, and residuals inside the colons.

Keywords: active contour, modified level-set model, gradient vector flow, edge flow vector, multi-scale gradient* vector flow.

1 Introduction

Edge detection is a fundamental method in image processing, especially in segmentation. Several methods have been presented in the literatures, such as active contours, gradient vector flow, edge flow vector, and each of them has its own strengths and weaknesses. Snakes or active contours are curves defined within an image domain that can move under the influence of internal forces within the curve itself and external forces derived from the image data [1]. The weaknesses of a snake are that its result depends on the initial contour and its difficulties

---

* Corresponding author.

T.-h. Kim et al. (Eds.): SIP/MulGraB 2010, CCIS 123, pp. 49–57, 2010.
© Springer-Verlag Berlin Heidelberg 2010

in progressing into concave boundary regions[2]-[4]. On the other hand, gradient vector flow fields are dense vectors derived from images by minimizing an energy function by solving a pair of decoupled linear partial differential equations[5],[6]. This method can solve the initialization problem, however, the weakness is that it is guaranteed to work properly if an image contains only one single object. Another method called the edge flow vector utilizes a predictive coding model to identify the direction of change in color and texture at each image location at a given scale, and constructs an edge flow vector field[7]. The drawback of this method is the possibility that the denominator might be equal to zero.

The main objective of this paper is to propose a new method that can detect the boundaries of all objects in an image with only one initial curve by integrating the MGVF into a modified level-set model to generate a vector flow field that can effectively pull the initial curve toward the boundary of each object.

2   Multi-Scale Gradient* Vector Flow

The MGVF is a modified vector flow of the edge flow vector(EFV)[8] whose magnitude and direction are estimated using the intensity gradient magnitude(E_{grad}) and the prediction error(E_{pred}) at a particular location, $r = (x, y)$ along the orientation θ. The equations to compute E_{grad} and E_{pred} can be expressed as

$$E_{grad}(r, \sigma, \theta) = |\nabla_\theta I_\sigma(x, y)| \tag{1}$$

$$E_{pred}(r, \sigma, \theta) = |I_\sigma(x + d\cos\theta, y + d\sin\theta) - I_\sigma(x, y)| \tag{2}$$

where ∇_θ is the gradient in the direction of θ, I_σ is the Guassian blurred(scaled) image, σ is a blurring scale, d(offset parameter) is the distance of the prediction and is usually set to 4σ. The probability function used to search the orientation $\hat\theta$ for the flow direction can be computed from

$$P(r, \sigma, \theta) = \frac{E_{pred}(r, \sigma, \theta)}{E_{pred}(r, \sigma, \theta) + E_{pred}(r, \sigma, \theta + \pi)} \tag{3}$$

Finally, the EFV field is calculated as the vector sum

$$E(r, \sigma) = \int_{\hat\theta - \frac{\pi}{2}}^{\hat\theta + \frac{\pi}{2}} E_{grad}(r, \sigma, \theta) e^{i\theta} d\theta \tag{4}$$

where $i = \sqrt{-1}$ and $\hat\theta$ is a probable flow direction at r[7]. Morever, the obtained result E (for a given scale, σ) from the above equation is a complex number whose magnitude and angle represent the resulting edge energy and the flow direction at a particular location r.

Instead, our proposed MGVF uses the gradient of gradient magnitude of a scaled image to compute a vector flow field, S, as

$$G_\sigma(x, y) = \frac{1}{\sqrt{2\pi}\sigma} e^{\frac{-x^2 - y^2}{2\sigma^2}} \tag{5}$$

where G_σ is the Gaussian distribution with a blurring scale, σ, and the window size of the Gaussian filter is varied with respect to σ, i.e., $(2*(4*\sigma)+1) \times (2*(4*\sigma)+1)$.

$$I_\sigma = G_\sigma * I_{org} \qquad (6)$$

where I_{org} is an original image, I_σ is a scaled image or a blurred image for the purpose of noise suppression at a blurring scale, σ.

$$\|\nabla I_\sigma\| = \sqrt{\left(\frac{\partial I_\sigma}{\partial x}\right)^2 + \left(\frac{\partial I_\sigma}{\partial y}\right)^2} \qquad (7)$$

where $\|\nabla I_\sigma\|$ is gradient magnitude of a scaled image. Then the MGVF can be defined as

$$S = \nabla \|\nabla I_\sigma\| \qquad (8)$$

where S is the gradient of gradient magnitude of a scaled image. This vector is then used as an initial vector in a multi-scale gradient* vector flow algorithm(MGF) to iteratively find the best vector flow field that will be substituted into a modified level-set model.

Algorithm 1. Algorithm for computing multi-scale gradient* vector flow where $\sigma_1{=}0.25, \sigma_2{=}2, \Delta\sigma{=}0.25, \theta{=}\frac{\pi}{4}$, and $\gamma{=}1/15$

Function $MultiScaleMGF(I, \sigma_1, \sigma_2, \Delta\sigma)$
% initialize the vector;
$S = \text{MGF}(I, \sigma_1)$;
Set $\sigma = \sigma_1$;
% find the best vector flow field ;
while $\sigma < \sigma_2$ **do**
    Set $\sigma = \sigma + \Delta\sigma$;
    $T = \text{MGF}(I, \sigma)$;
    M $= \max \|S\|$;
    forall the *pixels in* $S(x, y)$ **do**
        if $\|S(x, y)\| < \gamma \cdot M$ **then**
            | $S(x, y) = T(x, y)$
        end
        else if $Angle(S(x, y), T(x, y)){<}\theta$ **then**
            | $S(x, y) = S(x, y) + T(x, y)$
        end
        else
            | $S(x, y)$ is kept the same.
        end
    end
end
return S

The above multi-scale gradient* vector flow algorithm updates the MGVF, S, by first initializing the blurring scale, σ, to be equal to the lower bound, σ_1, and then iteratively increasing it until it reaches the upper bound, σ_2. While γ is a positive constant that is used to adjust the threshold value, and θ defines the maximum angle between S and T where S and T are two vectors with different blurring scales. At the small blurring scale, the vector field appears only on a thin line along the edges. As the blurring scale gets larger, the edges become thicker thus the vector field also gets thicker as shown in Fig. 1. From our experiments on several kinds of image data, the range to search for the optimal blurring scale, σ, is $[0.25, 2.0]$ and the directions of two vectors with different blurring scales are considered to be the same if the angle between them is less than $\frac{\pi}{4}$. Since we prefer to keep edges that appear at multiple scales and suppress edges that exist at finer scales, we thus check the vector directions from larger and finer scales and if they have the same direction, we sum the vectors up to increase the strength of the edge.

  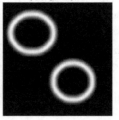

(a) Original image   (b) Gradient magnitude at $\sigma = 0.25$ (c) Gradient magnitude at $\sigma = 2$

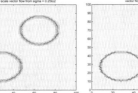 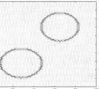 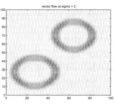

(d) MGVF from $\sigma = 0.25$ to $\sigma = 2$ (e) Vector flow field at $\sigma = 0.25$ (f) Vector flow field at $\sigma = 2$

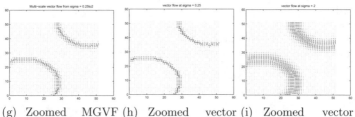

(g) Zoomed MGVF from $\sigma = 0.25$ to $\sigma = 2$ (h) Zoomed vector flow field at $\sigma = 0.25$ (i) Zoomed vector flow field at $\sigma = 2$

Fig. 1. Examples of MGVF at different blurring scales

Algorithm 2. MGF method

Function $S = MGF(I, \sigma)$
set window size to $(2 * (4 * \sigma) + 1) \times (2 * (4 * \sigma) + 1)$
generate Gaussian filter, G_σ, according to Eq.(5) with size as calculated above
calculate I_σ using Eq.(6)
calculate gradient magnitude of I_σ using Eq.(7)
calculate MGVF,S, using Eq.(8)

3   Modified Level-Set Model

The level-set model was first proposed by Osher and Sethian [9] which uses zero-crossing of evolution function value where points inside a contour have positive ϕ values, points outside a contour have negative ϕ values and points on a contour have zero values. An active contour in the level-set model is embedded as a zero set in a function called an embedded function, ϕ where $C = \{x | \phi(x) = 0\}$ The PDE for the level-set model [10] can be written as

$$\phi_t = |\nabla\phi| \left[\alpha(I - \mu) + \beta \left\langle \nabla g, \frac{\nabla\phi}{|\nabla\phi|} \right\rangle + \beta g div \left(\frac{\nabla\phi}{|\nabla\phi|} \right) \right] \tag{9}$$

where I is an image, $g = \frac{1}{1 - c|\nabla I|^2}$ is an edge stopping function with c controlling the slope, α and β are set as weights to balance the terms in Eq. (9), μ is the lower bound of the gray-level of an object. The first term is the propagation term describing a movement, the second term describes the movement in a vector field induced by the gradient of g to attract the curve to the boundary of an object[10]. The third term describes a curvature flow weighted by the gradient feature map g [7].

In this section, we present a modified level-set model which can pull the initial curve toward the edges of all objects including concave objects. The modified level-set model substitutes the second term of the hybrid level-set method [10] with our proposed multi-scale gradient* flow vector and discards the third term from the above equation. The new PDE for our modified level-set can be written as

$$\phi_t = |\nabla\phi| \left[\alpha(I - \mu) + \beta k \left\langle S, \frac{\nabla\phi}{|\nabla\phi|} \right\rangle \right] \tag{10}$$

where I is an image to be segmented, μ is a predefined parameter indicating the lower bound of the gray-level of an object, k is the curvature [9], and S is our MGVF. It can be seen that the first term forces an active contour to attract the regions with gray-levels greater than μ. The global minimum of this term alone is the boundary of a binary image resulting from simple thresholding of an image I with a threshold value μ.

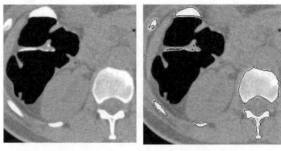

(a) Original image   (b) Segmentation result from the proposed method

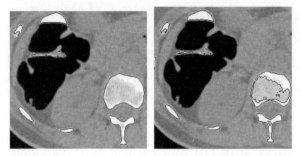

(c) Segmentation result from hybrid level-set method   (d) Segmentation result from thresholding method

Fig. 2. Comparison of edge detection and segmentation results from three different methods

4   Experimental Results

The proposed method was evaluated on a set of 1600 CT images and the results were compared with the results from the same data using the hybrid level-set method[10], and the conventional thresholding method. Let us assume that the target objects to be segmented have high intensity values, we can notice from Fig. 2(b)-(c) that the proposed method and the hybrid level-set method yielded similar results, i.e., they could effectively and acurrately pull the curve to enclose the boundaries of all objects while the thresholding method might not be able to enclose all objects as can be seen from the lower right corner of Fig. 2(d) where the bone is located. The thresholding method resulted in two separated regions instead of one region. Moreover, our proposed method can perform better than the hybrid level-set method for the cases when the target objects are concave, when the objects have low gradient magnitude as shown in Fig. 3(a)-(d), and when the boundaries of objects are very close to each other as shown in Fig. 4(a)-(c) and Fig. 5(a)-(c). Figure 3(a) reveals that our proposed method can automatically select the appropriate value for σ that gives optimal segmentation

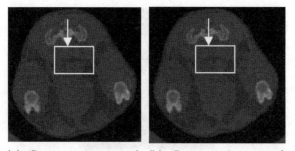

(a) Segmentation result from the proposed method with σ_1=0.25, σ_2=2, and $\Delta\sigma$=0.25

(b) Segmentation result from hybrid level-set method with σ=0.25

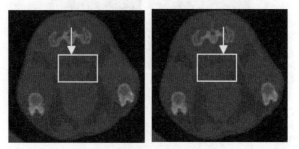

(c) Segmentation result from hybrid level-set method with σ=0.1

(d) Segmentation result from hybrid level-set method with σ=0.5

Fig. 3. Comparison between hybrid level-set method and the proposed method where $\mu = 28$, $\theta=\frac{\pi}{4}$, and γ=1/15

result while Fig. 3(b)-(d) reveal that it is quite difficult to manually select the appropriate σ. Figures 4 and 5 show that the proposed method can accurately detect two objects that are close to each other while the hybrid level-set method can detect only one object.

5   Disscussion

We propose a new method for medical image segmentation that adapts the concept of multi-scale edge flow vector and hybrid level-set model. Instead of using the edge flow vector to attract the curve to the edges of each object, we propose to use the multi-scale gradient* vector flow. The proposed method can solve the problem of having zero denominator in EFV, can detect the edges of concave objects, and can automatically select the optimal blurring scale. The multiple scaling is performed by blurring an image with Gaussian filter of different sizes and computing the MGVF, iteratively. The final MGVF is the summation of

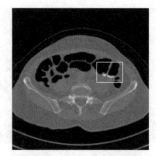

(a) Original image

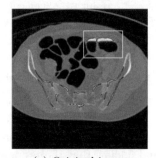

(b) Segmentation re- (c)   Segmentation
sult from the pro- result from hybrid
posed method   level-set method

Fig. 4. Comparison between hybrid level-set method and the proposed method

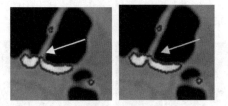

(a) Original image

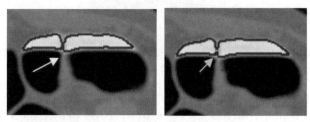

(b) Segmentation result from (c)   Segmentation   result
the proposed method   from   hybrid   level-set
                              method

Fig. 5. Comparison between hybrid level-set method and the proposed method

multi-scale MGVFs with the same direction. This MGVF is applied to the modified level-set model which is used to detect the edges of all objects in an image. The main advantages of the proposed method are that it can detect and segment the boundaries of all objects effectively and accurately regardless of their size, orientation, intensity, and shape.

Acknowledgments. This research has been partially supported by grant under the A1B1 Scholarships from Faculty of Science, Chulalongkorn University, Thailand. The Abdominal CT-scan image data set is supported by Chulalongkorn Hospital. We would also like to give a special thank for the invaluable advices from Associate Professor Laddawan Vajragupta, M.D. and Bandit Chaopathomkul, M.D. from Department of Radiology, Faculty of Medicine, Chulalongkorn University and Associate Professor Wallapak Tavanapong from Department of Computer Science, Iowa State University.

References

1. Kass, M., Witkin, A., Terzopoulos, D.: Snakes: Active ContourModels. International Jouranl of Computer Vision 1(4), 321–331 (1987)
2. Davatzikos, C., Prince, J.L.: An Active Contour Model for Mapping the Cortex. IEEE Transactions on Medical Imaging 14(1), 65–80 (1995)
3. Chan, T.F., Vese, L.A.: Active contours without edges. IEEE Transactions on Image Processing 10(2), 266–277 (2001)
4. Caselles, V., Kimmel, R., Sapiro, G.: Geodesic Active Contours. International Journal of Computer Vision 22(1), 61–79 (1997)
5. Xu, C., Prince, L.J.: Gradient Vector Flow: A New External Force for Snakes. In: IEEE Conference on Computer Vision and Pattern Recognition, pp. 66–71. IEEE Computer Society, San Juan (1997)
6. Xu, C., Prince, L.J.: Snakes, Shapes, and Gradient Vector Flow. IEEE Transactions on Image Processing 7(3) (1998)
7. Ghosh, P., Bertelli, L., Sumengen, B., Manjunath, B.S.: A Nonconservative Flow Field for Robust Variational Image Segmentation. IEEE Transactions on Image Processing 19(2), 478–488 (2010)
8. Ma, W.Y., Manjunath, B.S.: Edgeflow: A Technique for Boundary Detection and Image Segmentation. IEEE Transactions on Image Processing, 1375–1388 (2000)
9. Osher, S., Sethian, J.A.: Fronts Propagating with Curvature-Dependent Speed: Algorithms Based on Hamilton-Jacobi Formuations. Journal of Computational Physics 79(1), 12–49 (1988)
10. Zhang, Y., Matuszewski, B.J., Shark, L., Moore, C.J.: Medical Image Segmentation Using New Hybrid Level-Set Method. In: Fifth International Conference BioMedical Visualization: Information Visualization in Medical and Biomedical Informatice, pp. 71–76. IEEE Computer Society, Los Alamitos (2008)

Generalized Hypersphere d-Metric in Rough Measures and Image Analysis

Dariusz Małyszko and Jarosław Stepaniuk

Department of Computer Science
Bialystok University of Technology
Wiejska 45A, 15-351 Bialystok, Poland
{d.malyszko,j.stepaniuk}@pb.edu.pl

Abstract. Dynamic advances in data acquisition methods accompanied by widespread development of information systems that depend upon them, require more sophisticated data analysis tools. In the last decades many innovative data analysis approaches have been devised in order to make possible deeper insight into data structure. **R**ough **E**xtended **F**ramework - REF - presents recently devised algorithmic approach to data analysis based upon inspection of the data object assignment to clusters. In the paper, in the Rough Extended Framework, a new family of generalized hypersphere d-metrics has been presented. The introduced hypersphere d-metrics have been applied in the rough clustering (entropy) measures setting. The combined analysis based upon hyper-sphered d-metric type and rough (entropy) measures is targeted as a robust medium for development of image descriptors.

1   Introduction

Rough Extended (Entropy) Framework presents extensively developed method of data analysis. The introduced solutions are targeted into incorporation into multimedia standards as a general family of image descriptors. The introduced REF framework combined with standard and generalized hypersphere d-metrics by means of inclusion into rough measures seems to be robust technique in construction of image descriptors. The resultant image descriptors based upon these innovative metrics and rough measures possibly are adequate into incorporation into multimedia standards as a general family of image descriptors. Rough Extended Framework in image segmentation has been primarily introduced in [7] in the domains of image thresholding routines. In [7], rough set notions, such as lower and upper approximations have been put into image thresholding. Further, rough entropy measures have been employed in image data clustering setting in [3], [5] described as **R**ough **E**ntropy **C**lustering **A**lgorithm.

By means of selecting the proper rough entropy measures, different properties of the analyzed are possible to capture. In this context, the notion and definition of the generalized hypersphere d-metrics is a valuable tool during specialized analysis of image properties. The introduced generalized hypersphere metrics are seamlessly incorporated into Rough Entropy Framework giving the measure

T.-h. Kim et al. (Eds.): SIP/MulGraB 2010, CCIS 123, pp. 58–67, 2010.
© Springer-Verlag Berlin Heidelberg 2010

of distance between image points. Distance measure is required during data analysis based upon rough measures. The main paper contribution consists in

1. introducing generalized hypersphere d-metric,
2. incorporating generalized hypersphere d-metric into RECA measures,
3. giving some illustration of the proposed d-metrics.

This paper is structured in the following way. Section 2 presents standard hypersphere metrics and introduces and describes innovative generalized hypersphere metric. In Section 3 RECA concepts have been described. Presentational and experimental materials have been included and discussed in Section 4 followed by concluding remarks in Section 5.

2   Standard and Generalized Hypersphere Metrics

In the subsequent material, the standard scalar product between two points x and y in R^d is denoted as

$$S(x * y) = \Sigma_{i=1}^{d} (x_i - y_i)^2 \tag{1}$$

the p-norm between two d-dimensional points x and y is defined as

$$||x - y||_p = \left(\Sigma_{i=1}^{d} (x_i - y_i)^p \right) \tag{2}$$

For point x the coordinates are denoted as $x_1, \ldots, x_d$. In the same manner, for point x_i the coordinates are denoted as $x_{i1}, \ldots, x_{id}$

2.1   Voronoi d-Metric

Let $\mathcal{P} = \{p_1, \ldots, p_n\}$ denotes a set of points of R^d. For each p_i it is possible to assign its Voronoi region $V(p_i)$

$$V(p_i) = \{x \in R^d : ||x - p_i|| \leq ||x - p_j||, 1 \leq j \leq n\} \tag{3}$$

In this way, this distance type metric is most often used in many theoretical and practical domains, referred to as Voronoi d-metric (diagram metric).

2.2   Power d-Metric

The power of a point x to a hypersphere σ of center c and radius r is defined as the real number

$$\sigma(x) = (x - c)^2 - r^2 \tag{4}$$

In this way, an extension of the standard approach to data analysis emerges by means of assignment to the clusters additional weighing properties. In typical applications, during clustering process, cluster centers are represented only by cluster centers in R^s metric space. In the clustering process, cluster centers are extended into hypershperes. Now, incorporation of the notion of power of a point to the hypersphere gives way to more detailed data analysis by means of appropriate cluster weighing.

Let $\mathcal{S} = \{\sigma_1, \ldots, \sigma_n\}$ represents a set of hyperspheres of R^d. In the subsequent material, c_i denotes the center of σ_i, its radius is referred to as r_i. In this way, the power function to hypersphere σ_i is defined as

$$\sigma_i(x) = (x - c_i)^2 - r_i^2 \tag{5}$$

For each hypersphere σ_i, it is possible to define the region $L(\sigma_i)$ consisting of the points of R^d whose power to σ_i do not exceeds than their power to the other hyperspheres of $\mathcal{S}$

$$L(\sigma_i) = \{x \in R^d : \sigma_i(x) \le \sigma_j(x), 1 \le j \le n\} \tag{6}$$

The radical hyperplane of σ_i and σ_j, denoted as π_{ij} represents the set of points that have equal power to two hyperspheres σ_i and σ_j. The region $L(\sigma_i)$ contains the intersection of all halfspaces π_{ij}^i.

The power diagram of $\mathcal{S}$, denoted as $PD(\mathcal{S})$ represents all power regions and their faces.

2.3   Mobius d-Metric

Let $\otimes = \{\omega_1, \ldots, \omega_n\}$ denotes a set of so-called Mobius sites of R^d with ω_i meaning triple (p_i, λ_i, μ_i) that is formed of a point p_i of R^d, and two real numbers λ_i and μ_i. For a point $x \in R^d$, the distance $\delta_i(x)$ from x to the Mobius site ω_i is defined as

$$\delta_i(x) = \lambda_i(x - p_i)^2 - \mu_i \tag{7}$$

The Mobius region of the Mobius site $\omega_i, i = 1, \ldots, n$ is

$$M(\omega_i) = \{x \in R^d : \omega_i(x) \le \omega_j(x), 1 \le j \le n\} \tag{8}$$

Mobius (diagram) metric represents more advanced extension of a standard cluster definition or cluster definition based on power distance metric. Now, clusters are defined by means of hyperspheres and additionally scaling factor λ.

2.4   Additively Weighted d-Metric

Given a finite set of weighted points $\{\sigma_0, \sigma_1, \ldots, \sigma_n\}$ with $\sigma_i = (p_i, r_i), p_i \in R$, the distance from x to σ_i is defined as follows

$$\delta_i(x) = ||x - c_i|| - r_i \tag{9}$$

This distance is equivalently referred to as additively weighted distance from x to the weighted point δ_i

The Apollonius region $A(\sigma_i)$ of σ_i represents

$$A(\sigma_i) = \{x \in R^d : \sigma_i(x) \le \sigma_j(x), 1 \le j \le n\} \tag{10}$$

2.5   Anisotropic d-Metric

Let consider a finite set of anisotropic sites $\mathcal{S} = \{s_1, \ldots, s_n\}$. Each site $s_i, i = 1, \ldots, n$ is denoted as (p_i, M, π_i), where p_i represents point $p_i \in R^d$, M is a $d \times d$ symmetric positive definite matrix, and π_i denotes a scalar weight.

The anisotropic distance $\delta_i(x)$ of a point $x \in R^d$ to site s_i is defined by

$$\delta_i(x) = (x - p_i)^T M_i (x - p_i) - \pi_i \tag{11}$$

The anisotropic region of site s is defined as follows

$$AV(s_i) = \{x \in R^d : \sigma_i(x) \leq \sigma_j(x), 1 \leq j \leq n\} \tag{12}$$

The anisotropic diagram is the minimalization diagram of the functions δ_i.

2.6   Generalized Hypersphere d-Metric

In the paper, the following definition of the generalized hypersphere metric - or cluster in standard setting - is defined

$$\sigma(x) = \lambda(x - c)^p - \mu \tag{13}$$

Let $\oplus = \{\phi_1, \ldots, \phi_n\}$ denotes a set of so-called generalized (Mobius) sites of R^d with ϕ_i meaning

$$C_i^h = \phi_i = (p_i, \lambda_i, \mu_i, r, s, M_i) \tag{14}$$

that is formed of a point p_i of R^d, and four real numbers λ_i, μ_i, r and s. Additionally, M represents matrix. For a point $x \in R^d$, the distance $\delta_i(x, \phi_i)$ from x to the generalized (Mobius) site ϕ_i is defined as

$$\delta_i(x, \phi_i) = \delta_i(x, C_i^h) = \lambda_i \left[(x - p_i)^r \right]^T M_i (x - p_i)^s - \mu_i \tag{15}$$

For $r = s = 1.0$ the formulae is

$$\delta_i(x, \phi_i) = \delta_i(x, C_i^h) = \lambda_i (x - p_i)^T M_i (x - p_i) - \mu_i \tag{16}$$

$$M(\omega_i) = \{x \in R^d : \phi_i(x) \leq \phi_j(x), 1 \leq j \leq n\} \tag{17}$$

3   Rough Entropy Measures

3.1   General REF and RECA Concepts

Rough measures considered as an image descriptor gives possible robust theoretical background for development of high-quality clustering schemes. These clustering algorithms incorporate rough set theory, fuzzy set theory and entropy measure. Three basic rough properties that are applied in clustering scheme include: selection of the threshold metrics (crisp, fuzzy, probabilistic, fuzzified probabilistic) - tm , the threshold type (thresholded or difference based) - tt - and the measure for lower and the upper approximations - ma - (crisp, fuzzy, probabilistic). Data objects are assigned to lower and upper approximation on the base of the following threshold type definition: assignment performed on the basis of the distance to cluster centers within given threshold value (threshold based) or assignment performed on the basis of the difference of distances to the cluster centers within given threshold value (difference based).

3.2   Standard Metrics for RECA Measures

Introduced rough measures are primarily based on the similar pattern - it means data points closest to the given cluster center relative to the selected threshold metrics (crisp, fuzzy, probabilistic, fuzzified probabilistic) are assigned to this cluster lower and upper approximations. The upper approximations are calculated in the specific, dependant upon threshold type and measure way presented in the subsequent paragraphs. Standard crisp distance most often applied in many working software data analysis systems depends upon Euclidean distance or Minkowsky distance, calculated as follows

$$d_{cr}(x_i, C_m) = \left(\sqrt{\Sigma_{j=1}^{d}(x_{ij} - C_{mj})^p} \right)^{\frac{1}{p}} \tag{18}$$

Fuzzy membership value $d_{fz}(x_i, C_m) = \mu_{fz}C_m(x_i) \in [0,1]$ for the data point $x_i \in U$ in cluster C_m is given as

$$d_{fz}(x_i, C_m) = \mu_{C_m}(x_i) = \frac{d(x_i, C_m)^{-2/(\mu-1)}}{\sum_{j=1}^{k} d(x_i, C_j)^{-2/(\mu-1)}} \tag{19}$$

where a real number $\mu > 1$ represents fuzzifier value and $d(x_i, C_m)$ denotes distance between data object x_i and cluster (center) C_m.

Probability distributions in RECA measures are required during measure calculations of probabilistic distance between data objects and cluster centers. Gauss distribution has been selected as probabilistic distance metric for data point $x_i \in U$ to cluster center C_m calculated as follows

$$d_{pr}(x_i, C_m) = (2\pi)^{-d/2}|\Sigma_m|^{-1/2}exp\left(-\frac{1}{2}(x_i - \mu_m)^T \Sigma_m^{-1}(x_i - \mu_m)\right) \tag{20}$$

where $|\Sigma_m|$ is the determinant of the covariance matrix Σ_m and the inverse covariance matrix for the C_m cluster is denoted as Σ_m^{-1}. Data dimensionality is denoted as d. In this way, for standard color RGB images $d = 3$, for gray scale images $d = 1$. Mean value for Gauss distribution of the cluster C_m has been denoted as μ_m.

Fuzzified membership value of probabilistic distance $\mu_{C_m}(x_i) \in [0,1]$ for the data point $x_i \in U$ in cluster C_m is given as

$$d_{fp}(x_i, C_m) = \mu_{C_m}(x_i) = \frac{d_{pr}(x_i, C_m)^{-2/(\mu-1)}}{\sum_{j=1}^{k} d_{pr}(x_i, C_j)^{-2/(\mu-1)}} \tag{21}$$

3.3   Hypersphered Metrics for RECA Measures

The introduced rough measures are primarily based on the similar pattern as in case of crisp and fuzzy RECA measures. The probability rough measures require selecting adequate probability measure. Data points closest to the given cluster center relative to the selected threshold metrics (crisp, fuzzy, probabilistic) are assigned to its lower and upper approximation. The upper approximations are calculated in the specific, dependant upon threshold type and measure way presented in the subsequent paragraphs.

Standard crisp distance most often applied in many working software data analysis systems depends upon Euclidean distance or Minkowsky distance, calculated as follows

$$d_{crh} = h_{cr}(x_i, C_m^h) = \delta_m(x, \phi_m) = \delta_m(x, C_m^h) \qquad (22)$$

Fuzzy membership value $d_{fzh}(x_i, C_m) = \mu_{C_m}(x_i) \in [0,1]$ for the data point $x_i \in U$ in cluster C_m is given as

$$d_{fzh} = h_{fz}(x_i, C_m^h) = \frac{\delta_m(x_i, C_m^h))^{-2/(\mu-1)}}{\sum_{j=1}^{k} \delta_j(x_i, C_j^h))^{-2/(\mu-1)}} \qquad (23)$$

where a real number $\mu > 1$ represents fuzzifier value and $\delta_m(x_i, C_m^h)$ denotes distance between data object x_i and hypersphered cluster (center) C_m^h.

Probability distributions in RECA measures are required during measure calculations of probabilistic distance between data objects and cluster centers. Gauss distribution has been selected as probabilistic distance metric for data point $x_i \in U$ to cluster center C_m^h calculated as follows

$$d_{prh}(x_i, C_m^h) = (2\pi)^{-d/2} |\Sigma_m|^{-1/2} exp\left(-\frac{1}{2}(\delta_m(x_i, C_m^h))^T \Sigma_m^{-1}(\delta_l(x_i, C_m^h)\right) \qquad (24)$$

In fact, in this distance, cluster center C_m^h represents cluster mean value, as calculated during cluster mean, and standard deviation calculation.

Where $|\Sigma_m|$ is the determinant of the covariance matrix Σ_m and the inverse covariance matrix for the C_m cluster is denoted as Σ_m^{-1}. Data dimensionality is denoted as d. In this way, for standard color RGB images $d = 3$, for gray scale images $d = 1$. Mean value for Gauss distribution of the cluster C_m has been denoted as μ_m.

Fuzzified membership value of probabilistic distance $\mu(pr)_{C_l^h}(x_i) \in [0,1]$ for the data point $x_i \in U$ in cluster C_m is given as

$$d_{fph}(x_i, C_m) = \frac{d_{prh}(x_i, C_m)^{-2/(\mu-1)}}{\sum_{j=1}^{k} d_{pr-h}(x_i, C_j)^{-2/(\mu-1)}} \qquad (25)$$

The distance d_{prh} is calculated in the same way as in Equ. 24.

4   Experimental Setup and Results

Image data - in the subsequent material, the image 27059 from Berkeley image database [6] has been selected for experiments. The original color RGB image as given in Figure 1 (a) have been preprocessed and 2D data for bands G and B have been obtained. In Figure 1 (a) image in 0GB bands has been presented. **Cluster centers** - for the presentation purposes, 7 cluster centers have been created as shown in Figure 1 (c) and displayed in blue numbered circles. Data points assignment and border of data points that are the closest to the given clusters depends heavily on the metric. Additionally, in Figure 1 (c) cluster borders and cluster assignment (it means data points belonging to the nearest cluster center) have been given in case of crisp or fuzzy metric.

In Figures 2, 3, 4, 6, 7 boundary regions and equidistances for different RECA measures and metrics (standard and hyperspered d-metrics) have been presented.

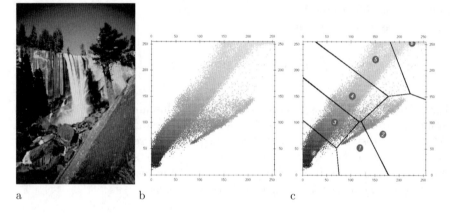

a   b   c

Fig. 1. Berkeley dataset image: (a) 27059 - 0GB bands, (b) image 27059, 0GB attributes, (c) image 27059, 0GB attributes with 7 cluster centers

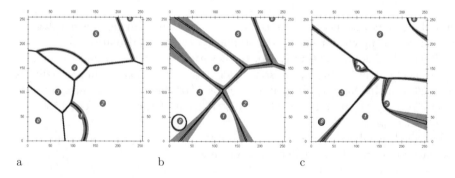

a   b   c

Fig. 2. Berkeley dataset image 27059 - 0GB bands: (a), (b), (c) CDRECA - $\epsilon = 10^{-14}$ with three different λ parameters during hypershpere d-metric calulations

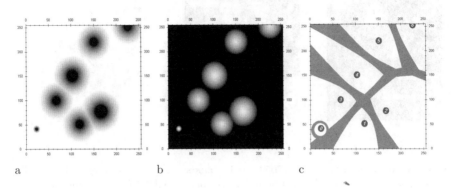

Fig. 3. Berkeley dataset image 27059 - 0GB bands: CDRECA - $\epsilon = 10^{-14}$ with low value of λ parameter for the first hypercluster hypershpere d-metric calulations

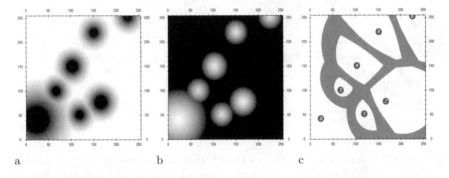

Fig. 4. Berkeley dataset image 27059 - 0GB bands: CDRECA - $\epsilon = 10^{-14}$ with high value of λ parameter for the first hypercluster hypershpere d-metric calulations

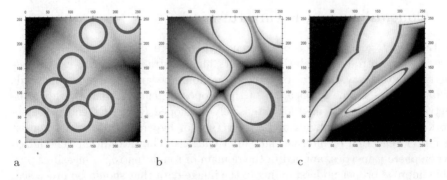

Fig. 5. Berkeley dataset image 27059 - 0GB bands: equidistances displayed in blue color for (a) crisp, (b) fuzzy and (c) probabilistic metrics

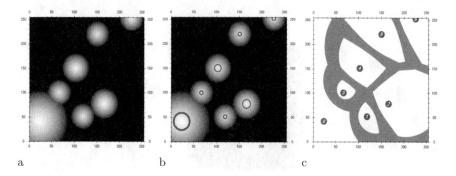

a   b   c

Fig. 6. Berkeley dataset image 27059 - 0GB bands: (a) distance image, (b) equidistances displayed in blue color, (c) boundary regions for CDRECA for d-metric

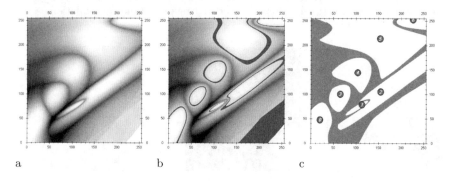

a   b   c

Fig. 7. Berkeley dataset image 27059 - 0GB bands: (a) distance image, (b) equidistances displayed in blue color, (c) boundary regions for PDRECA in standard metric

5   Conclusions and Future Research

In the study, generalization of different hypershpere d-metrics has been presented. The introduced solution presents high abstraction of the existing d-metrics. In this way, the area of application should be suitable in many methods of data analysis. The application of the proposed generalized hypershpere d-metrics has been performed for extensively developed Rough Extended Framework. The main advantages of the new type of hypershere d-metrics consist in better reflection of internal data interrelations. The introduced generalized hypersphere d-metrics, applied in the domain of rough (entropy) measures possible improve proper understanding of the image data that should be processed. By means of selecting the proper rough (entropy) measures and hypersphere distances, different properties of the analyzed are possible to capture.

On the other hand, the combination of crisp, fuzzy, probabilistic and fuzzified probabilistic rough measures together with application of different distance hypersphere d-metrics seems to be suitable for image segmentation and analysis.

In this context, further research directed into extension and incorporation of the introduced hypershpere *d*-metrics and rough (entropy) measures in the area of image descriptors for image analysis multimedia applications seems to be a reasonable task.

Acknowledgments

The research is supported by the grant N516 0692 35 from the Ministry of Science and Higher Education of the Republic of Poland.

References

1. Boissonnat, J.D., Teillaud, M.: Effective Computational Geometry for Curves and Surfaces (Mathematics and Visualization). Springer, New York (2006)
2. Malyszko, D., Stepaniuk, J.: Granular Multilevel Rough Entropy Thresholding in 2D Domain. In: 16th International Conference Intelligent Information Systems, IIS 2008, Zakopane, Poland, June 16-18, pp. 151–160 (2008)
3. Malyszko, D., Stepaniuk, J.: Standard and Fuzzy Rough Entropy Clustering Algorithms in Image Segmentation. In: Chan, C.-C., Grzymala-Busse, J.W., Ziarko, W.P. (eds.) RSCTC 2008. LNCS (LNAI), vol. 5306, pp. 409–418. Springer, Heidelberg (2008)
4. Malyszko, D., Stepaniuk, J.: Adaptive multilevel rough entropy evolutionary thresholding. Information Sciences 180(7), 1138–1158 (2010)
5. Malyszko, D., Stepaniuk, J.: Adaptive Rough Entropy Clustering Algorithms in Image Segmentation. Fundamenta Informaticae 98(2-3), 199–231 (2010)
6. Martin, D., Fowlkes, C., Tal, D., Malik, J.: A database of human segmented natural images and its application to evaluating segmentation algorithms and measuring ecological statistics. In: ICCV 2001, vol. (2), pp. 416–423. IEEE Computer Society, Los Alamitos (2001)
7. Pal, S.K., Shankar, B.U., Mitra, P.: Granular computing, rough entropy and object extraction. Pattern Recognition Letters 26(16), 2509–2517 (2005)
8. Skowron, A., Stepaniuk, J.: Tolerance Approximation Spaces. Fundamenta Informaticae 27(2-3), 245–253 (1996)
9. Stepaniuk, J.: Rough–Granular Computing in Knowledge Discovery and Data Mining. Springer, Heidelberg (2008)

Semantic Supervised Clustering Approach to Classify Land Cover in Remotely Sensed Images

Miguel Torres, Marco Moreno, Rolando Menchaca-Mendez,
Rolando Quintero, and Giovanni Guzman

Intelligent Processing of Geospatial Information Laboratory,
Computer Research Center, National Polytechnic Institute, Mexico City, Mexico
{mtorres,marcomoreno,rmen,quintero,jguzmanl}@cic.ipn.mx
http://piig-lab.cic.ipn.mx/geolab/

Abstract. GIS applications involve applying classification algorithms to remotely sensed images to determine information about a specific region on the
Earth's surface. These images are very useful sources of geographical data
commonly used to classify land cover, analyze crop conditions, assess mineral
and petroleum deposits and quantify urban growth. In this paper, we propose a
semantic supervised clustering approach to classify multispectral information in
satellite images. We use the maximum likelihood method to generate the clustering. In addition, we complement the analysis applying spatial semantics to
determine the training sites and refine the classification. The approach considers
the a priori knowledge of the remotely sensed images to define the classes related to the geographic environment. In this case, the properties and relations
that involve the geo-image define the spatial semantics; these features are used
to determine the training data sites. The method attempts to improve the supervised clustering, adding the intrinsic semantics of multispectral satellite images
in order to establish the classes that involve the analysis with more precision.

1 Introduction

The integration of remote sensing and geographic information systems (GIS) in environmental applications has become increasingly common in recent years. Remotely
sensed images are an important data source for environmental GIS-applications, and
conversely, GIS capabilities are being used to improve image analysis approaches. In
fact, when image processing and GIS facilities are combined in an integrated vector
data system, they can be used to assist in image classification and raster image statistics. In addition, vectors are used as criteria for spatial queries and analysis [1].

Moreover, remotely sensed images are very useful sources of geographical data
commonly used to classify land cover, analyze crop conditions, assess mineral and
petroleum deposits and quantify urban growth [2].

A common GIS application involves applying classification algorithms to remotely
sensed images to determine information about a specific region on the Earth's surface.
For instance, classification of image pixels is an important procedure used to segment
a satellite image, according to ground cover type (e.g., water, forest, desert, urban,

T.-h. Kim et al. (Eds.): SIP/MulGraB 2010, CCIS 123, pp. 68–77, 2010.

cropland, rangeland, etc.). Classification is also a common task applied to other forms of scientific imagery [3].

This paper proposes a semantic supervised clustering algorithm applied to Landsat TM images in order to classify the land cover, according to the *a priori* knowledge. The semantics of remotely sensed images is used to detect adequate *training sites* considering the geometrical and topological properties and relations. This approach attempts to overcome some of the limitations associated with computational issues from previous supervised clustering methods, such as high number of classes, lack of processing of quantitative values, generation of classes that may not be useful for users, lack of post-refinement classes treatment before they can be used in the classification and so on. The approach is based on the *spatial semantics*[1], which is used to determine the behavior of a certain geographic environment by means of spatial properties and relations [4].

The rest of the paper is organized as follows. Section 2 presents related work that describes supervised clustering methods applied to different fields. In section 3 we describe the semantic supervised clustering approach to classify land cover in remotely sensed images. Section 4 shows the results obtained by applying the approach to Landsat TM images. Our conclusions and future work are outlined in section 5.

2 Related Work

Several works have been published with classical methods, as well as approaches using supervised clustering to process remotely sensed images (e.g., [5] and [6]).

In [7] a semi-supervised classifier based on the combination of the expectation-maximization algorithm for Gaussian mixture models and the mean map kernel is described. It improves the classification accuracy in situations where the available labeled information does not properly describe the classes in the test image. In [8] Markov random fields are used to model the distribution of points in the 2-dimensional geometrical layout of the image and in the spectral grid. The mixture model of noise and appropriate Gibbs densities yield the same approach and the same efficient iterated conditional modes (ICM) for filtering and classifying.

In [9] a method for segmenting multiple feature images is presented. It builds on two well-established methodologies: Bayes' decision rule and a multi-resolution data representation. The method is aimed at image analysis applications requiring routine processing. The goal is to improve the classification reliability of conventional statistical-based pixel classifiers by taking into account the spatial contextual information conveyed in an image via multi-resolution structure.

In [10] three different statistical approaches to automatic land cover classification from satellite images are described: maximum likelihood classification, support vector machines and iterated conditional modes. These three approaches exploit spatial context by using a Markov random field. These methods are applied to Landsat 5 Thematic Mapper (TM) data, after using an unsupervised clustering method to identify subclasses. In [11] automated approaches for identifying different types of

---

[1] To define *spatial semantics* [4], we use some features that involve the remote sensing imagery. Our definition is based on providing a description of the geo-image. This description is composed of relations and properties that conceptually define the behavior of the raster.

glaciated landscapes using digitally processed elevation data were evaluated. The tests were done over geomorphic measures derived from digital elevation models to differentiate glaciated landscapes using maximum likelihood classification and artificial neural networks. In [12] the benefits and the pitfalls of including spatial information in clustering techniques are studied. Spatial information is taken into consideration in the initialization of clustering parameters, during cluster iterations by adjusting the similarity measure or at a post-processing step. Homogeneous regions in the image are better recognized and the amount of noise is reduced by this method, because multivariate image data provide detailed information in variable and image space.

On the other hand, the combination of neural and statistical algorithms for supervised classification of remote sensing images is proposed in [13] as a method to obtain high accuracy values after much shorter design phases and to improve the accuracy–rejection tradeoff over those allowed by algorithms that are purely statistical or purely based on neural networks. In [14] a supervised algorithm for vessel segmentation in red-free images of the human retina is proposed. The algorithm is modular and composed of two fundamental blocks. The optimal values of the two algorithm parameters are found by maximizing measures of performance, which can be used to evaluate from a quantitative point of view the results provided by the algorithm. In [15] an approach for incorporating a spatial weighting into a supervised classifier for remote sensing applications is described. The classifier modifies the feature-space distance-based metric with a spatial weighting. This is facilitated by the use of a non-parametric (k-nearest neighbor) classifier in which the spatial location of each pixel in the training data set is known and available for analysis.

All the previous approaches are strictly based on numerical analysis. This is true for both feature selection and the classification itself. On the other hand the algorithm proposed here takes advantage of the intrinsic semantic information involved in any geo-image to enhance a supervised clustering algorithm.

3 Semantic Supervised Clustering Approach

3.1 Supervised Clustering Method

Supervised clustering is a procedure used for quantitative analysis of remotely sensed image data [16]. In this work, we used the classical *Maximum Likelihood Classification* method. In many cases, supervised clustering is defined through training sites, which are determined by pixels according to *a priori* semi-automatic selection. When the clustering process is finished, a set of M classes is obtained. We have considered six steps to generate a supervised clustering framework: (I) Determine the number and type of classes to be used in the analysis. (II) Choose training regions (sites) for each of the classes according to the intrinsic semantic information. (III) Use these regions to identify the spectral characteristics of each specific class. (IV) Use these training regions to determine the parameters of the supervised clustering. (V) Classify all the pixels from the multispectral satellite image, assigning them to one of the classes defined by the training regions. (VI) Summarize the results of the supervised clustering.

An important assumption in supervised clustering that is usually adopted in remote sensing is that each spectral class is described by a probability distribution in a multispectral space. A multidimensional normal distribution is defined by a function of a vector location in this space, denoted by Eqn 1.

$$p(x) = \frac{1}{(2\pi)^{N/2} |\Sigma|^{1/2}} \exp\left\{ -\frac{1}{2} (x - \mu)^t \Sigma^{-1} (x - \mu) \right\},$$ (1)

where: x is a vector location in the N-dimensional pixel space, N is the number of components of vector x, μ is the mean position of the spectral class, and Σ is the covariance matrix of the distribution.

The multidimensional normal distribution is completely specified by its *mean vector* and its *covariance matrix* (Σ). In fact, if the mean vectors and covariance matrices are known for each spectral class, it is possible to compute the set of probabilities that describe the relative *likelihoods* of a pattern at a particular location belonging to each of these classes.

From these likelihoods, we can compute the maximum probability to belong to a class and the element with this maximum value is considered a member of the class. Since μ and Σ are known for each spectral class in an image, also each pixel is examined and labeled according to the probabilities computed for the particular location of each pixel. Before the classification, μ and Σ are estimated for each class from a representative set of pixels, commonly called *training sites*.

3.2 Semantic-Based Detection of Training Sites Algorithm

The main issue of this research is to semantically obtain the *training sites* (classes) to determine the areas that are relevant to the classification. We define these areas by means of the spatial semantics taking into account the a priori knowledge of the specific area. For this paper, the number and type of classes were defined by a set of experts in land cover. The semantic clustering algorithm is described as follows:
[Step 1].   Let O be the sampling set of objects. See Eqn. 2.

$$O = \left\{ o_i \mid o_i \in Z^3 \right\},$$ (2)

Each vector o_i is a 3-tuple, which contains the RGB values[2] at the position (x, y) of a randomly selected object in the geo-image. To perform this random selection we define the function *seeds(i)* that returns the set of reflectivity values related with the relevant properties of the geo-image. According to our experiments, the number of *seeds(i)* that provides good results is defined by Eqn. 3.

$$k_1(m \cdot n) \leq i \leq k_2(m \cdot n),$$ (3)

where: m is the number of rows in the geo-image, n is the number of columns in the geo-image, $k_1 = 0.01$ is a constant used to compute the lower bound of the proposed value of seeds for each class (ω) and $k_2 = 0.03$ is a constant used to compute the upper bound of the proposed value of seeds for each class (ω).

---

[2] In other types of images, we can use other spectral components.

[Step 2]. Transform each vector o_i from the spatial domain Z^3 to the semantic domain S by means of Eqn. 4. This function maps the numerical values to a set of labels that represents concepts. These concepts are associated with a range of values defined by the spectral signatures of each element in a geo-image. A table of spectral signatures can be found in [17].

$$S_O = \{s_i \mid s(o_i) \neg o_i \in O\},\tag{4}$$

$s(o_i)$ is the semantic characteristic vector, which is obtained by applying Eqn 5.

$$s = <s_1, s_2, s_3, s_4, s_5, s_6>,\tag{5}$$

where: s_1, s_2, s_3 are the mean of the original seed and their 8-neighbors[3] in the spatial domain Z^3, according to the RGB spectral bands respectively.

Additionally s_4, s_5, s_6 are the values of the standard deviation of the original seed and their 8-neighbors in the same domain. Standard deviation represents the variability of the reflectance in every selected spectral band.

[Step 3]. Obtain the dissimilitude coefficient (DSC) $d(s_i, s_j)$ for s_i, s_j that belongs to S_o, which is defined by Eqn. 6.

$$d(s_x, s_y) = \sqrt{\left(s_{x1}^2 - s_{y1}^2\right) + \left(s_{x2}^2 - s_{y2}^2\right) + \dots + \left(s_{x6}^2 - s_{y6}^2\right)}\tag{6}$$

[Step 4]. Let $D = \{d(s_i, s_j) \mid s_i, s_j \in S_o \forall i \neq j\}$ be the set of DSCs, and $d_{max} = Max(D)$ and $d_{min} = Min(D)$. The threshold of the highest similarity U is obtained with Eqn. 7.

$$U = \lambda(d_{max} - d_{min}),\tag{7}$$

where: λ is the discrimination coefficient[4].

[Step 5]. Let s_i, s_j and s_k be vectors that belong to S_o, with $s_k = f(s_i, s_j)$ and in which S_o is the merging process of s_i and s_j. The process can be done if Eqn. 8 is satisfied.

$$s_k = \left\langle \frac{s_{i1} + s_{j1}}{2}, \frac{s_{i2} + s_{j2}}{2}, \dots, \frac{s_{i6} + s_{j6}}{2} \right\rangle\tag{8}$$

Let $g : S_o \rightarrow S_o$ be the minimal dissimilitude function (MDS) in which $g(s_i)$ is the most similar vector to s_i. Then an evolution of the set S_o is defined by Eqn. 9.

$$S_o' = \{s_i \mid g(g(s_i)) \neq s_i \forall s_i \in S_o\} \cup \tag{9}$$
$$\{s_f = f(s_i, s_j) \mid g(s_i) = s_j, g(s_j) = s_i \forall s_i, s_j \in S_o\}$$

If the dissimilitude distance of a vector s_i is greater than the threshold U with respect to the rest of the vectors, then s_i goes to the next generation of vectors. Otherwise, the

---

[3] In this view, it is important to define the topological property of the geo-image, because this property is considered the semantics of the raster data.

[4] The initial value for λ is 0.5. This value provides good results according to our experiments.

vector s_i must be merged with other vectors, whose dissimilitude distance is less than the threshold U.

[Step 6]. If $card(S'_o) < M$ (in which M is the number of desired classes), then repeat from step 4, with $\lambda = \lambda / 2$.

[Step 7]. Determine the proportional ratio of the classes in the semantic domain S, which is given by Eqn. 10.

$$P_r = \frac{d_{\min}}{d_{\max}} \tag{10}$$

While the proportional ratio of the classes is closer to 1, the partition of the semantic domain will be more accurate; that means that the dissimilitude distances between vectors are closer.

[Step 8]. Since the process is iterative, it should be repeated with the new generation of vectors, which are obtained; that is, with $S_o = S_o'$.

[Step 9]. Repeat the process from step 3, until $card(S_o) = M$.

3.3 Semantic Supervised Clustering (SSC) Algorithm

The result of the previously described algorithm (section 3.2) is the set Ω of semantic characteristic vectors ω_i of the classification. From the Ω, it is necessary to compute the mean and covariance matrix for each $\omega_i \in \Omega$. In order to determine the class or category for each pixel at a location x, it is necessary that each pixel contains a conditional probability, denoted by Eqn. 11:

$$p(\omega_i \mid x), i = 1,...,M \tag{11}$$

The constraint represented by Eqn. 12 is used in the classification algorithm, since the probabilities $p(x)$ are known from the training data, we assume that it is conceivable that the $p(\omega_i)$ is the same for each ω_i.

$$x \in \omega_i \text{ if } p(x \mid \omega_i)p(\omega_i) > p(x \mid \omega_j)p(\omega_j) \text{ for all } j \neq i \tag{12}$$

In this analysis, the classes have multidimensional normal distributions and each pixel can assign a probability of being a member of each class. After computing the probabilities of a pixel being in each of the available classes, we assign the class with the highest probability. The algorithm consists of the following steps:

[Step 1]. Determine the number of classes ω_i by means of the semantic training sites algorithm (section 3.2).

[Step 2]. Compute the maximum likelihood distribution and the covariance matrix for each of the generated classes ω_i.

[Step 3]. For each image pixel, determine its semantic vector by applying Eqn. 5.

[Step 4]. Compute the probability of vector s to know if it belongs to each class ω_i.

[Step 5]. Obtain the coordinates (x, y) of the pixel, if the constraint (see Eqn. 13) is satisfied, then the pixel belongs to the class ω_i.

$$p(x \mid \omega_i)p(\omega_i) > p(x \mid \omega_j)p(\omega_j) \text{ for all } j \neq i \tag{13}$$

[Step 6]. Repeat from step 3, until all pixels in the geo-image are classified.

4 Results

We perform a semantic supervised clustering in Landsat TM images. The algorithm has been implemented in C++ Builder. We carried out a series of experiments over different Landsat TM images. However, due to space limitations, in this paper we only present a small but representative sample of them. The segment of the Mexican state of Tamaulipas Landsat TM image is shown in Fig. 1. Additionally, Fig. 1a shows the Landsat TM image, which is composed of seven multispectral bands, using the combination (4 3 2) for the processing. We overlapped vector data to identify the urban and land areas. Fig. 1b depicts the results of the classification, using the semantic supervised clustering approach.

We have considered that to obtain a good estimation of class statistics, it is necessary to choose trainings fields for the one-cover type, located in different regions of the geo-image. The band signatures for the classes are obtained from the training fields, which are given in Table 1.

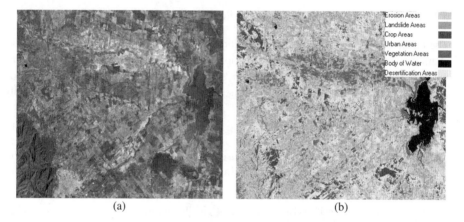

(a)   (b)

Fig. 1. (a) Landsat image of Tamaulipas basin[5]. (b) SSC of Landsat image.

Table 1. Class signatures generated from the training areas in Fig. 1b

Class	Mean vector	Standard deviation	Class	Mean vector	Standard deviation
Body of Water (blue)	39.16 34.30 21.38	1.35 1.74 2.43	Vegetation areas (green)	44.83 49.26 62.79	1.30 2.32 5.36
Urban areas (yellow)	52.27 58.40 58.50	5.11 7.61 7.31	Desertification areas (pink)	44.08 43.08 83.87	1.27 2.44 5.89
Crop areas (brown)	46.89 54.74 62.52	1.01 1.64 2.55	Erosion areas (gray)	41.66 42.81 54.45	0.85 1.28 4.84
Landslide areas (red)	39.68 37.07 63.65	0.93 1.50 5.14			

[5] A basin is the entire geographical area drained by a river and its tributaries.

We implemented other supervised clusterings using PCI Geomatics software to compare the values of each class in the classifications. We use the *Parallelepiped* method (PM) depicted in Fig 2a, and the *Minimum Distance* method (MDM) shown in Fig 2b. In the three supervised methods, we use the same number of training sites. The PM method does not classify several quantity pixels of the geo-image. On the other hand, the MDM method classifies the pixels with some mistakes. For instance, the areas of vegetation and desertification classes are less than the areas generated with the SSC method. Our approach improves the classification process using the semantics to compute the training sites. To assess the accuracy of the proposed method we use the overall accuracy metric, which is simply defined as the ratio of the pixels that were correctly classified over the total number of pixels contained in the raster. The values attained by the three supervised clustering methods are described in Table 2.

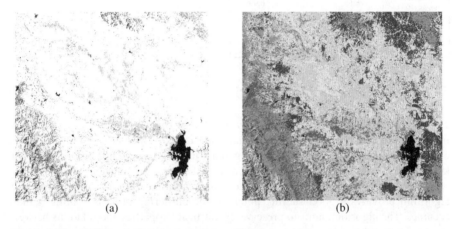

(a)   (b)

Fig. 2. (a) Parallelepiped supervised clustering. (b) Minimum Distance supervised clustering.

Table 2. Comparative results of three supervised clustering methods

Class	Overall Accuracy	Mean vector	Standard deviation	Class	Overall Accuracy	Mean vector	Standard deviation
Body of Water (SSC)	97.1	39.16 34.30 21.38	1.35 1.74 2.43	Vegetation areas (SSC)	97.7	44.83 49.26 62.79	1.30 2.32 5.36
Body of Water (PM)	97.3	23.53 22.50 17.36	1.23 1.14 1.05	Vegetation areas (PM)	96.8	29.20 37.46 58.77	1.18 1.72 3.98
Body of Water (MDM)	92.5	32.62 30.27 9.78	1.99 3.17 2.78	Vegetation areas (MDM)	91.5	38.29 45.23 51.19	1.94 3.75 5.71
Urban areas (SSC)	96.2	52.27 58.40 58.50	5.11 7.61 7.31	Desertification areas (SSC)	96.4	44.08 43.08 83.87	1.27 2.44 5.89
Urban areas (PM)	97.1	44.11 49.34 48.95	5.04 7.31 6.29	Desertification areas (PM)	95.2	28.45 31.28 79.85	1.15 1.84 4.51

Table 2. (*continued*)

Urban areas (MDM)	90.3	47.35	5.28	Desertification areas (MDM)	92.7	37.54	1.91
		53.33	8.98			39.05	3.87
		51.67	7.17			72.27	6.24
Crop areas (SSC)	93.7	46.89	1.01	Erosion areas (SSC)	95.8	41.66	0.85
		54.74	1.64			42.81	1.28
		62.52	2.55			54.45	4.84
Crop areas (PM)	98.1	40.12	0.67	Erosion areas (PM)	96.1	26.03	0.73
		42.64	1.02			31.01	0.68
		58.54	1.76			50.43	3.46
Crop areas (MDM)	89.7	49.33	1.41	Erosion areas (MDM)	95.1	35.12	1.49
		61.81	3.39			38.78	2.71
		40.19	2.22			42.85	5.19
Landslide areas (SSC)	96.9	39.68	0.93				
		37.07	1.50				
		63.65	5.14				
Landslide areas (PM)	96.1	23.45	1.23				
		25.78	0.90				
		59.16	4.23				
Landslide areas (MDM)	90.9	34.67	1.89				
		37.07	2.34				
		58.32	6.07				

5   Conclusions and Future Work

In the present work the *semantic supervised clustering* approach for Landsat TM images is proposed. We use *spatial semantics* to improve the classification by means of *a priori* knowledge that consists of geometrical and topological properties. This knowledge is considered into the selection criteria of the training sites.

The set of properties obtained by the semantic-based detection of training sites algorithm provides a good starting point for the clustering process making it more accurate. The algorithm aims to preserve the intrinsic properties and relations between the semantic information and the spectral classes of the geo-image. The semantic clustering algorithm consists of transforming the problem from the spatial domain to the semantic domain. In this context, the most similar semantic classes will be merged according to their semantic information. The process is based on computing the dissimilitude coefficient (DSC) and merging pairs of classes. On the other hand, however one of the main limitations of the method is that it relies on the expertise of a human being to select the training sites.

We are working towards the definition of an *automatic* methodology to classify geo-images by means of *semantic unsupervised clustering*. This approach is also being successfully used to detect flooding and landslide areas by using conceptual structures such as ontologies.

Acknowledgments

Work partially sponsored by the Mexican National Polytechnic Institute (IPN), by the Mexican National Council for Science and Technology (CONACyT) and by the SIP-IPN under grants 20101282, 20101069, 20101088, 20100371, and 20100417.

References

1. Unsalan, C., Boyer, K.L.: Classifying land development in high-resolution Satellite imagery using hybrid structural-multispectral features. IEEE Transactions on GeoScience and Remote Sensing 42(12), 2840–2850 (2004)
2. De Vleeschouwer, C., Delaigle, J.F., Macq, B.: Invisibility and application functionalities in perceptual watermarking – an overview. Proceedings of the IEEE 90(1), 64–77 (2002)
3. Heileman, G.L., Yang, Y.: The Effects of Invisible Watermarking on Satellite Image Classification. In: Proceedings of the ACM Workshop on Digital Rights Management, Washington, DC, USA, pp. 120–132 (2003)
4. Torres, M., Levachkine, S.: Ontological representation based on semantic descriptors applied to geographic objects. Revista Iberoamericana Computacióny Sistemas 12(3), 356–371 (2009)
5. Nishii, R., Eguchi, S.: Supervised image classification by contextual AdaBoost based on posteriors in neighborhoods. IEEE Transactions on GeoScience and Remote Sensing 43(11), 2547–2554 (2005)
6. Bandyopadhyay, S.: Satellite image classification using genetically guided fuzzy clustering with spatial information. International Journal of Remote Sensing 26(3), 579–593 (2005)
7. Gómez-Chova, L., Bruzzone, L., Camps-Valls, G., Calpe-Maravilla, J.: Semi-Supervised Remote Sensing Image Classification based on Clustering and the Mean Map Kernel. In: Proceedings of Geoscience and Remote Sensing Symposium (IGARSS), Boston, MA, USA, vol. 4, pp. 391–394 (2008)
8. Granville, V., Rasson, J.P.: Bayesian filtering and supervised classification in image remote sensing. Computational Statistics & Data Analysis 20(2), 203–225 (1995)
9. Neg, I., Kittler, J., Illingworth, J.: Supervised segmentation using a multiresolution data representation. Signal Processing 31(2), 133–163 (1993)
10. Keuchel, J., Naumann, S., Heiler, M., Siegmund, M.: Automatic land cover analysis for Tenerife by supervised classification using remotely sensed data. Remote Sensing of Environment 86(4), 530–541 (2003)
11. Brown, D.G., Lusch, D.P., Duda, K.A.: Supervised classification of types of glaciated landscapes using digital elevation data. Geomorphology 21(3-4), 233–250 (1998)
12. Krooshof, P., Postma, G.J., Melssen, W.J., Buydens, L.M., Tranh, T.N.: Effects of including spatial information in clustering multivariate image data. Transactions of Trends in Analytical Chemistry 25(11), 1067–1080 (2006)
13. Giacinto, G., Roli, F., Bruzzone, L.: Combination of neural and statistical algorithms for supervised classification of remote-sensing images. Pattern Recognition Letters 21(5), 385–397 (2000)
14. Anzalone, A., Bizzarri, F., Parodi, M., Storace, M.: A modular supervised algorithm for vessel segmentation in red-free retinal images. Computers in Biology and Medicine 38, 913–922 (2008)
15. Atkinson, P.M.: Spatially weighted supervised classification for remote sensing. International J. of Applied Earth Observation and Geoinformation 5, 277–291 (2004)
16. Bandyopadhyay, S., Maulik, U., Pakhira, M.K.: Clustering using simulated annealing with probabilistic redistribution. International Journal of Pattern Recognition and Artificial Intelligence 15(2), 269–285 (2001)
17. Chuvieco, E., Huete, A.: Fundamentals of Satellite Remote Sensing. CRC Press, Boca Raton (2010)

Fuzzified Probabilistic Rough Measures in Image Segmentation

Dariusz Małyszko and Jarosław Stepaniuk

Department of Computer Science
Bialystok University of Technology
Wiejska 45A, 15-351 Bialystok, Poland
{d.malyszko,j.stepaniuk}@pb.edu.pl

Abstract. Recent advances witnessed during widespread development of information systems that depend upon detailed data analysis, require more sophisticated data analysis procedures and algorithms. In the last decades, deeper insight into data structure has been made more precise by means of many innovative data analysis approaches. **R**ough **E**xtended (Entropy) **F**ramework presents recently devised algorithmic approach to data analysis based upon inspection of the data object assignment to clusters. Data objects belonging to clusters contribute to cluster approximations. Cluster approximations are assigned measures that directly make possible calculation of cluster roughness. In the next step, total data rough entropy measure is calculated on the base of the particular roughness. In the paper, in the Rough Extended (Entropy) Framework, a new family of the probabilistic rough (entropy) measures has been presented. The probabilistic approach has been extended into fuzzy domain by fuzzification of the probabilistic distances. The introduced solution seems to present promising area of data analysis, particulary suited in the area of image properties analysis.

1 Introduction

Data analysis based on the fuzzy sets depends primarily on the assumption, stating that data objects may belong in some degree not only to one concept or class but may partially participate in other classes. Rough set theory on the other hand assigns objects to class lower and upper approximations on the base of complete certainty about object belongingness to the class lower approximation and on the determination of the possible belongingness to the class upper approximation. Probabilistic approaches have been developed in several rough set settings, including decision-theoretic analysis, variable precision analysis, and information-theoretic analysis. Most often, probabilistic data interpretation depends upon rough membership functions and rough inclusion functions.

Rough Extended (Entropy) Framework presents extensively developed method of data analysis. In the paper, a new family of the probabilistic rough (entropy) measures has been introduced. The probabilistic approach has been expanded by the operation of the fuzzification of the probabilistic distances. The introduced solution has been supported and visualized by mean od several images of

T.-h. Kim et al. (Eds.): SIP/MulGraB 2010, CCIS 123, pp. 78–86, 2010.

the boundary regions in case of standard probabilistic and fuzzified probabilistic measures. The operation of the fuzzification of the probabilistic measures seems to present promising area of data analysis, particulary suited in the area of image properties analysis.

This paper is structured in the following way. In Section 2 the introductory information about rough sets, rough sets extensions and rough measures has been presented. In Section 3 RECA concepts and probabilistic RECA measures have been described. Presentation and experimental material has been included and discussed in Section 4 followed by concluding remarks.

2   Rough Set Data Models

2.1   Rough Set Theory Essentials

An information system is a pair (U, A) where U represents a non-empty finite set called the universe and A a non-empty finite set of attributes. Let $B \subseteq A$ and $X \subseteq U$. Taking into account these two sets, it is possible to approximate the set X making only the use of the information contained in B by the process of construction of the lower and upper approximations of X and further to express numerically the roughness $R(AS_B, X)$ of a set X with respect to B by assignment

$$R(AS_B, X) = 1 - \frac{card(LOW(AS_B, X))}{card(UPP(AS_B, X))}. \tag{1}$$

In this way, the value of the roughness of the set X equal 0 means that X is crisp with respect to B, and conversely if $R(AS_B, X) > 0$ then X is rough (i.e., X is vague with respect to B). Detailed information on rough set theory is provided in [8,10]. During last decades, rough set theory has been developed, examined and extended in many innovative probabilistic frameworks as presented in [11]. Variable precision rough set model VPRS improves upon rough set theory by the change of the subset operator definition, designed to analysis and recognition of statistical data patterns. In the variable precision rough set setting, the objects are allowed to be classified within an error not greater than a predefined threshold. Other probabilistic extensions include decision-theoretic framework and Bayesian rough set model.

2.2   Rough Extended (Entropy) (Clustering) Framework - REF

In general Rough Extended Framework data object properties and structures are analyzed by means of their relation to the selected thresholds or cluster centers. In this context, Rough Extended Framework basically consists of two interrelated approaches, namely thresholding Rough Extended Framework and clustering Rough Extended Framework. Each of these approaches gives way development and calculation of rough measures. Rough measures based upon entropy notion are further referred to as rough entropy measures.

Cluster centers are regarded as representatives of the clusters. The main assumption made during REF based analysis consists on the remark that the way data objects are distributed in the clusters determines internal data structure. In the process of the inspection of the data assignment patterns in different parametric settings it is possible to reveal or describe properly data properties.

Rough Extended Entropy Framework in image segmentation has been primarily introduced in [6] in the domains of image thresholding routines, it means forming Rough Extended Entropy Thresholding Framework. In [6], rough set notions - lower and upper approximations have been applied into image thresholding. This thresholding method has been extended into multilevel thresholding for one-dimensional and two-dimensional domains - [3] rough entropy notion have been extended into multilevel granular rough entropy evolutionary thresholding of 1D data in the form of 1D MRET algorithm. Additionally, in [1] the authors extend this algorithm into 2D MRET thresholding routine of 2D image data. Further, rough entropy measures have been employed in image data clustering setting in [2], [4] described as **R**ough **E**ntropy **C**lustering **A**lgorithm.

3 Rough Entropy Measures

3.1 General REF and RECA Concepts

Rough (entropy) measures, considered as a measure of quality for data clustering gives possibility and theoretical background for development of robust clustering schemes. These clustering algorithms incorporate rough set theory, fuzzy set theory and entropy measure. The procedure for calculation of rough entropy value has been given in Algorithm 1. Three basic rough properties that are applied during rough entropy calculation and further in clustering scheme include

1. selection of the threshold metrics (crisp, fuzzy, probabilistic, fuzzified probabilistic) - tm,
2. the threshold type (thresholded or difference based) - tt,
3. the measure for lower and the upper approximations - crisp, fuzzy, probabilistic, fuzzified probabilistic - ma.

Data objects are assigned to lower and upper approximation on the base of the following criteria

1. assignment performed on the basis of the difference of distances to the cluster centers within given threshold value, see Table 1.
2. assignment performed on the basis of the distance to cluster centers within given threshold value, see Table 2,

Measures for lower and upper approximations have been presented in Table 3. In General Rough Entropy Framework, data objects are analyzed by means of their relation to the selected number of cluster centers. Cluster centers are regarded as representatives of the clusters. The main assumption made during **REF** based analysis consists on the remark that the way data objects are distributed in the clusters determines internal data structure. In the process of the inspection of the data assignment patterns in different parametric settings it is possible to reveal or describe properly data properties. In the following subsections, the RECA measures are presented and explained in detail.

Algorithm 1. Rough_Entropy Calculation

Data: Rough Approximations
Result: R - Roghness, RE - Rough Entropy Value

for $l = 1$ **to** k *(number of data clusters)* **do**
| **if Upper**(C_l) **!= 0 then   R**$(C_l) = 1$ **- Lower**(C_l) **/ Upper**(C_l)
end
RE $= 0$
for $l = 1$ **to** k *(number of data clusters)* **do**
| **if R**(C_l) **!= 0 then   RE = RE -** $\frac{exp}{2} \cdot$ **R**$(C_l) \cdot log($**R**$(C_l))$;
| (see **Eq. 1**)
end

Table 1. Difference metric

Metric	Symbol	Condition
Crisp - CD	$d_{cr}(x_i, C_m)$	$|d_{cr}(x_i, C_m) - d_{cr}(x_i, C_l)| \leq \epsilon_{cr}$
Fuzzy - FD	$d_{fz}(x_i, C_m)$	$|(\mu_{C_m}(x_i) - \mu_{C_l}(x_i)| \leq \epsilon_{fz}$
Pr - PD	$d_{pr}(x_i, C_m)$	$|d_{pr}(x_i, C_m) - d_{pr}(x_i, C_l)| \leq \epsilon_{pr}$
FP - FPD	$d_{fp}(x_i, C_m)$	$|d_{fp}(x_i, C_m) - d_{fp}(x_i, C_l)| \leq \epsilon_{fp}$

Table 2. Threshold metric

Metric	Symbol	Condition
Crisp - CT	$d_{cr}(x_i, C_m)$	$d_{cr}(x_i, C_m) \leq \epsilon_{cr}$
Fuzzy - FT	$d_{fz}(x_i, C_m)$	$(\mu_{C_m}(x_i) \geq \epsilon_{fz}$
Pr - PT	$d_{pr}(x_i, C_m)$	$d_{pr}(x_i, C_m) \geq \epsilon_{pr}$
FP -FPT	$d_{fp}(x_i, C_m)$	$d_{fp}(x_i, C_m) \geq \epsilon_{fp}$

Table 3. Approximation measures

Approximation	Distance	Value
C - Crisp	$m_{cr}(x_i, C_m)$	1
F - Fuzzy	$m_{fz}(x_i, C_m)$	μ_{C_m}
P - Pr	$m_{pr}(x_i, C_m)$	$|d_{pr}(x_i, C_m)$
FP	$m_{fp}(x_i, C_m)$	$|d_{fpr}(x_i, C_m)$

3.2 Crisp RECA Measures

In crisp setting, RECA measures are calculated on the base of the crisp metric.
In rough clustering approaches, data points closest to the given cluster center
or sufficiently close relative to the selected threshold type, are assigned to this
cluster lower and upper approximations. The upper approximations are calculated in the specific, dependant upon threshold type and measure way presented

in the subsequent paragraphs. Standard crisp distance most often applied in many working software data analysis systems depends upon Euclidean distance or Minkowsky distance, calculated as follows

$$d_{cr}(x_i, C_m) = \left(\sqrt{\Sigma_{j=1}^d (x_{ij} - C_{mj})^p} \right)^{\frac{1}{p}} \tag{2}$$

3.3 Fuzzy RECA Measures

Fuzzy membership value $\mu_{C_l}(x_i) \in [0,1]$ for the data point $x_i \in U$ in cluster C_m is given as

$$d_{fz}(x_i, C_m) = \mu_{C_l}(x_i) = \frac{d(x_i, C_m)^{-2/(\mu-1)}}{\sum_{j=1}^k d(x_i, C_j)^{-2/(\mu-1)}} \tag{3}$$

where a real number $\mu > 1$ represents fuzzifier value and $d(x_i, C_m)$ denotes distance between data object x_i and cluster (center) C_m.

3.4 Probabilistic RECA Measures

Probability distributions in RECA measures are required during measure calculations of probabilistic distance between data objects and cluster centers. Gauss distribution has been selected as probabilistic distance metric for data point $x_i \in U$ to cluster center C_m calculated as follows

$$d_{pr}(x_i, C_m) = (2\pi)^{-d/2} |\Sigma_m|^{-1/2} exp\left(-\frac{1}{2}(x_i - \mu_m)^T \Sigma_m^{-1}(x_i - \mu_m) \right) \tag{4}$$

Table 4. Threshold and difference based measures and related RECA algorithms, for condition for the difference metric see Table 1 and Table 2 for threshold metric

Algorithm	Measure	Threshold	Condition
CC-TRECA	$m_{cr}(x_i, C_m)$	crisp	$d_{cr}(x_i, C_m)$
CF-TRECA	$m_{cr}(x_i, C_m)$	fuzzy	$d_{fz}(x_i, C_m)$
CP-TRECA	$m_{cr}(x_i, C_m)$	pr	$d_{pr}(x_i, C_m)$
C-FP-TRECA	$m_{cr}(x_i, C_m)$	fuzzy-pr	$d_{fp}(x_i, C_m)$
FC-TRECA	$m_{fz}(x_i, C_m)$	crisp	$d_{cr}(x_i, C_m)$
FF-TRECA	$m_{fz}(x_i, C_m)$	fuzzy	$d_{fz}(x_i, C_m)$
FP-TRECA	$m_{fz}(x_i, C_m)$	pr	$d_{pr}(x_i, C_m)$
F-FP-TRECA	$m_{fz}(x_i, C_m)$	fuzzy-pr	$d_{fp}(x_i, C_m)$
PC-TRECA	$m_{pr}(x_i, C_m)$	crisp	$d_{cr}(x_i, C_m)$
PF-TRECA	$m_{pr}(x_i, C_m)$	fuzzy	$d_{fz}(x_i, C_m)$
PP-TRECA	$m_{pr}(x_i, C_m)$	pr	$d_{pr}(x_i, C_m)$
P-FP-TRECA	$m_{pr}(x_i, C_m)$	fuzzy-pr	$d_{fp}(x_i, C_m)$
FP-C-TRECA	$m_{fp}(x_i, C_m)$	crisp	$d_{cr}(x_i, C_m)$
FP-F-TRECA	$m_{fp}(x_i, C_m)$	fuzzy	$d_{fz}(x_i, C_m)$
FP-P-TRECA	$m_{fp}(x_i, C_m)$	pr	$d_{pr}(x_i, C_m)$
FP-FP-TRECA	$m_{fp}(x_i, C_m)$	fuzzy-pr	$d_{fp}(x_i, C_m)$

where $|\Sigma_m|$ is the determinant of the covariance matrix Σ_m and the inverse covariance matrix for the C_m cluster is denoted as Σ_m^{-1}. Data dimensionality is denoted as d. In this way, for standard color RGB images $d = 3$, for gray scale images $d = 1$. Mean value for Gauss distribution of the cluster C_m has been denoted as μ_m. The summary of the parameters for the probabilistic RECA measures has been given in Table 4.

3.5   Fuzzified Probabilistic RECA Measures

In Fuzzified probabilistic RECA measures, the probabilistic distances to all clusters are fuzzified by means of the following formulae applied to the $d_{pr}(x_i, C_1)$, $\ldots$, $d_{pr}(x_i, C_k)$ distances. Fuzzified membership value of probabilistic distance $d_{fp}(x_i, C_m) \in [0, 1]$ for the data point $x_i \in U$ in cluster C_m is given as

$$d_{fp}(x_i, C_m) = \frac{d_{pr}(x_i, C_m)^{-2/(\mu-1)}}{\sum_{j=1}^{k} d_{pr}(x_i, C_j)^{-2/(\mu-1)}} \tag{5}$$

The main parameters during calculating fuzzified probabilistic distance are as follows

1. probability distribution function, in the experiments, Gauss distribution have been selected as the pdf
2. crisp distance metric, in the experiments, Euclidean distance has been selected as the metric
3. fuzzifier value, the higher fuzzifier the data objects in higher degree participate in all clusters
4. transfer function, because probabilities that the object belongs to the clusters are comparatively low, especially when data dimensionality is high, most often during fuzzification procedure, the probabilities should be given in logarithmic domain.

4   Experimental Setup and Results

Image data - in the subsequent material, the image 27059 from Berkeley image database [5] has been selected for experiments. The original color RGB image as given in Figure 1 (a) have been preprocessed and 2D data for bands G and B have been obtained. In Figure 1 (a) image in 0GB bands has been presented.

Cluster centers - for the presentation purposes, 7 cluster centers have been created as shown in Figure 1 (c) and displayed in blue numbered circles. Data points assignment and border of data points that are the closest to the given clusters depends heavily on the metric. Additionally, in Figure 1 (c) cluster borders and cluster assignment (it means data points belonging to the nearest cluster center) have been given in case of crisp or fuzzy metric.

In order to make the presented concepts more understandable, two types of boundary regions have been illustrated. The first boundary regions are calculated with difference based metric and the second with threshold based metric.

The boundary regions for probability difference based metric are presented in Figure 2 PDRECA, and in Figure 2 are displayed boundary regions for Fuzzified PDRECA.

The boundary regions for probability threshold based metric are presented in Figure 4 PTRECA, and in Figure 5 are displayed boundary regions for Fuzzified PTRECA. In all images, boundary regions, understood as attribute data that belong to the upper approximations but not belong to the lower approximations of the same cluster are presented in gray color. In the material, in case of PDRECA measures, the threshold value decreases from $\epsilon = 10^{-4}$ (a) to $\epsilon = 10^{-12}$ (c). At the same time, the boundary regions are becoming smaller.

The same situation is for Fuzzified PDRECA measures. In this case, threshold values are different (the data is fuzzy) and the boundary regions have different shapes. For PDRECA measures and Fuzzified PDRECA measure the similar tendency is also observable.

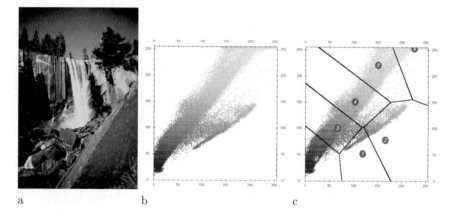

a   b   c

Fig. 1. Berkeley dataset image: (a) 27059 - 0GB bands, (b) image 27059, 0GB attributes, (c) image 27059, 0GB attributes with 7 cluster centers

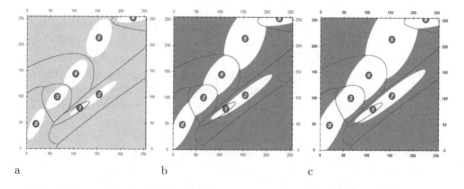

a   b   c

Fig. 2. Berkeley dataset image, image 27059, R-B bands, PDRECA: (a) $\epsilon = 10^{-4}$, (b) $\epsilon = 10^{-8}$, (c) $\epsilon = 10^{-12}$

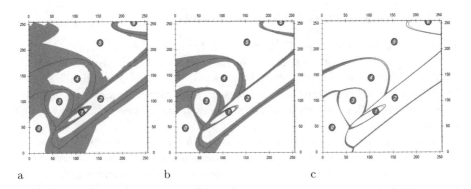

Fig. 3. Berkeley dataset image, image 27059, R-B bands, Fuzzified PDRECA: (a) $\epsilon = 10^{-4}$, (b) $\epsilon = 5 * 10^{-4}$, (c) (a) $\epsilon = 10^{-6}$

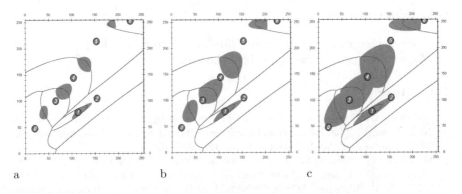

Fig. 4. Berkeley dataset image, image 27059, 0RB bands, TDRECA: (a) $\epsilon = 5 * 10^{-8}$, (b) $\epsilon = 10^{-11}$, (c)$\epsilon = 10^{-16}$

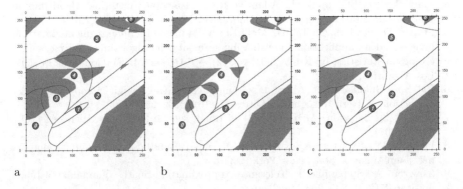

Fig. 5. Berkeley dataset image, image 27059, 0RB bands, Fuzzified TDRECA: (a) $\epsilon = 0.22$, (b) $\epsilon = 0.23$, (c) $\epsilon = 0.242$

5   Conclusions and Future Research

In the study, an extension of the probabilistic rough (entropy) measures has been introduced and presented. The introduced solution consists of the fuzzification operation of the probabilistic measures. In this way, probabilistic and fuzzified probabilistic rough (entropy) measures are capturing and revealing data structure properties that seem to be complementary to crisp and fuzzy rough measures.

Unification and integration of crisp, fuzzy, probabilistic and fuzzified probabilistic rough entropy measures presents promising emerging area of image segmentation and analysis. The main advantage of this new approach consist in better reflection of internal data structure. The introduced fuzzification of the probabilistic rough (entropy) measures improves proper understanding of the data that should be processed meaning better data analysis robustness.

Acknowledgments

The research is supported by the grant N516 0692 35 from the Ministry of Science and Higher Education of the Republic of Poland.

References

1. Malyszko, D., Stepaniuk, J.: Granular Multilevel Rough Entropy Thresholding in 2D Domain. In: 16th International Conference Intelligent Information Systems, IIS 2008, Zakopane, Poland, June 16-18, pp. 151–160 (2008)
2. Malyszko, D., Stepaniuk, J.: Standard and Fuzzy Rough Entropy Clustering Algorithms in Image Segmentation. In: Chan, C.-C., Grzymala-Busse, J.W., Ziarko, W.P. (eds.) RSCTC 2008. LNCS (LNAI), vol. 5306, pp. 409–418. Springer, Heidelberg (2008)
3. Malyszko, D., Stepaniuk, J.: Adaptive multilevel rough entropy evolutionary thresholding. Information Sciences 180(7), 1138–1158 (2010)
4. Malyszko, D., Stepaniuk, J.: Adaptive Rough Entropy Clustering Algorithms in Image Segmentation. Fundamenta Informaticae 98(2-3), 199–231 (2010)
5. Martin, D., Fowlkes, C., Tal, D., Malik, J.: A database of human segmented natural images and its application to evaluating segmentation algorithms and measuring ecological statistics. In: ICCV 2001, vol. (2), pp. 416–423. IEEE Computer Society, Los Alamitos (2001)
6. Pal, S.K., Shankar, B.U., Mitra, P.: Granular computing, rough entropy and object extraction. Pattern Recognition Letters 26(16), 2509–2517 (2005)
7. Pawlak, Z., Skowron, A.: Rudiments of rough sets. Information Sciences 177(1), 3–27 (2007)
8. Pedrycz, W., Skowron, A., Kreinovich, V. (eds.): Handbook of Granular Computing. John Wiley & Sons, New York (2008)
9. Skowron, A., Stepaniuk, J.: Tolerance Approximation Spaces. Fundamenta Informaticae 27(2-3), 245–253 (1996)
10. Stepaniuk, J.: Rough–Granular Computing in Knowledge Discovery and Data Mining. Springer, Heidelberg (2008)
11. Slezak, D., Ziarko, W.: The investigation of the Bayesian rough set model. International Journal of Approximate Reasoning 40, 81–91 (2005)

JHUF-5 Steganalyzer: Huffman Based Steganalytic Features for Reliable Detection of YASS in JPEG Images

Veena H. Bhat[1,3], S. Krishna[1], P. Deepa Shenoy[1],
K.R. Venugopal[1], and L.M. Patnaik[2]

[1] Department of Computer Science and Engineering,
University Visvesvaraya College of Engineering, Bangalore, India
[2] Vice Chancellor, Defense Institute of Advanced Technology, Pune, India
[3] IBS-Bangalore, Bangalore, India
{veena.h.bhat,krishna.somandepalli}@gmail.com,
shenoypd@yahoo.com

Abstract. Yet Another Steganographic Scheme (YASS) is one of the recent steganographic schemes that embeds data at randomized locations in a JPEG image, to avert blind steganalysis. In this paper we present JHUF-5, a statistical steganalyzer wherein J stands for JPEG, HU represents Huffman based statistics, F denotes FR Index (ratio of file size to resolution) and 5 - the number of features used as predictors for classification. The contribution of this paper is twofold; first the ability of the proposed blind steganalyzer to detect YASS reliably with a consistent performance for several settings. Second, the algorithm is based on only five uncalibrated features for efficient prediction as against other techniques, some of which employs several hundreds of predictors. The detection accuracy of the proposed method is found to be superior to existing blind steganalysis techniques.

Keywords: Statistical Steganalysis, Huffman Coding, YASS.

1 Introduction

Steganography is the art of hiding the very presence of communication by embedding a secret message in an innocent looking cover object that includes digital images, audio or video. Yet Another Steganographic Scheme (YASS) is a recent embedding scheme for JPEG images proposed in 2007 that has successfully eluded most of the popular blind steganalyzers in vogue. Several attacks proposed to detect steganography thrive on their sensitivity to variations in the higher order statistical properties when images are subjected to steganography - YASS is designed to sustain such attacks successfully.

Based on the notion of calibration of features extracted proposed in [4], blind steganalyzers can be classified as those techniques that use self calibrated features extracted from stego-images as predictors and those that do not work on these features. Calibration refers to the difference between a specific (non-calibrated) functional calculated from a stego image and the same functional obtained from the corresponding cover image [4]. Several attacks employ this method for efficient steganalysis.

T.-h. Kim et al. (Eds.): SIP/MulGraB 2010, CCIS 123, pp. 87–97, 2010.

Steganalyzers that do not use calibration adopt Markov process to monitor the magnitudes of the quantized Block DCT (BDCT) coefficients of the JPEG images before and after embedding data in the cover image [6].

The idea of steganography in YASS is simple yet effective. YASS does not embed data in JPEG coefficients directly. Instead, it uses the strategy of Quantization Index Modulation (QIM) to hide information in the quantized Discrete Cosine Transform (DCT) coefficients in randomly chosen 8×8 host blocks, whose locations may not coincide with the 8×8 grids used in JPEG compression. After data embedding, images are compressed back to JPEG format; this process of randomized embedding successfully evades detection by calibrated steganalytic features [1]. Several techniques, both blind and targeted attacks have been proposed for steganalysis of YASS, though no blind steganalysis method is promising and most do not give reliable detection over the vagary of block sizes used during embedding using YASS. Targeted steganalysis techniques aimed at attacking YASS specifically have shown an efficient accuracy [1, 10].

This work presents a blind steganalyzer that works on uncalibrated statistical features (extracted from both cover and stego-images) for reliable detection of YASS.

2 Related Work

YASS, a JPEG steganographic technique, hides data in the non-zero DCT coefficients of randomly chosen image blocks [8]. YASS robustly embeds messages in digital images in a key dependent transform domain. This random embedding scheme evades detection by most blind steganalyzers which employ analysis of the higher order statistical features. YASS is further strengthened against statistical detection by the introduction of randomization, using a mixture based approach as proposed in [9] and is referred to as Extended YASS (EYASS) in [17]. YASS and EYASS (explained in section 3) thus pose a new challenge for steganalysts to ponder on the predictor features extracted and algorithms designed for steganalysis.

The original proposers of YASS [8] evaluated its detectability against six popular blind steganalyzers [3, 5, 13, 14, 15, 16] using Support Vector Machine (SVM) as the classifier and demonstrated its robustness, however with a low embedding capacity. The improved YASS algorithm (EYASS) introduces more calculated randomness into the original YASS algorithm besides improving the embedding capacity. This work also proves the non-detectablility of EYASS by evaluating against two blind steganalyzers [3, 5]. The notion of using self calibration features for statistical detection of YASS and its failure has been explored in [4].

Steganalysis YASS / EYASS using four state-of-the-art steganalytic feature sets is proposed in [10]; CC-PEV, a 548 dimensional Cartesian-Calibrated Pevny feature set [10], 486 dimensional Markov process features (MP) [7], the Subtractive Pixel Adjacency Model (SPAM) feature set consisting of 686 features [16] and a combination of SPAM and CC-PEV to derive a 1,234 dimensional Cross-Domain Feature (CDF) set. These feature sets are used to evaluate YASS / EYASS techniques against twelve different settings of YASS and resulted in prediction with an error probability of less than 15% even for payloads as small as 0.03 bpac and in small images. Huang et al., in a study of the security performance of YASS proves that the Markov Process feature set [5, 7] is the most accurate in detecting YASS [17]. A targeted steganalysis approach that successfully predicts YASS with a high accuracy has been proposed in the work, Steganalysis of YASS [1]. [1] demonstrates two weakness of YASS - the insufficient randomization of the locations of the embedding host blocks and the introduction of zero coefficients by the Quantization Index Modulation embedding. Fisher Linear Discriminant analysis is used for classification.

The proposed work is focused on deploying the blind steganalytic features introduced in our previous work [2] to attack YASS and thereby analyze the performance of this proposed steganalytic model as a universal blind steganalyzer. The JHUF-5 steganalyzer proposed here, works on only 5 features for prediction - four are Huffman based statistics and the fifth being FR Index, the ratio of file size to resolution of the image.

3 YASS – Yet Another Steganographic Scheme

Given an input image of resolution M×N, the embedding process of YASS consists of the following steps [10]:

1) An image is divided into non-overlapping consecutive B×B (B>8) blocks so as to get $M_B \times N_B$ blocks in the image where $M_B = M/B$ and $N_B = N/B$. In the rest of this paper, B is referred to as 'Big block size'.
2) In each B-block, an 8×8 sub-block is randomly selected using a secret key shared only with the receiver.
3) Two quality factors - design quality factor Q_{Fh} and advertised quality factor Q_{Fa} of the final JPEG compression, are noted.
4) For each sub-block selected in step 2, a two-dimensional DCT is computed and these coefficients are further divided by quantization steps specified by Q_{Fh}. This results in an output block with unrounded coefficients.
5) A QIM scheme [18] is employed for data hiding in predetermined non-zero low frequency alternate current (AC) DCT coefficients (called *candidate embedding bands*). The unrounded coefficients whose rounding values are zeros and unrounded coefficients which are not in the candidate embedding bands are unaltered thereby preventing unnecessary visual as well as statistical artifacts being introduced. The resulting output blocks from step 4 are embedded with data and are referred to as *data embedded blocks*.
6) The data embedded blocks are multiplied with the quantization steps specified by Q_{Fh} and further the 2-D inverse DCT is performed (termed as *modified blocks*).
7) Using the advertised quality factor Q_{Fa}, the whole image is compressed using the standard JPEG format. Thus resulting in a 'stegged' image. A Repeat Accumulate (RA) encoding framework is used to correct the errors that are caused during JPEG image compression as described in [8].

EYASS introduces further randomness in two stages. The first by randomly selecting the 8×8 embedding blocks from each of the big block size. Next, by the attack aware iterative embedding strategy, referred to as M1 [9], which lowers the error rate while increasing the embedding capacity as compared to the embedding capacity of YASS.

4 Feature Extraction – Huffman Based Statistics and FR Index

- **Huffman Based Statistics**

Huffman coding, a data compression technique employed in JPEG image format, encodes DCT coefficients that are computed during JPEG compression with variable length codes assigned on statistical probabilities. A grayscale image employs 2 Huffman tables, one each for AC and DC portions. The number of particular lengths of the Huffman codes is unique for a given image of certain size, quality and resolution. The numerical statistics of the DC portion of the Huffman table is referred as Huffman based statistics (H). Almost 90% of the DCT coefficients of an image are encoded using Huffman codes of lengths ranging from 2 to 5 bits. The number of codes of lengths 6 to 16 bits is negligible. Hence we use only statistics of codes of length 2 to 5 bits denoted as H_2, H_3, H_4 and H_5. These features are generated using a Huffman decoder on decompression of the JPEG bit-stream using Matlab as a tool. A considerable variation can be observed in the Huffman code length statistics before and after embedding as illustrated in table 1.

Some of the scoring features of the proposed Huffman based statistics as steganalytic features for detection of YASS could be as follows:

1) The pseudo random number generator used to locate the 8×8 block in the B×B block successfully confuses a steganalyzer looking for anomalies related to synchronous blocks, however the proposed Huffman based statistics reflect the extra bits that are embedded by YASS irrespective of the complexity it uses to distort the steganalyzer's perception of synchronous blocks.

2) The randomized embedding strategy of YASS can successfully evade detection when calibrated features used as predictors. In contrast to this the Huffman based statistics identified by [2], are non-calibrated and are computed from both cover and stego images and are used as predictors for supervised learning.

3) One of the important characteristics of the Huffman coding algorithm is its 'unique prefix property' that is no code is a prefix to any other code, making the codes assigned to the symbols unique. This fact further supports our choice of Huffman based statistics as predictor features.

Table 1. Huffman based statistics before and after YASS embedding for a 507×788 image

Huffman Based Statistics	Notation	Before embedding	After embedding
No. of codes of length 1 bit	H_1	0	0
No. of codes of length 2 bits	H_2	**1287**	**3275**
No. of codes of length 3 bits	H_3	**549**	**2514**
No. of codes of length 4 bits	H_4	**360**	**318**
No. of codes of length 5 bits	H_5	**158**	**125**
No. of codes of length 6 bits	H_6	187	104
No. of codes of length 7 bits	H_7	6	0
.	.	.	.
No. of codes of length 16 bits	H_{16}	0	0

- **FR Index: Ratio of file size to resolution of an image**

When an image is compressed to the JPEG format, based on the resolution of the image, its quality, the compression ratio and few other factors, the resulting JPEG file takes up a particular file size. This indicates that the file size and the resolution of an image and further its quality are interrelated. Thus the ratio of file size of the image in bytes to that of its resolution is found to be unique and in a certain range for a given resolution, this functional is termed '*FR Index*'. In our cover image database, the FR index ranges from 0.034 to 0.921.

5   Implementation

5.1   Image Database

One of the important aspects of performance evaluation of any steganalyzer is the nature of image database employed in implementation. JPEG images are a popular format for storing, presenting and exchanging images. Grayscale JPEG images are selected for our study as it is harder to detect hidden data in grayscale images as compared to color images where steganalysis can utilize dependencies between color channels. The images used span across a wide vagary of sizes, resolutions and textures. The entire image database used in the experiments consists of over 20,000 images among which over 2,000 are used as cover images. The cover images are taken from the database provided by Memon et al [20]. A subset of 1000 cover images is embedded with 16 different combinations of YASS settings. Embedding with YASS was carried out in Matlab using the code provided by Anindya Sarkar [9]. This code implements EYASS which is an improvement to YASS.

Three parameters are used in these 16 settings namely; the design quality factor (Q_{Fh}), the advertised quality factor (Q_{Fa}) and big block size (B). Big block sizes of 9, 10, 12 and 14 are tested against two sets of quality factor settings. The first setting is where Q_{Fh} and Q_{Fa} are chosen randomly from a combination of 50, 60 and 70, the second set being combinations of quality factors 50 and 75 for Q_{Fh} and Q_{Fa} respectively. For convenience, in this work we denote these settings of YASS as YASS1, YASS2 upto YASS16. The embedding parameters and its corresponding notation are detailed in table 2. In table 2, 'rand(50,60,70)' implies that the quality factors are chosen randomly amongst these three numbers.

Table 2. Embedding setting of YASS and notations used (Q_{Fh} - Quality Factor, Q_{Fa} - Advertised Quality Factor and B the big block size)

Q_{Fh} / Q_{Fa}	B	Notation	Q_{Fh} / Q_{Fa}	B	Notation
rand(50,60,70) / rand(50,60,70)	9	YASS1	50 / 75	9	YASS9
	10	YASS2		10	YASS10
	12	YASS3		12	YASS11
	14	YASS4		14	YASS12
75 / 75	9	YASS5	50 / 50	9	YASS13
	10	YASS6		10	YASS14
	12	YASS7		12	YASS15
	14	YASS8		14	YASS16

5.2 Model

Fig. 1 illustrates the Huffman feature extraction model used by the proposed stegana-lyzer. The model is designed to extract the four Huffman based statistics; H_2, H_3, H_4 and H_5 from a grayscale JPEG image using a Huffman decoder. A fast Huffman de-coder is used to extract the two Huffman tables, one each of AC and DC portions, the code buffer holds all the codes decoded from which the required statistics (number of 2 to 5 bit code lengths from DC portion only) are selectively separated and fed to the functional space of the classifier.

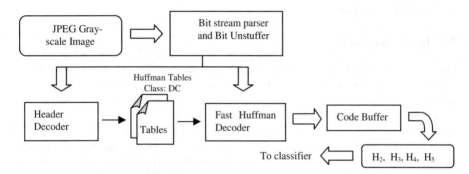

Fig. 1. JHUF-5 Steganalyzer model illustrating Huffman feature extraction

5.3 Classification

In any blind statistical steganalyzer, the classifier used for pattern recognition plays a pivotal role. In this work we use three different classifiers to evaluate the features extracted; Artificial Neural Networks (ANN), Support Vector Machines (SVM) and Random Forests (RF). The functional space for these classifiers consists of five vari-ables, the four Huffman based statistics; H_2, H_3, H_4 and H_5 and FR Index and is de-signed for binary classification to distinguish stego images from genuine ones.

To train and test each classifier for each of the 16 combinations of YASS settings, the following steps are followed:

- The image database is divided into several combinations of training and testing sets for each of the 16 YASS settings tested. In each trial, 60% of the data is used for training the classifier.
- To evaluate the accuracy of the model, the minimal total average probability of error (P_e) is computed, given by:

$$P_e = (P_{FP} + P_{FN})/2. \tag{1}$$

where P_{FP} and P_{FN} are the probability of false positives and false negatives of the test images respectively.

Artificial Neural Network (ANN) classifier: we implement a feed forward back-propagation neural network with a single hidden layer of 3 neurons with radial basis as the activation function, for this work. **Support Vector Machines (SVM):** In this work we employ a C type binary class SVM with a Gaussian kernel; the two hyper-parameters of the C-SVMs; penalty parameter C and kernel width γ are estimated using a 10-fold cross-validation employing grid-search over the following multiplicative grid.

$$(C, \gamma) \in [(2_i, 2_j)| \, i \in \{1, ..., 10\}, j \in \{-10, ..., 2\}]. \tag{2}$$

Random Forest (RF) is an ensemble classifier that combines Breiman's "bagging" idea and the random selection of features to construct an ensemble of decision trees. The number of trees is kept constant at 500 and the maximum number of nodes is varied from 3 to 9 for a sufficiently low Out-of-Bag (OOB) error with 2 variables tried at each split, these values are computed and assigned by a tune function proposed in R [19].

These techniques are chosen as they can classify non-linear data with a huge number of instances effectively. The performance of each of these classifiers and the error probability (P_e), across different payloads are described in section 6.

6   Results and Performance Analyses

Table 3 illustrates the error probability P_e for all the 16 settings of YASS tested across the classifiers - ANN, SVM and RF. It can be observed that ANN Classifier gives the best and consistent performance of detecting YASS. Over all we conclude that the proposed model detects YASS with an accuracy of more than 99% for both Q_{Fh} and Q_{Fa} set to 50 and for all four big block sizes; 9, 10, 12 and 14. Further, it can be observed the error probability is consistent for several settings of YASS that shows the reliability of the JHUF-5 steganalyzer.

Though the image database is implemented using EYASS, we evaluate against the YASS too as the features extracted in this work do not depend on the further randomization attributed by EYASS. We adopt a four-fold methodology to evaluate the performance of the proposed JHUF-5 blind steganalyzer against existing methods. First, the proposed model is compared with the set of 6 blind steganalyzers tested in [8]; secondly the 2 steganalyzers reported EYASS [9] are used to evaluate. The results are further compared against steganalysis using the four schemes described in [10] for the twelve settings of YASS. Finally, we compare the proposed method against the work in [17] where security performance of YASS against four state-of-the-art blind JPEG steganalyzers is reported.

6.1   Comparison of JHUF-5 against Blind Steganalyzers Tested in YASS [8]

The original authors of YASS evaluate its steganographic security against the following 6 blind steganalysis schemes [8]. The numbers following the name of the scheme indicate the number of features used in the prediction model; we adopt the same notation as used in [8]. The comparison results are shown in table 4.

1. **Farid-72:** uses wavelet based features for steganalysis [11].
2. **PF-23:** uses DCT based steganalytic feature vectors [12].
3. **PF-274:** uses a combination of Markov and DCT features [3].

4. **DCT hist:** Histogram of DCT coefficients from a low-frequency band [13].
5. **Xuan-39:** uses spatial domain features for steganalysis [14].
6. **Chen-324:** Steganalysis based on statistical moments [15].

Table 3. Error probability for various YASS settings using ANN, SVM and RF

YASS Setting	Error Probability P_e (%) for Classifiers			YASS Setting	Error Probability P_e (%) for Classifiers		
	ANN	SVM	RF		ANN	SVM	RF
YASS1	4.83	5.16	4.81	YASS9	4.16	7.35	6.55
YASS2	3.83	6.67	5.62	YASS10	4.16	7.22	6.78
YASS3	5.00	6.76	5.56	YASS11	4.33	6.89	6.55
YASS4	4.50	6.76	5.53	YASS12	4.33	7.01	6.55
YASS5	4.33	6.89	6.66	YASS13	0.50	2.07	0.92
YASS6	4.66	6.76	6.19	YASS14	0.16	2.07	0.92
YASS7	4.83	6.65	5.96	YASS15	0.16	2.07	1.03
YASS8	4.83	6.65	6.08	YASS16	0.16	2.07	1.03

Table 4. Comparison of JHUF-5 with blind steganalyzers tested in [8]

YASS Settings	Steganalyzer Detection Accuracy						
	Farid-72	PF-23	PF-274	DCT hist	Xuan-39	Chen-324	JHUF-5 (ANN)
YASS5	0.52	0.53	0.59	0.55	0.52	0.54	0.913
YASS6	0.51	0.57	0.60	0.54	0.54	0.55	0.907
YASS7	0.52	0.53	0.62	0.55	0.53	0.53	0.903
YASS8	0.52	0.52	0.54	0.53	0.52	0.53	0.903
YASS9	0.55	0.59	0.77	0.64	0.63	0.75	0.917
YASS10	0.55	0.59	0.79	0.64	0.64	0.65	0.917
YASS11	0.54	0.56	0.74	0.60	0.57	0.60	0.913
YASS12	0.51	0.60	0.65	0.54	0.53	0.55	0.913
YASS13	0.52	0.56	0.58	0.53	0.54	0.57	0.990
YASS14	0.51	0.55	0.56	0.53	0.56	0.51	0.997
YASS15	0.52	0.54	0.53	0.51	0.54	0.55	0.997
YASS16	0.51	0.54	0.55	0.53	0.51	0.54	0.997

6.2   Comparison of JHUF-5 against Blind Steganalyzers Tested in EYASS [9]

EYASS evaluates the settings for YASS1 using the steganalyzers, PF-274 [3] and Chen-324 [15]. The detection accuracy using PF-274 is found to be 59% and Chen-324 to be 58% whereas the proposed JHUF-5 steganalyzer yields a detection accuracy of 90.34% which is superior to the compared methods.

6.3   Comparison of JHUF-5 against Steganalytic Feature Sets Tested in [10]

A 1,234 dimensional Cross Dimensional Feature (CDF) set for steganalysis is proposed in [10]. Table 5 shows the comparative analysis. The CDF feature set is a combination of three other steganalytic feature sets - MP-486 (uses Markov process features for steganalysis) [5], CC-PEV-548 (uses Cartesian-calibrated Pevny feature set) [10] and SPAM-686 (uses second order Markov chain based features) [16].

Table 5. Comparison of JHUF-5 with blind steganalyzers tested in [10]

Steganalyzer	MP-486	CC-PEV-548	SPAM-686	CDF-1234	JHUF-5
YASS5 P_e (%)	15.5	16.4	15.2	9.7	4.33
YASS6 P_e (%)	27	26	14.5	12.4	4.66

6.4 Comparison of JHUF-5 against Steganalytic Feature Sets Tested in [17]

YASS is tested against four state-of-the-art steganalyzers in [17]. Chen-324, MP-486, PF-274 are tested besides a fourth steganalyzer that uses the same features of PF-274 but without calibration, we denote this steganalyzer as NonClbPF-274. The comparative results are shown in table 6. The classification results obtained using only the ANN classifier only is tabulated.

Table 6. Comparison of JHUF-5 with blind steganalyzers tested in [17]

YASS Setting	Chen-324	MP-486	PF-274	NoClbPF-274	JHUF-5 (ANN)
YASS5	79.1	80.4	87.2	83.2	91.3
YASS6	80.8	80.4	86.3	85.4	90.7
YASS7	69.3	71.7	77.8	73.8	90.3
YASS8	61.5	63.4	68.9	66.3	90.3
YASS9	96.3	97.2	96.6	97.2	91.7
YASS10	98.5	98.7	98.6	98.7	91.7
YASS11	93.2	94.4	95.3	94.4	91.3
YASS12	82.1	83.0	88.0	85.5	91.3
YASS13	83.9	85.8	87.7	88.0	99.0
YASS14	84.1	86.0	89.5	89.4	99.7
YASS15	72.4	75.0	81.1	79.7	99.7
YASS16	64.4	67.2	72.5	70.6	99.7

7 Conclusions and Future Work

The proposed work evaluates the JHUF-5 blind steganalyzer to attack YASS; the performance analyses show that the detection accuracy is consistent over a wide range of YASS settings.

The features used in this statistical steganalyzer are unique when evaluated against several classifier techniques; moreover the proposed model employs a 5-dimensional feature vector only as compared against several reliable attacks that use several hundreds of features.

Our future work includes analyzing settings of the variance parameter in Extended YASS. Further we would attempt to extend the existing model to a multi-class classification problem to predict block size.

References

1. Bin, L., Jiwu, H., Yun, Q.S.: Steganalysis of YASS. IEEE Transaction on Information Forensics and Security 3, 369–382 (2009)
2. Bhat, V.H., Krishna, S., Shenoy, P.D., Venugopal, K.R., Patnaik, L.M.: JPEG Steganalysis using HBCL Statistics and FR Index. In: Chau, M. (ed.) PAISI 2010. LNCS, vol. 6122, pp. 105–112. Springer, Heidelberg (2010)
3. Pevn'y, T., Fridrich, J.: Merging Markov and DCT Features for Multi-class JPEG Steganalysis. In: Proc. of SPIE, San Jose, CA, vol. 6505, pp. 3–4 (2007)
4. Kodovsk'y, J., Fridrich, J.: Calibration Revisited. In: Proc. of the 11th ACM workshop on Multimedia and Security, New York, USA, pp. 63–74 (2009)
5. Shi, Y.Q., Chen, C., Chen, W.: A Markov Process based Approach to Effective Attacking JPEG Steganography. In: Camenisch, J.L., Collberg, C.S., Johnson, N.F., Sallee, P. (eds.) IH 2006. LNCS, vol. 4437, pp. 249–264. Springer, Heidelberg (2007)
6. Fu, D., Shi, Y.Q., Zou, D., Xuan, G.: JPEG Steganalysis using Empirical Transition Matrix in Block DCT Domain. In: International Workshop on Multimedia Signal Processing, Victoria, BC, Canada (2006)
7. Chen, C., Shi, Y.Q.: JPEG Image Steganalysis Utilizing both Intrablock and Interblock Correlations. In: Proc. of Intl. Symposium on Circuits and Systems, pp. 3029–3032 (2008)
8. Solanki, K., Sarkar, A., Manjunath, B.S.: YASS: Yet Another Steganographic Scheme that Resists Blind Steganalysis. In: Proc. of 9th International Workshop on Information Hiding, pp. 16–31 (2007)
9. Solanki, K., Sarkar, A., Manjunath, B.S.: Further Study on YASS: Steganography based on Randomized Embedding to Resist Blind Steganalysis. In: Proc. of SPIE, San Jose, CA, vol. 6819, pp. 16–31 (2008)
10. Kodovsky, J., Pevny, T., Fridrich, J.: Modern Steganalysis can Detect YASS. In: Proc. SPIE, Electronic Imaging, Media Forensics and Security XII, San Jose, CA (2010)
11. Lyu, S., Farid, H.: Detecting Hidden Messages using Higher-order Statistics and Support Vector Machines. In: Petitcolas, F.A.P. (ed.) IH 2002. LNCS, vol. 2578, pp. 340–354. Springer, Heidelberg (2002)
12. Pevny, T., Fridrich, J.: Multi-class Blind Steganalysis for JPEG Images. In: Proc. of SPIE, San Jose, CA, vol. 6072, pp. 1–13 (2006)
13. Solanki, K., Sullivan, K., Madhow, U., Manjunath, B.S., Chandrasekaran, S.: Probably Secure Steganography: Achieving Zero K-L Divergence using Statistical Restoration. In: Proc. ICIP, pp. 125–128 (2006)
14. Xuan, G., et al.: Steganalysis based on Multiple Features Formed by Statistical Moments of Wavelet Characteristic Functions. In: Barni, M., Herrera-Joancomartí, J., Katzenbeisser, S., Pérez-González, F. (eds.) IH 2005. LNCS, vol. 3727, pp. 262–277. Springer, Heidelberg (2005)
15. Chen, C., Shi, Y.Q., Chen, W., Xuan, G.: Statistical Moments based Universal Steganalysis using JPEG-2D Array and 2-D Characteristic Function. In: Proc. ICIP, Atlanta, GA, USA, pp. 105–108 (2006)
16. Pevný, T., Bas, P., Fridrich, J.: Steganalysis by Subtractive Pixel Adjacency Matrix. In: Proc. of the 11th ACM Multimedia & Security Workshop, Princeton, pp. 75–84 (2009)
17. Huang, F., Huang, J., Shi, Y.Q.: An Experimental Study on the Security Performance of YASS. IEEE press, Los Alamitos

18. Chen, B., Wornell, G.W.: Quantization Index Modulation: A class of Provably Good Methods for Digital Watermarking and Information Embedding. IEEE Trans. on Information Theory 47, 1423–1443 (2001)
19. Random Forests,
    `http://debian.mc.vanderbilt.edu/R/CRAN/web/packages/`
    `randomForest/randomForest.pdf`
20. Kharrazi, M., Sencar, H.T., Memon, N.: A Performance Study of Common Image Steganography and Steganalysis Techniques. Journal of Electronic Imaging 15(4) (2006)

Region Covariance Matrices for Object Tracking in Quasi-Monte Carlo Filter

Xiaofeng Ding, Lizhong Xu*, Xin Wang, and Guofang Lv

College of Computer and Information Engineering, Hohai University,
Nanjing, 210098, China
xiaoqi.ding@gmail.com, lzhxu@hhu.edu.cn

Abstract. Region covariance matrices (RCMs), categorized as a matrix-form feature in a low dimension, fuse multiple different image features which might be correlated. The region covariance matrices-based trackers are robust and versatile with a modest computational cost. In this paper, under the Bayesian inference framework, a region covariance matrices-based quasi-Monte Carlo filter tracker is proposed. The RCMs are used to model target appearances. The dissimilarity metric of the RCMs are measured on Riemannian manifolds. Based on the current object location and the prior knowledge, the possible locations of the object candidates in the next frame are predicted by combine both sequential quasi-Monte Carlo (SQMC) and a given importance sampling (IS) techniques. Experiments performed on different type of image sequence show our approach is robust and effective.

Keywords: Object tracking, quasi-Monte Carlo filter, sample area, region covariance matrices.

1   Introduction

Object tracking is always an important task in computer vision, it has various application in visual surveillance, human-computer interaction, vehicle navigation and so on.

In general, object tacking can be performed in two different ways [1]. One is the bottom-up process such as target representation and location, the other is top-down process such as filtering and data association. In either of the tracking approaches, because of the variations of target appearance and environment, object tracking is often non-linear and non-Gaussian. To handle the non-linear and non-Gaussian problems in objet tracking, the non-parametric techniques based on Monte Carlo simulations are introduced. Among these non-parametric techniques, the particle filter (PF) [2,3,9] is the most typical one and is firstly applied for visual tracking in [9], also known as Condensation. In recent years, PFs are widely applied in visual object tracking.

In the beginning of the particle filter tracker, the raw pixel values of several image statics such as color, gradient, shape and filter responses are used as

---

* Corresponding author.

T.-h. Kim et al. (Eds.): SIP/MulGraB 2010, CCIS 123, pp. 98–106, 2010.
© Springer-Verlag Berlin Heidelberg 2010

target appearance [9], which are easy to change in the presence of illumination changes and non-rigid motions. Then, histogram, a nature extension of raw pixel values, is used [3], such as color histogram and Histogram of Oriented Gradient (HOG). However color histograms are exponential with its numbers and in many applications it is not discriminative enough to detect objects. Objects' interest point-based matching is another way to achieve object tracking where the point features are used for target representation, such as Harris feature, Kanade-Lucas-Tomasi (KLT) feature, scale-invariant feature transform (SIFT) feature. The performance of interest point-based matching is poor with noisy low resolution images. Recently, region covariance matrices (RCMs) [6,7] was presented to outperform color histogram. RCMs, categorized as a matrix-form feature, are denoted as symmetric positive define matrices. The RCMs fuse multiple different image features which might be correlated such as intensity, color, gradients, filter responses. It describes how features vary together.

Wu et al. [12] and Palaio, Batista [13] proposed the covariance-based particle filter for single and multiple objects tracking with update mechanisms based on means on Riemannian manifolds, respectively. In PF, particle impoverishment is inevitable. To overcome this problem, various improved strategies are proposed to design better proposal distributions, to employ complex sampling strategies, or to employ specific prior knowledge [2]. However, the pure random sampling in PF is apt to lead to "gaps and clusters" in the particle support, especially in high dimension spaces. In order to solve this problem, Guo and Wang [8] proposed the sequential quasi-Monte Carlo (SQMC) filter, where the deterministic random sampling is instead of the pure random sampling.

In this paper, we present a RCMs-based object tracking algorithm in SQMC filter. RCMs provide a natura tool for integrating multiple features, and are scale and illumination independent. RCMs-based target appearance model can hold strong discriminating ability and robustness. In the proposed algorithm, a target is tracked with SQMC filter by comparing its RCM with the RCMs measured on a Riemannian manifold of the sampling positions. The novelties of the proposed algorithm is that the sampling area is predicted by the object's last location, then the IS is used. It will decrease the number of the particles and improve the accuracy of the prediction.

This paper is organized as follows. Section 2 reviews the generic PF and introduces the SQMC filter. In Section 3, the details of the proposed RCMs-based SQMC filter are presented. Experiments are performed to verify the algorithm and compare the new algorithm with existing methods in section 4, which is followed by conclusion in Section 5.

2   Quasi-Monte Carlo filter

In this section, we first briefly review the generic particle filter algorithm. Then the quasi-Monte Carlo filter is introduced.

2.1   Generic Particle Filter

Particle filters solve non-linear and non-Gaussian state estimation problems and have become popular for visual tracking. In general, tracking problem is considered as a nonlinear dynamic system model

$$x_k = f(x_k) + w_k \qquad (1)$$

and an observation model is given as

$$z_k = g(x_k) + v_k, \qquad (2)$$

where x_k and z_k denote the state and observation variables, respectively; w_k and v_k denote the state noise and observation noise which follow certain know distributions. The sequence $\{w_k\}$ and $\{v_k\}$ are assumed to be independent. Let x_0 be the initial state, $z_{1:k} \doteq \{z_1, \cdots, z_k\}$ be the observation up to time k. Using the given data $z_{1:k}$, we can estimate the value of the hidden state variable x_k. The posterior state distribution $p(x_k/z_{1:k})$ at each time k can be obtained recursively as follows:

$$p(x_k/z_{1:k-1}) = \int p(x_k/x_{1:k-1})p(x_{k-1}/z_{1:k-1}) \qquad (3)$$

$$p(x_k/z_{1:k}) = \frac{p(z_k/x_k)p(x_k/z_{1:k-1})}{c_k} \qquad (4)$$

where $c_k \triangleq \int p(z_k/x_k)p(x_k/z_{1:k-1})dx_k$, $k = 1, 2, \cdots$; $p(z_k/x_k)$ and $p(x_k/x_{k-1})$ correspond to an observation model and a dynamic model, respectively.

2.2   Sequential Quasi-Monte Carlo Filter

Particle filter make use of Monte Carlo (MC) integration to approximate the integrals in (3) and (4). Quasi-Monte Carlo (QMC) integration techniques have been proven asymptotically more efficient than the MC integration [10].

Suppose we want to evaluate the integral of a real-valued function f defined over d-dimension unit cube $c^d = [0, 1)^d$

$$I = \int_{c^d} f(u)du \qquad (5)$$

In either MC integration and QMC integration, we sample N points $P_N = \{u^{(1)}, u^{(2)}, \cdots, u^{(N)}\}$ in the integral region, then using (6) to compute the approximation of I.

$$\hat{I} = \frac{1}{N}\sum_{j=1}^{N} f(u^{(j)}) \qquad (6)$$

In MC method, the P_N is a set of N independent and uniformly distributed vectors over $[0, 1)^d$. In QMC methods, P_N is chosen from a low-discrepancy point set and P_N is more regular distributed than the random point set P_N in MC methods. There are different low-discrepancy sequences, such as Halton sequence, Sobol sequence, Niederreiter sequence. Specifically, for a given integral in d-dimension, if the number of sampling points is N, then the error bounds of the best known low-discrepancy sequence have the asymptotic order $\mathcal{O}(\frac{\log_d N}{N})$, which is better asymptotic error rate than the $\mathcal{O}(\frac{1}{\sqrt{N}})$ rate associated with MC.

The SQMC filter was given by Guo and Wang [8] for the nonlinear dynamic systems. By employing the QMC integration techniques in (3) and (4), they get the generic SQMC filter. Then, SQMC and importance sampling (SQMC/IS) are combined to further improve the performance of the Bayesian nonlinear filters. The general SQMC/IS recurrence is:

$$p(x_k/z_{1:k}) \cong \sum_{j=1}^{N} w_{k-1}^{(j)} p(x_k|x_{k-1}^{(j)}), \tag{7}$$

$$w_k^{(j)} \triangleq \frac{p(x_k^{(j)}|z_{1:k})}{q(x_k^j)} \propto \frac{1}{\hat{c}_k} p(z_k|x_k^{(j)}) \frac{p(x_k^{(j)}|z_{1:k-1})}{q(x_k^{(j)})}, \tag{8}$$

with

$$\hat{c}_k = \sum_{j=1}^{N} p(z_k|x_k^{(j)}) \frac{p(x_k^{(j)}|z_{1:k-1})}{q(x_k^{(j)})}, \tag{9}$$

where $q(x_k)$ is a proposed density.

A general procedure of SQMC/IS algorithm is given in Fig. 1. The readers are referred to [8] for more details.

Initialization: Generate QMC points $\{u^{(j)}, j = 1, \cdots, N\}$ in $[0,1)^d$ and transform them to $\{x_0^{(j)}, j = 1, \cdots, N\}$ by $x^{(j)} = [a + (b - a) \circ u^{(j)}]$, where $\circ$ denotes the point-wise vector product. Then compute the prior density $p(x_0^{(j)}), j = 1, \cdots, N$.

The following steps are implemented at the kth recursion ($k = 1, 2, \cdots$):

Prediction:

----Compute the sampling density $q(x_k)$.

----Generate QMC points $\{u^{(j)}, j = 1, \cdots, N\}$ in $[0,1)^d$ and transform them to $\{x_0^{(j)}, j = 1, \cdots, N\}$ by $q^{-1}(u^{(j)})$.

----Compute the predictive density values $p(x_k^{(j)}/z_{1:k-1}), j = 1, \cdots, N$, by (8).

Filtering: Compute the posterior density values $p(x_k^{(j)}/z_{1:k}), \; j = 1, \cdots, N$, by (9).

Fig. 1. A General Procedure of SQMC/IS Algorithm

3 Sequential Quasi-Monte Carlo Tracking with Region Covariance Matrices

SQMC techniques for filtering nonlinear dynamic systems [8,11] have been proven more accuracy than sequential Monte Carlo techniques. This section gives a detailed description of the proposed algorithm. First, the construction of RCMs are presented. Then, the proposed RCMs-based SQMC tracker is discussed.

3.1 Region Covariance Matrices

Let I be a three dimensional color image. Let F be the $W \times H \times d$ dimensional feature image extracted from I

$$F(x, y) = \phi(I, x, y) \tag{10}$$

where the function ϕ can be any mapping such as intensity, color, gradients, filter responses, etc. For a given rectangular region R in image I, let $\{f_k\}_{k=1..n}$ be the d-dimensional feature points inside R. The region covariance can be expressed as

$$C_R = \frac{1}{n-1} \sum_{k=1}^{n} (f_k - \mu)(f_k - \mu)^T, \tag{11}$$

where C_R is a $d \times d$ positive matrix, μ is the mean of all the points in R.

Differen composite of different sub-feature extracted from the image represents differen property. To have a correct object representation we need to choose what kind of features to extract from the image I. In this paper, we define ϕ as

$$\begin{aligned} \phi = \begin{bmatrix} x & y & R(x,y) & G(x,y) & B(x,y) \\ & I(x,y) & I_x(x,y) & I_y(x,y) \end{bmatrix}, \end{aligned} \tag{12}$$

where x and y are the pixel location; $R(x,y)$, $G(x,y)$ and $B(x,y)$ are the red, green and blue color values; I is the intensity; $I_x(x,y)$ and $I_y(x,y)$ are the image derivatives calculated through the filter $[-1 \quad 0 \quad 1]$ in x, y direction. With the define feature vector ϕ, the RCM of a region is a 8×8 matrix, and due to symmetry, only upper triangular part is stored, which has only 36 different values. Let X and Y denote the RCMs of target model and target candidate, respectively. Based on the Riemannian geometry, the distance between them can be measured by

$$d(X, Y) = \sqrt{\operatorname{tr}(\log^2(X^{-\frac{1}{2}} Y X^{-\frac{1}{2}}))}, \tag{13}$$

where $\log()$ is the ordinary matrix logarithm operator; $\operatorname{tr}()$ is the trace operator.

3.2 RCMs-Based SQMC Tracker

The computational complexity of the generic SQMC filter in [8] is $\mathcal{O}(N^2)$, which is greater than PFs' $\mathcal{O}(N)$. In general IS of SQMC filter, sampling at time k is depending on those large particles at time $k - 1$. However, some large weight particles are illusive. Sampling from these particles will result in a wrong prediction. To handle this problem, we need to find the exact sample area before IS. In this paper, we randomly sample 20 particles from the object location at time $k - 1$ after iteration. These particles are used to search likelihood area. The weights of the particles are given by the dissimilarity between the target model and target candidate. Then, apply IS base on these particles and corresponding weights.

To obtain the most similar region to the target, we can use (13) to measure the distance between the RCMs corresponding to the target object window and the candidate regions. Theoretically, a single covariance matrix extracted from a region is usually enough to matching the region in different views and poses. So we can find the best object matching by finding the minimum distance with (13). However, during object tracking, occlusions often happen. Tuzel et al. [6] represented an object with five covariance matrixes of the image features computed inside the object region as show in Fig. 2. Then, a different dissimilarity [13] of the two objects is given as follows.

$$\rho(O_1, O_2) = \sum_{i=1}^{5} d^2(C_{O_1}^i, C_{O_2}^i), \tag{14}$$

where $C_{O_1}^i$ and $C_{O_2}^i$ $(i = 1, ..., 5)$ are the five covariance matrixes of object 1 and object 2, respectively, which integrates all the distances of the five regions of two object. It is robust when the target is with partial occlusions. Therefore, (14) is used to measure the dissimilarity between the target model and target candidate.

Let X and Y are the last and current RCMs. Then, our target appearance model is updated by

$$C = X^{\frac{1}{2}} \exp(X^{-\frac{1}{2}}(\alpha d(X,Y))X^{-\frac{1}{2}})X^{\frac{1}{2}}, \tag{15}$$

where α is the rate of model update, and when $d(X,Y)$ is more than a given threshold the model will de updated.

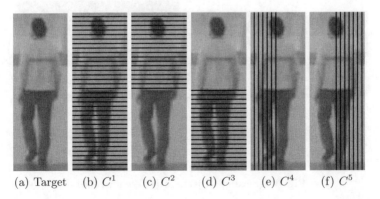

(a) Target   (b) C^1 (c) C^2 (d) C^3 (e) C^4 (f) C^5

Fig. 2. Target representation. Construction of the five RCMs from overlapping regions of an object feature image.

4   Experiments

To demonstrate the performance of the proposed approach, experiments were carried on several video sequences, see Fig. 3. The proposed algorithm is coded in Matlab and implemented on a Core2 2.4 GHz CPU computer with 2-G RAM and Windows XP. The experiments show that our approach is effective, feasible and robust.

(a) Street A   (b) Street B

(c) Room   (d) Subway

(e) Running dog   (f) Motorcycle

Fig. 3. Tracking Objects in Different Scenes

For each sequence, we used hybrid tracker [4], covariance-based probabilistic tracker [12], SQMC tracker and our tracker. To quantify the performance of these tracking algorithms, we adopt the criteria proposed in [14] is decribed as:

$$ S = \frac{1}{M} \sum_{m=1}^{M} \left(\frac{1}{|L|} \sum_{t \in L} \frac{2A_{O,t}^{(m)}}{A_{R,t} + A_{\hat{x},t}^{(m)}} \right) \in [0,1], \qquad (16) $$

where L is the set of indices for the labelled frames; $A_{R,t}$ and $A_{\hat{x},t}^{(m)}$ are the areas of the labelled bounding box and the bounding box corresponding to the state estimate in frame t, respectively; $A_{O,t}^{(m)}$ is the area of the overlap between the labelled and estimated bounding boxes in frame t, and m is an index ranging over the number of independent experiments. The score is a performance measure that ranges between 0 (no overlap in any of the labelled frames) and 1 (perfect overlap in all of the labelled frames).

The average results of all these 6 sequences of 4 tracking methods are shown in Table 1.

Table 1. The performance of the four algorithms. M1: color-based particle filter. M2: covariance-based probabilistic tracker. M3: SQMC tracker. M4: our tracker.

	M1	M2	M3	M4
Score	0.38	0.69	0.78	0.80

5   Conclusion

The RCMs are embedded in the sequential quasi-Monte Carlo framework to track objects. In the proposed algorithm, we do not re-sample particles from the particles at time $t - 1$. We give a sample area, then the IS is done in this area, which could reduce the number of particles and improve the accuracy. Experimental results show that the performance of the proposed algorithm is robust and can track the object under temporal occlusions, and the proposed method take more time than M1 and M2 tracking method and less time than M3, when the tracking results are better than M1, M2 and M3.

Acknowledgment

The authors would like to thank the anonymous reviewers for their valuable comments on this paper.

References

1. Comaniciu, D., Ramesh, V., Meer, P.: Kernel-based Object Tracking. IEEE Trans. Pattern Anal. Machine Intell. 25, 564–577 (2003)
2. Sanjeev Arulampalam, M., Maskell, S., Gordon, N., Clapp, T.: A Tutorial on Particle Filters for Online Nonlinear/Non-Gaussian Bayesian Tracking. IEEE Trans. Signal Process. 50, 174–188 (2002)
3. Perez, P., Hue, C., Vermaak, J., Gangnet, M.: Color-based Probabilistic Tracking. In: Heyden, A., Sparr, G., Nielsen, M., Johansen, P. (eds.) ECCV 2002. LNCS, vol. 2350, pp. 661–675. Springer, Heidelberg (2002)
4. Maggio, E., Cavallaro, A.: Hybrid Particle Filter and Mean Shift Tracker with Adaptive Transition Model. In: IEEE Conf. Signal Processing Society International Conference on Acoustics, Speech, and Signal Processing (ICASSP 2005), pp. 221–224. IEEE Press, New York (2005)
5. Dalal, N., Triggs, B.: Histograms of Oriented Gradients for Human Detection. In: IEEE Conf. Computer Vision and Pattern Recognition (CVPR 2005), pp. 886–893. IEEE Press, New York (2005)
6. Tuzel, O., Porikli, F., Meer, P.: Region Covariance: A Fast Descriptor for Detection and Classification. In: Leonardis, A., Bischof, H., Pinz, A. (eds.) ECCV 2006. LNCS, vol. 3952, pp. 589–600. Springer, Heidelberg (2006)
7. Tuzel, O., Porikli, F., Meer, P.: Pedestrian Detection via Classification on Riemannian Manifolds. IEEE Trans. Pattern Anal. Machine Intell. 30, 1713–1727 (2008)
8. Guo, D., Wang, X.: Quasi-Monte Carlo filtering in nonlinear dynamic systems. IEEE Tans. Signal Proc. 54, 2087–2098 (2006)
9. Isard, M., Blake, A.: CONDENSATION-Conditional Density Propagation for Visual Tracking. Int. J. Comput. Vis. 29, 5–28 (1998)
10. L'Ecuye, P., Lemieux, C.: Recent Advance in Randomized Quasi-Monte Carlo Methods. Kluwer Academic, Boston (2000)
11. Mathias, F., Patrick, D., Frederic, L.: Quasi Monte Carlo Partitioned Filtering for Visual Human Motion Capture. In: 16th IEEE Conf. Image Processing (ICIP 2009), pp. 2553–2556. IEEE Press, New York (2009)
12. Wu, Y., Wu, B., Liu, J., et al.: Probabilistic Tracking on Riemannian Manifolds. In: 19th International Conf. Pattern Recognition (ICPR 2008). IEEE Press, New York (2008)
13. Palaio, H., Batista, J.: Multi-object Tracking using An Adaptive Transition Model Particle Filter with Region Covariance Data Association. In: 19th International Conf. Pattern Recognition (ICPR 2008). IEEE Press, New York (2008)
14. Vermaak, J., Lawrence, N.D., Perez, P.: Variational inference for visual tracking. In: IEEE Conf. Computer Vision and Pattern Recognition (CVPR 2003), pp. 773–780. IEEE Press, New York (2003)

Target Detection Using PCA and Stochastic Features

Suk-Jong Kang[1] and Hyeon-Deok Bae[2]

[1] Agency for Defense Development, Yuseong
P.O. Box 35, Daejeon, Korea
luckybill007@yahoo.co.kr
[2] Department of Electrical Engineering, Chungbuk National University,
Jeungdeok, Cheongju, Korea
hdbae@chungbuk.ac.kr

Abstract. Automatic target detection (ATD) system which uses forward-looking infrared (FLIR) consists of two stages: image signal processing and clutter rejection. Images from electro-optical sensors are processed to express target well in signal processing stage. And true targets are well identified in clutter rejection stages. However, it is difficult to process target express well and to identify target from target candidates because they are obscure and there are many target-like objects. We propose new target detection algorithm using PCA and stochastic features. The proposed algorithm consists of two stages; image processing and clutter rejection. Image erosion, dilation and reconstruction is applied to eliminate multiple target candidates that are actually the same, single target and to remove small clutters in image processing stage. Linear Discriminant Analysis (LDA) using principal component analysis (PCA) and stochastic features is applied to clutter rejection. Several FLIR images are used to prove the performance of the proposed algorithm. The experimental results show that the proposed algorithm accurately detects targets with a low false alarm rate.

Keywords: target detection, clutter rejection, PCA and stochastic feature.

1 Introduction

Automatic target detection (ATD) generally refers to the autonomous or aided target detection by computer processing of data from an electro-optical sensor such as FLIR in military vehicle, such as tanks, reconnaissance vehicles and aircraft. It is an extremely important capability for targeting and surveillance missions of defense systems operating from a variety of platform.

ATD is aimed at reducing the workload for human operators who are tasked with the battlefield situation cover mission areas to targeting individual targets using unmanned vehicles or weapon systems. But automatic target detection using FLIR images are very challenging tasks because a number of factors can affect thermal images, including range, location, aspect angle, meteorological conditions and time of day.

A typical ATD system consists of two stages; signal processing and clutter rejection. The target candidates are well expressed in signal processing stage and true targets are identified in the clutter rejection stages. This paper focuses on signal processing and clutter rejection stages.

T.-h. Kim et al. (Eds.): SIP/MulGraB 2010, CCIS 123, pp. 107–114, 2010.
© Springer-Verlag Berlin Heidelberg 2010

Many researchers have developed target detection algorithm for FLIR images. Typical methods are thresholding [1]-[3], hit-miss transform [4], morphological wavelet [5], and directional wavelet [6]. Recently, Rizvi et el, proposed a neural network approach using region based principal component analysis (PCA) [7] and Chan el el. used eigenspace separation transform (EST) to reject potential targets [8]. These methods are used only one feature such as thresholding level, PCA feature and shape. In our experience with variety detection methods, one weakness of the above approaches is that they can miss targets with low contrast or warn false alarms with target-like clutters. So we propose new target detection method that use grayscale reconstruction technique [9] to eliminate small clutters in signal processing stage and LDA with PCA features in raw images and stochastic features of edge images.

This paper organized as follows. In section 2, algorithm for signal processing is proposed. In signal processing stage, images are processed for target images to eliminate multiple target candidates that are actually the same, single target. and to remove small clutters. In section 3, clutter rejection algorithm using PCA of region of interest (ROI) and stochastic features of edge images is proposed. In section 4, experimental results show that the proposed algorithm works well in low contrast and in a cluttered environment. Finally, concluding remarks are given in section 5.

2   Paper Preparation

As shown in Fig. 1, the proposed target detection algorithm consists of two stages; signal processing and clutter rejection. In the signal processing stage, candidate targets are extracted, while real targets are selected in clutter rejection stage.

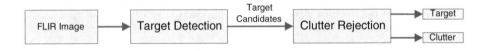

Fig. 1. Block diagram of target detection

2.1   Signal Processing Stage

Target candidates are detected in the signal processing stage, which consists of four steps; 1) image processing of input FLIR using grayscale reconstruction 2) segmentation of target candidates using experimental method 3) a closing and opening operation to prevent target separation in the binary image 4) the rejection of target candidates larger and smaller target candidate than expected size.

1) The processed image using Grayscale reconstruction is obtained using Eq. (1) and Eq. (2).

$$\sigma_{IO}(J) = (J \oplus B) \wedge I \tag{1}$$

where $\sigma_{IO}(J)$: reconstruct image using dilation

$\quad\quad J \oplus B$: dilation of J by disk structuring element B

$\quad\quad \wedge$: pointwise minimum

$$\sigma_{IC}(J) = (J \ominus B) \vee I \tag{2}$$

where $\sigma_{IC}(J)$: reconstruct image using erosion

$J \ominus B$: erosion of J by disk structuring element B

$\vee$: pointwise maximum

2) Thresholding values for segmentation are computed by mean (μ_I) and standard deviation (σ_I)

$$Th = \mu_I + \alpha \times \sigma_I \tag{3}$$

where Th : threshold value , μ_I : mean of input image

σ_I : standard deviation of input image, α : weighting factor

3) As a single target candidate may be detected as multiple targets, a closing and opening operation is used to eliminate multiple target candidates that are actually the same, single target.

4) Target candidates larger or smaller than expected target size are removed.

3   Proposed Clutter Rejection Method

Clutter rejection is needed to identify true targets by discarding the clutters from the target candidate provided in the signal processing stage. PCA features of targets and clutters are calculated using Mahalanobis distance, which is obtained from projection to eigenvectors of targets. Stochastic features are calculated using variance and standard deviation of edge image in target candidate. And optimal discrimination solution to identify is obtained from LDA using PCA and stochastic features.

3.1 PCA Feature Extraction

PCA features of targets and clutters are calculated with Mahalanobis distance when target candidate is projected to target eigenvectors.

PCA Features. PCA features are calculated using Euclidian distance by projecting target candidate to target eigenvectors. The procedures of PCA feature extraction is as follows.

1) Normalize N-target images using Eq. (4), which reduce the effect of background brightness.

$$I_{Nor} = \frac{(I_{TC} - m) - \mu_{std}}{std} + \mu_m \tag{4}$$

where I_{TC}: target chip image

m   : mean of target image

std : standard deviation of target image

μ_{std} : desired standard deviation for pre-defined

μ_m : desired mean for pre-defined

2) Calculate covariance matrix using Eq. (5)

$$\Sigma = \frac{1}{N}\sum_{n=1}^{N}\Phi_n\Phi_n^T \tag{5}$$

where $\Phi_i = \Gamma_i - \frac{1}{N}\sum_{n=1}^{N}I_{Nor\,n}$ and $\Gamma_i = [I_{Nor\,1}, I_{Nor\,2}, I_{Nor\,3}, ..., I_{Nor\,N}]$

3) Obtain eigenvalues using Eq. (5).

4) Calculate eigenvector using Eq. (6)

$$u_i = \frac{1}{N}\sum_{n=1}^{N}\lambda_{in}\Phi_n \quad (i = 1, 2, 3, ..., N) \tag{6}$$

where λ_{in} : eingenvalues of target image

Calculate PCA Feature using Euclid Distance. To obtain discrimination equation using LDA, Euclidian distance is calculated from train images. The procedure for Euclidian distance is as follow.

1) Normalize train image same as Eq. (4)
2) Project train images to eigenvectors of target as Eq. (6)

$$P_{pj} = u_i^T \cdot \varphi \tag{7}$$

where φ : normalized target image

3) Obtain Euclidian distance using Eq. (8)

$$E_{dis} = \sum_{n=1}^{N}|P_{pj}|^2 \tag{8}$$

3.2 Stochastic Feature Extraction

Edge image of man-made objects are much stronger than that of natural objects. So, stochastic features of edge images are also used for calculating discrimination equation to distinguish targets from clutters. To obtain stochastic features in same condition, the coefficients of variance (CV) of images for train are used

1) Calculate edge image for train image using Sobel operator.
2) Stochastic features are calculated using CV expressed as Eq. (9).

$$CV = \frac{std_{eg}}{m_{eg}} \times \alpha \tag{9}$$

where std_{eg} : standard deviation of edge image,

$\qquad m_{eg}$: mean of edge image, α : weighting factor

3.3 Clutter Separation Using LDA

To obtain threshold level to remove clutters from target candidates, optimal solution w which maximize the object function J(w) is calculated using PCA and stochastic features of target image.

$$J(w) \quad = \quad \frac{w^T S_b w}{w^T S_w w} \tag{10}$$

$$w^* \quad = \quad \underset{w}{\mathrm{argmax}} \{J(w)\} \quad = S_w^{-1}(\mu_1 - \mu_2)$$

where S_b : variance of interclass

S_w: variance of between class

And S_b, S_w is expressed as Eq. (11).

$$S_b = \frac{1}{c}\sum_{k=1}^{c}(\mu_k - \mu)(\mu_k - \mu)^T \tag{11}$$

$$S_w = \frac{1}{c}\sum_{k=1}^{c}\sum_{i|x_i=1}(\mu_k - \mu)(\mu_k - \mu)^T$$

where c: the number of classes, k: the number of train images

μ: mean of train image, μ_k : mean of target and clutter classes

3.4 Clutter Rejection from Target Candidates

Clutter rejection is performed using Eq. (7) – Eq. (11) to eliminate clutters from target candidates.

4 Simulation Results

4.1 Image Precessing

To demonstrate the performance of the proposed algorithm, various FLIR images obtained from a military infrared thermal sight installed on an army vehicle were tested. The IR images of targets were taken under different conditions, such as a riverside with a stone dike and natural field with a small stream and mountain, in different seasons year round. The image measured 720×480 pixels and the range between target and FLIR sensor was 1200m~1700m. We selected 202 images for the simulation. Fig. 2(a) shows raw IR images with three targets. Fig. 2(b) shows image with processed using grayscale reconstruction filters. Fig. 2(c) and Fig. 2(d) show binary images after thresholding of images Fig 2(a) and Fig 2(b). As you see in Fig. 2(c) and Fig. 2(d), Fig. 2(d) is better than Fig. 2(c) in that segmented binary images are much fewer and small targets are not included. As a result of effectiveness, target candidates are much well expressed in Fig. 2(d).

(a) input image   (b) image processed using grayscale reconstruction

(c) binary image using original image   (d) binary image using grayscale reconstruction

Fig. 2. (a) shows the input image from FLIR sensor. (b) shows that grayscale reconstruction filter is processed in input image. (c) shows that binary images obtained withdout image processing. (d) shows that binary image with image processing with grayscale reconstruction and small clutters are eliminated.

4.2 Clutter Rejection

The proposed clutter rejection technique operates on potential target images extracted from database for target chips and clutter chips. The size of target and clutter chips are 80×40 pixels. Target images for simulation are 5 targets, each with 8 different views (at a 45 degree separation in aspect angle). These target images are selected manually from database for calculating eigenvectors of targets. Fig. 3(a) shows target images of 5 targets with 8 different views. Fig. 3(b) shows eigenvectors of Fig. 3(a). Fig. 3(c) and Fig. 3(d) show some of target and clutter chips for PCA features (144 target and clutter chips are used for training). PCA features for targets are calculated using sum of each Euclid distance when target chips are projected to target eigenvectors. And PCA features for clutters are calculated using same method as target chips.

Fig. 4(a) and Fig. 4(b) show edge images of target and clutter chips. As in Fig. 4, edge images of man-made objects have stronger than those of nature. And stochastic features (CV of targets and clutters edge image) are calculated using edge images of target and clutter. Fig. 4 shows the target and clutter features. An x-axis shows the

PCA features, and y axis shows the stochastic features (CV of edge images). CV features scaled considering the magnitude of PCA features. Optimal solutions for LDA are calculated using Eq. 10.

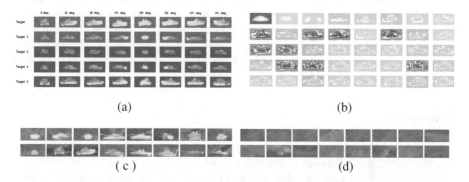

(a)   (b)

(c)   (d)

Fig. 3. (a) shows 5target image with 8 different views (at a 45 degree separation in aspect angle), (b) eigenvectors of targets (a), (c) and (d) some images of targets and clutters for calculating PCA features

(a) edge images of targets   (b) edge images of clutters

Fig. 4. These figures show edge images of targets and clutters. As they are shown, edge images of targets have stronger than those of clutters.

4.3 Simulation Results of Clutter Rejection

Target detection is simulated using 158 image including 177 targets. The result of LDA is applied to target candidates from input images. Simulation is performed in 3 cases: Only using PCA features, only using stochastic features and using PCA and stochastic features together. The results of simulation are shown in Table 1. In case of using PCA features, targets are detected well compared with using stochastic features. But clutters are rejected well when CV features used. The simulation results show that the performance of target detection and clutter rejection is more effective when PCA and stochastic features are used.

Table 1. Summarizations of simulation results

section	Number of test image	PCA features	CV features	PCA + CV features
Target	177	153	132	151
clutter	124	107	119	115
Detection rate(%)		86.4	83.4	88.4

5 Conclusion

This paper proposed a target detection algorithm using signal processing with gray-scale reconstruction filters and clutter rejection with PCA features and stochastic features. Gray scale reconstruction filters are used for small target rejection and a closing and opening operation is used to eliminate multiple target candidates that are actually the same, single. And clutters are eliminated using PCA and stochastic features. The simulation results using MATLAB showed that the proposed algorithm works well in cluttered and low contrast environment.

References

1. Otsu, N.: A Threshold Selection Method from Gray-Level Histogram. IEEE Trans. On Systems, Man, and Cybernetics SMC-9(1), 62–66 (1979)
2. Weszka, J.S., Rosenfeld, A.: Threshold Evaluation Techniques. IEEE Trans. On Systems, Man, and Cybernetics SMC-8(8), 622–629 (1978)
3. Sahoo, P.K., Soltani, S., Wang, A.K.C.: A Survey of Thresholding Techniques. Computer Vision, Graphics, and Image processing 41, 233–260 (1988)
4. Casasent, D., Ye, A.: Detection Filters and Algorithm fusion for ATR. IEEE Trans. Image Processing 6(1), 114–125 (1997)
5. Pham, Q.H., Brosnan, T.M., Smith, M.J.T.: Sequential Digital Filters for Fast Detection of Targets in FLIR Image Data. In: Proc. SPIE, vol. 3069, pp. 62–73 (1997)
6. Murenzi, R., et al.: Detection of targets in low resolution FLIR Imagery using two-Dimensional directional Wavelets. In: Proc. SPIE, vol. 3371, pp. 510–518 (1998)
7. Rizvi, S.A., Nasrabadi, N.M., Der, S.Z.: A Clutter Rejection Technique for FLIR Imagery Using Region-Based Principal Component Analysis. In: Proc. SPIE, vol. 3718, pp. 139–142 (1999)
8. Chan, L.A., Nasrabadi, N.M., Torrieri, D.: Bipolar Eigenspace separation Transformation for Automatic Clutter Rejection. In: Proc. IEEE Int. Conf., Image Processing, vol. 1, pp. 139–142 (1999)
9. Vincent, L.: Morphological Grayscale Reconstruction. IEEE Transactions on Image Processing 2(1), 176–201 (1993)
10. Hackyong, H.: Pattern Recognition Handbook, Hanbit Media, seoul (2005)

Cloud Image Resolution Enhancement Method Using Loss Information Estimation

Won-Hee Kim and Jong-Nam Kim

Div. of Electronic Computer and Telecommunication Engineering,
Pukyong University
{whkim,jongnamr}@pknu.ac.kr

Abstract. Image resolution enhancement techniques are required in various multimedia systems for image generation and processing. The main problem is an artifact such as blurring in image resolution enhancement techniques. Specially, cloud image have important information which need resolution enhancement technique without image quality degradation. To solve the problem, we propose error estimation and image resolution enhancement algorithm using low level interpolation. The proposed method consists of the following elements: error computation, error estimation, error application. Our experiments obtained the average PSNR 1.11dB which is improved results better than conventional algorithm. Also we can reduce more than 92% computation complexity. The proposed algorithm may be helpful for applications such as satellite image and cloud image.

Keywords: Cloud Image, Interpolation, Resolution Enhancement, Loss Information Estimation.

1 Introduction

Image resolution enhancement is a technique to increase image quality of low resolution image [1]. The technique is useful for applications in various multimedia systems such as image restoration, reconstruction, compression, and re-sampling [2]. But perfect image reconstruction is a very difficult work [3]. Some transformation processes cause information loss of images. It means error which is a hard to calculate correctly. Reconstruction images surely include some artifact such as blurring [4]. Specially, cloud image have important information which need resolution enhancement technique without image quality degradation.

To solve the problems, we propose an image resolution enhancement algorithm. First step, we calculated error using low level interpolation. Next step, proposed method estimated interpolation error using interpolation of the calculated error. Lastly, the estimated interpolation error to applied weighting is added the interpolated image. We confirmed improved image quality and reduced computation complexity.

This paper is organized as follows. In Section 2, the conventional image interpolation algorithms are surveyed. The proposed error estimation and image resolution enhancement methods are presented in Section 3. In Section 4, we discuss our experimental results. Finally, the concluding remarks are given in Section 5.

T.-h. Kim et al. (Eds.): SIP/MulGraB 2010, CCIS 123, pp. 115–120, 2010.

2 Related Works

To date, researches on image resolution enhancement have been extensively reported. Interpolation algorithms have been studied largely in spatial or frequency domain [5].

For image re-sampling, the interpolation step must reconstruct a two dimensional continuous signal $s(x,y)$ from its discrete sample $s(k,l)$ with s, x, t and k, l IN. Thus the amplitude at the position (x,y) must be estimated from its discrete neighbors. This can be described formally as the convolution of the discrete image samples with the continuous 2-D impulse response of a 2-D reconstruction filter.

$$s(x,y) = \sum_k \sum_l s(k,l)h_{2D}(x-k,y-l) \tag{1}$$

Usually, symmetrical and separable interpolation kernels are used to reduce the computational complexity.

$$h_{2D}(x,y) = h(x)h(y) \tag{2}$$

Following the sampling theory, the scanning of a continuous image $s(x,y)$ yields infinite repetitions of its continuous spectrum $S(u,v)$ in the Fourier domain, which don't overlap since the *Nyquist* criterion is satisfied [6]. If it happens only then, the original image $s(x,y)$ can be reconstructed perfectly from its samples $s(k,l)$ by multiplication of an appropriate rectangular prism in the Fourier domain. The 1-D ideal interpolation equals the multiplication with a rect function in the Fourier domain and can be defines as shown Eq. (3).

$$Ideal_h(x) = \frac{\sin(\pi x)}{\pi x} = sinc(x) \tag{3}$$

The nearest neighbor interpolation is the easiest method to approximate the *sinc* function by a spatially limited kernel [7]. The value $s(x)$ at the location (x) is chosen as the next known value $s(k)$. Therefore, only $N=1$ supporting point is required for the nearest neighbor interpolation and is given by the following convolution with a rect function as shown Eq. (4).

$$h_1(x) = \begin{cases} 1, & 0 \leq |x| < 5 \\ 0, & elsewhere \end{cases} \tag{4}$$

For separated bilinear interpolation, the values of both direct neighbors are weighted by their distance to the opposite point of interpolation [8]. Therefore the linear approximation of the *sinc* function follows the triangular function as shown Eq. (5).

$$h_2(x) = \begin{cases} 1 - |x|, & 0 \leq |x| < 1 \\ 0, & elsewhere \end{cases} \tag{5}$$

In these conventional algorithms, nearest neighbor interpolation requires the least computation, but results in blockiness in the interpolated image. The bilinear interpolation, while providing better quality in the smooth portions of images, tends to smooth edges and other discontinuities. Finally, the *sinc* interpolator cannot be used in practice because its infinite filter length causes a lack of locality and high complexity [9].

3   The Proposed Cloud Image Resolution Enhancement Method

Conventional image enhancement methods have some problems which include blurring effect. To solve the problem, we propose error estimation and image resolution enhancement algorithm using low level interpolation.

The proposed algorithm consists of the following elements: error computation, error estimation, error application.

In error computation step, we calculate error of LR image using low level interpolation. In this paper, we assume to know HR image information as I_H. We can define LR image which is damaged by down sampling and noise addition.

$$I_L = S_X I_H + N \tag{6}$$

In Eq. (6), S cause information loses in the interpolation process. Accordingly, a perfect reconstruction is impossible by the error of information loses. Therefore, we proposed a method to estimate the error as shown Eq. (7) ~ (10). The process consists of two steps: error computation and error estimation. The error computation scheme is defined by Eq. (7) ~ (9). The error is calculated by down sampling and interpolation. The down sample and interpolation scale is 1/2. In the equations, I_{LL} is down sampled image, I_{RL} is interpolated image, I_{EL} is an error image.

$$I_{LL} = S_2 I_L \tag{7}$$

$$I_{RL} = R_2 I_{LL} \tag{8}$$

$$I_{EL} = I_L - I_{RL} \tag{9}$$

The calculated error I_{EL} is defined accuracy information by computation of LR image using low level interpolation. In second step, we estimate real error using calculated error I_{EL}. The real error is estimated by interpolation of calculated error as shown Eq. (10).

$$I_{EH} = R_X I_{EL} \tag{10}$$

Lastly, interpolated LR image is defined as shown Eq. (11).

$$I_{RH} = R_X I_L \tag{11}$$

I_{RH} is interpolated LR image which have error by down sampling and interpolation. Therefore, we can obtain more correct reconstruction image by the sum interpolated LR image and estimated error as shown Eq. (12).

$$I_{HH} = I_{RH} + I_{EH} \tag{12}$$

I_{HH} is an image that is a sum of interpolated image and estimated error image. The proposed algorithm can have more correct image than general interpolated image using estimated error. Estimated error includes most edges information of image as high frequency domain. Therefore, I_{HH} have more edges information of image than general interpolated image. So the proposed method can obtain interpolated image of clear resolution.

But I_{EH} include both edges information and noise information. So the edges region is clearer, but an average intensity of image is increased. It causes image quality deterioration. To solve this problem, the proposed algorithm utilize weighting factor as shown Eq. (13).

$$I_{HH} = I_{RH} + aI_{EH} \ (0 < a < 1) \tag{13}$$

In the equation, a is a weighting factor. We should be applied weighting factor to increase image quality. Empirically, the weighting factor a is suitable about 0.3~0.7. The proposed methods can obtain highest PSNR using these values.

4   Experimental Results and Analysis

In this section, we presented performance of the proposed method by experiments. The proposed method was tested in the following conditions. PC specifications are CPU Pentium Core Duo 2.4GHz and RAM 2GB. The algorithms are implemented with MATLAB 7.5. Eight cloud images were used which have 512×512 image size, PGM format, gray scale.

We obtained experimental results by two measurements: objective image quality and complexity of algorithm. The objective quality is evaluated a PSNR (peak signal to noise ration). The complexity of algorithm is evaluated a run-time of program.

The proposed method only uses bilinear function as interpolation kernel, so we compare the proposed method with bilinear interpolation algorithm. We use m-file of MATLAB as the interpolation kernel. Generally, MATLAB m-files are correct and trustable functions. We obtained LR test images as 256×256 size by bilinear function in MATLAB 7.5. Comparison methods are utilized following four conventional algorithms: ICBI [10], INEDI [11], Castro's method [12], Takeda's method [13].

In the experiment 1, we describe results of objective quality comparison as shown Table 1. In table, the proposed method is improved average 1.1dB BIL is a bilinear interpolation method, PRO1 is the proposed method, PRO2 is the proposed method applied weighting. PRO2 is improved 1.11dB than conventional algorithms. In all case and all image, the proposed method obtained highest PSNR values. So, we confirmed more effective than comparison algorithms for image resolution enhancement by experiment 1.

In the experiment 2, we represent results of complexity comparison as shown Table 2. We can know that the proposed method have minimum time. The proposed method reduces computation complexity over 92%.

In conclusion, we can confirm that the proposed method have high image quality and low computation complexity by two experiments.

Table 1. Comparison results of PSNR(unit : dB)

IMAGES	ICBI[10]	INEDI [11]	Castro's method[12]	Takeda's method[13]	Proposed method
image1	33.46	33.21	33.71	33.21	34.34
image2	33.43	33.17	33.69	33.18	34.30
image3	33.42	33.15	33.69	33.18	34.29
image4	33.39	33.10	33.68	33.15	34.26
image5	33.62	33.33	33.91	33.38	34.47
image6	33.43	33.12	33.73	33.20	34.29
image7	33.42	33.14	33.72	33.18	34.27
image8	33.37	33.10	33.67	33.13	34.21
average	33.44	33.16	33.73	33.20	34.31

Table 2. Comparison results of run-time(unit : second)

	ICBI[10]	INEDI [11]	Castro's method[12]	Takeda's method[13]	Proposed method
average	1.60	55.41	2.00	21.13	0.12
ratio	13.11	454.32	16.36	173.23	1.00

5 Conclusions

In this paper, we proposed image resolution enhancement algorithms. First step, we calculated error using low level interpolation. Next step, proposed method estimated interpolation error using interpolation of the calculated error. Lastly, the estimated interpolation error to applied weighting is added the interpolated image. We confirmed experimental results that objective image quality is improved. Also, we reduced complexity of algorithms more than 92%. The proposed methods may be useful for applications in various multimedia systems.

Acknowledgments. This research was financially supported by MEST and KOTEF through the Human Resource Training Project for Regional Innovation, and supported by LEADER.

References

1. Park, S.C., Park, M.K., Kang, M.G.: Super-Resolution Image Reconstruction: A Technical Overview. Signal Processing Magazine IEEE 20(3), 21–36 (2003)
2. Dai, S., Han, M., Wu, Y., Gong, Y.: Bilateral Back-Projection for Single Image Super Resolution. In: IEEE International Conference on Multimedia and Expo., pp. 1039–1042 (July 2007)
3. Hardie, R.: A Fast Image Super-Resolution Algorithm Using an Adaptive Wiener Filter. IEEE Transactions on Image Processing 16(12), 2953–2964 (2007)

4. Shen, H., Zhang, L., Huang, B., Li, P.: A MAP Approach for Joint Motion Estimation, Segmentation, and Super Resolution. IEEE Transactions on Image Processing 16(2), 479–490 (2007)
5. Bai, Y., Zhuang, H.: On the Comparison of Bilinear, Cubic Spline, and Fuzzy Interpolation Techniques for Robotic Position Measurements. IEEE Transactions on Instrumentation and Measurement 54(6), 2281–2288 (2005)
6. Farsiu, S., Robinson, D., Elad, M., Milanfar, P.: Advances and Challenges in Super-Resolution. International Journal of Imaging Systems and Technology 14, 47–57 (2004)
7. Hong, S.H., Park, R.H., Yang, S.J., Kim, J.Y.: Image Interpolation Using Interpolative Classified Vector Quantization. Image Vis. Comput. 26(2), 228–239 (2008)
8. Qing, W., Ward, R.K.: A New Orientation-Adaptive Interpolation Method. IEEE Transactions on Image Processing 16(4), 889–900 (2007)
9. Banerjee, S.: Low-Power Content-Based Video Acquisition for Super-Resolution Enhancement. IEEE Transactions on Multimedia 11(3), 455–464 (2009)
10. Giachetti, A., Asuni, N.: Fast Artifacts-free Image Interpolation. In: Proc. of the British Machine Vision Conf., pp. 123–132 (2008)
11. Asuni, N.: INEDI – Tecnica Adattativa Per l'interpolazione di Immagini. Master's thesis, Università degli Studi di Cagliari (2007)
12. http://www.mathworks.com/matlabcentral/fileexchange/21410-increase-image-resolution
13. Takeda, H., Farsiu, S., Milanfar, P.: Kernel Regression for Image Processing and Reconstruction. IEEE Transactions on Image Processing 16(2), 349–366 (2007)

Multiple Ship Detection and Tracking Using Background Registration and Morphological Operations

Nasim Arshad[1], Kwang-Seok Moon[1], and Jong-Nam Kim[2]

[1] Dept. of Electronics Engineering,
Pukyong National University, Busan, 608-737 Korea
[2] Dept. of IT Convergence and Application Engineering
Pukyong National University, Busan, 608-737 Korea
fn.arshad@gmail.com, moonks@pknu.ac.kr, jongnam@pknu.ac.kr

Abstract. This paper presents a method to accurately detect and monitor ships within the area of interest. It is an advanced version of the previous works done regarding moving ship detection and tracking. The proposed tracking scheme is based on the characteristics of both sea and ship, which includes: background information and local position of the ship. Background subtraction and registration is achieved using morphological 'Open' operation and the ships are located using their edge information. The experimental results demonstrate robust and real-time ship detection and tracking with 98.7% detection rate. The proposed algorithm will be useful in coastal surveillance and monitoring applications.

1 Introduction

Automatic detecting and tracking ships in video surveillance data is a very stimulating problem in computer vision with important practical applications [1]-[2]. Coastal Surveillance is one of the major applications of ship detection and vessel identification and it is used in VIDS systems. Manually reviewing the large amount of data cameras generate is often impractical. Thus, algorithms for analyzing video contents which require little or no human input are a good solution. Video surveillance systems focus on background modeling, moving object classification and tracking [1]-[4].

Automating ship detection is difficult, due to the unpredictable environment and high requirements for robustness and accuracy. Tracking is also a fundamental technology to extract regions of interest and video object layers as defined in JPEG-2000 and MPEG-4 standards. Even though it is essential to many applications, robust object tracking under uncontrolled conditions still poses a challenge [1]-[6]. The increasing availability of video sensors and high performance video processing hardware opens up exciting possibilities for tackling many video understanding problems, among which ship tracking and target classification are very important and challenging. This is because very less work and few attempts have been made to

T.-h. Kim et al. (Eds.): SIP/MulGraB 2010, CCIS 123, pp. 121–126, 2010.
© Springer-Verlag Berlin Heidelberg 2010

achieve this goal and most of the past work dealt mainly on images taken by the satellites or Temporal Sequences of Marine Radar Images.

This improved research is more robust in tracking ships within the area of interest. The system uses a single camera which is either mounted on a fixed platform or on a moving ship. It captures video sequences of the sea, and if there is a ship passing through the area of interest it is detected and tracked until it exits the area. This tracking scheme is based on characteristics of both sea and ship, which include: background information and local position of the ship [4]-[6].

The Morphological Opening operation is used for background subtraction. The ships are easily detected using Sobel edge detector and Morphological (thicken – dilate – bridge) operations. Locating the approximate position of the ships allows us to track them by marking the locations in the frames of the video sequences.

2   Proposed Algorithm

2.1   Background Subtraction and Foreground Detection

Background subtraction is a commonly used class of techniques for segmenting out objects of interest in a scene for applications such as surveillance. A reliable and robust background subtraction algorithm should handle sudden or gradual illumination changes, high frequency, repetitive motion in the background, and long-term scene changes [3]. The architecture and flowchart of the proposed method is represented in Fig. 1. At first a video clip is read and decomposed into a number of frames. Next, using these frames as inputs, the stationary background image is registered and identified [1]-[6], i.e. by applying the Morphology Open operation on each RGB frame, an intermediate image frame is generated, which is then subtracted from the original RGB frame, causing the unnecessary background to fade.

In order to identify the foreground dynamic objects and to detect the existence of the ship, we subtract the registered background image (open operation) from the given input video frame. In other words the position of each ship is obtained using frames of differential images between the background and the input sequence.

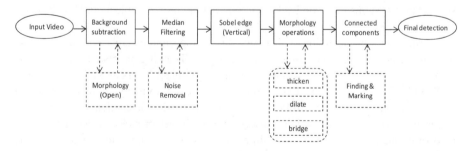

Fig. 1. System structure overview

2.2   Filtering and Morphological Processing

Usually due to the irregular movement of sea waves, there always exist some noise regions both in the foreground and background regions, so a Median filtering of 3x3 neighborhoods is applied to remove or minimize the noise and the unnecessary data. Median filtering is more effective than convolution when the goal is to simultaneously reduce noise and preserve edges.

All ships have vertical edges, where the sea region has mostly horizontal edges, so by extracting the vertical edges of each frame many unwanted data are eliminated.

This process detects outlines of the ship objects and boundaries between them and the background in the image [3]-[5], Most edge detection methods work on the assumption that the edge occurs where there is a discontinuity in the intensity function or a very steep intensity gradient in the image. Using this assumption, if one takes the derivative of the intensity value across the image and finds points where the derivatives are maximum then the edge can be located. We have employed the Sobel edge detection which is appropriate in our case.

Having extracted the edges we apply morphological operations to get the approximate position and the contour of the ship. The morphological operations used in this work are: *thicken*, *dilate* and *bridge* respectively.

> The "*thicken operation*" is used to thicken the objects by adding pixels to the exterior of the objects so it will result in previously unconnected objects being 8-connected. After Sobel edge detection many edges are disjoint.
> The "*dilate operation*" is used to thicken and enlarge the objects of interest in size.
> The "*bridge operation*", bridges unconnected pixels, that is, sets 0-valued pixels to 1 if they have two nonzero neighbors that are not connected.

The result of the morphological operations is a binary image where the presence of the ship is indicated as a white patch while the rest of the area is black.

2.3   Component Labeling and Counting

The approximate positions of the ships (white patches) are given as input to this phase. In order to track and detect the ships we should be able to count the number of objects in one image frame. We apply a counting and labeling algorithm onto the resultant images to count the number of ships, i.e. by finding the number of connected components we are able to find all the objects of interest in the frame. Due to occlusions, sometimes while performing the labeling algorithm, two components are merged together and treated as a single entity.

For detecting the ships each component on the original image frame is marked. Marking the consecutive frames of the videos generates the tracking movement.

3   Experimental Results

The proposed algorithm was executed on an Intel E2200 @ 2.20GHz CPU, with 1 GB RAM. All our algorithms were implemented in the MATLAB environment. The system was built on video-streams of YUV, AVI, MOV and MP4 formats taken by digital cameras, with input image resolutions of "1920 by 1080", "720 by 480", "640 by 480"and "320 by 240".

We tested the system on image sequences of 10 different videos, which were shot from the distance of 1~ 3 kilometers, with the zoom range of 25x ~ 100x with different angles. All the video sequences chosen for the ship tracking had different light intensity and were taken on different days with different backgrounds.

Table 1, shows the results of the proposed algorithm for the 10 different video sequences used.

The results reveal that the proposed algorithm is not affected by the shadows or the background of the object. Our test video sequences were comprised of minimum 100 to maximum 5000 frames, and the overall detection rate for the sequences was 98.7%. Fig. 2, shows the intermediate result of each phase of the proposed algorithm. Size of the images in the first row is 320*240, in the second row: 1920*1080, and the third row: 720*480. The images in the first column (a, e, i and m) are the original image frames; second column (b, f, j and n): after background subtraction; third column (c, g, k and o): after edge detection and morphology operation; and the forth column (d, h, l and p): Blue boxes indicate the detected ships. As observed from the figures, the background of each image frame is eliminated, and the ships are easily detected.

Table 1. Experimental results of the proposed algorithm

Input Video	Format	Dimensions	No. of frames	Actual moving objects	Detection Rate %
Video 1	YUV-AVI	720*480	2130	1	100
Video 2	YUV-AVI	720*480	3415	2	100
Video 3	YUV-MP4	720*480	5412	1	100
Video 4	YUV-MP4	1920*1080	6304	1	100
Video 5	MP4-AVI	1920*1080	1773	1	95.7
Video 6	MOV	320*240	120	2	93.5
Video 7	MOV	320*240	100	1	100
Video 8	MP4-AVI	720*480	2159	1	100
Video 9	MP4-AVI	1920*1080	3632	1	97.8
Video 10	MP4-AVI	640*480	100	1	100

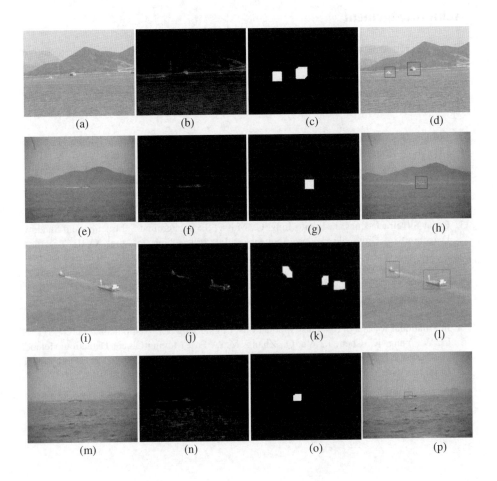

Fig. 2. Frames of different video sequences

4   Conclusions

As discussed above, ship detection and tracking are among the computationally most demanding object detection tasks. In this paper, we have developed and implemented a new algorithm for ship detection. The system can detect the ships within the region of interest (ROI) more accurately, and it is not affected by the shadows and background of the ship.

The implementation of this work has been successful with an overall correct identification rate of almost 98.7%, which was achieved in 0.375 seconds. The proposed system provides solutions to the problem of monitoring secured coastal areas. We expect hybrid solutions that can predict and apply the best combination of the detectors and trackers to be more popular in the future.

Acknowledgement

This research was financially supported by KOTEF through the Human Resource Training Project for Regional Innovation, and by LEADER.

References

1. Porikli, F.: Achieving Real-Time Object Detection and Tracking Under Extreme Conditions. Journal of Real-time Image Processing 1(1), 33–40 (2006) ISSN: 1861 8200
2. Vibha, L., Venkatesha, M., Rao, P.G., Suhas, N., Shenoy, P.D., Venugopal, K.R., Patnaik, L.M.: Moving Vehicle Identification using Background Registration Technique for Traffic Surveillance. In: International Multi-Conference of Engineers and Computer Scientists, Hong Kong, vol. 1, pp. 19–21 (2008)
3. Cezar Silveira Jacques, J., Rosito Jung, C., Musse, S.: Background Subtraction and Shadow Detection in Grayscale Video Sequences. In: Proceedings of the XVIII Brazilian Symposium on Computer Graphics and Image Processing, USA, pp. 189–197 (2005)
4. Wei, Z., Lee, D., Jilk, D., Schoenberger, R.: Motion Projection for Floating Object Detection. In: Proceedings of the 3rd International Conference on Advances in Visual Computing. LNCS, vol. 4842, pp. 152–161. Springer, Heidelberg (2007)
5. Khashman, A.: Automatic Detection, Extraction and Recognition of Moving Objects. International Journal of Systems Applications, Engineering & Development 2(1) (2008)
6. Li, W., Yang, K., Chen, J., Wu, Q., Zhang, M.: A Fast Moving Object Detection Method via Local Neighborhood Similarity. In: Proceedings of the 2009 IEEE, International Conference on Mechatronics and Automation, Changchun, China (August 9-12, 2009)

A Fast Motion Estimation Algorithm Using Unequal Search and Criterion for Video Coding

Jong-Nam Kim

Dept. of IT Convergence and Application Engineering
Pukyong National University, Busan, 608-737 Korea
jongnam@pknu.ac.kr

Abstract. In this paper, we propose an algorithm that reduces unnecessary computations, while keeping almost same prediction quality as that of the full search algorithm. In the proposed algorithm, we can reduce unnecessary computations efficiently by calculating initial matching error point from first partial errors. To do that, we use tighter elimination condition as error criterion than the conventional PDE algorithm. Additionally, we use different search strategy compared with conventional spiral search pattern. By doing that, we can increase the probability that hits minimum error point as soon as possible. Our algorithm decreases the computational amount by about 50% of the conventional PDE algorithm without any degradation of prediction quality. Our algorithm would be useful in real-time video coding applications using MPEG-2/4 AVC standards.

1   Introduction

In video compression, full search (FS) algorithm based on block matching algorithm (BMA) finds them optimal motion vectors which minimize the matching difference between reference block and candidate block. It has been widely used in video coding applications because of its simple and easy hardware implementation. However, heavy computational load of the full search with very large search range can be a significant problem in real-time video coding application. Many fast motion estimation algorithms to reduce the computational load of the full search have been studied in the last decades.

We classify these fast motion estimation methods into two main groups. One is lossy motion estimation algorithm with degradation of predicted images and the other is lossless one without any degradation of predicted images compared with the conventional FS algorithm. The former includes following subgroups: unimodal error surface assumption (UESA) techniques, multi-resolution techniques, variable search range techniques with spatial/temporal correlation of the motion vectors, half-stop techniques using threshold of matching distortion, integral projection technique of matching block, low bit resolution techniques, sub-sampling techniques of matching block, and so on [1]. The latter as fast full search technique contains following several algorithms: successive elimination algorithm (SEA) using sum of reference block and candidate block and its modified algorithms, partial distortion elimination (PDE) method and its modified algorithms, and so on [2].

T.-h. Kim et al. (Eds.): SIP/MulGraB 2010, CCIS 123, pp. 127–133, 2010.

In lossless motion estimation algorithms, PDE is very efficient algorithm to reduce unnecessary computation for matching error calculation. To further reduce unnecessary computation in calculating matching error, J.N. Kim and et al [3] proposed fast PDE algorithms based on adaptive matching scan, which requires additional computation to get matching scan order. But the additional computation for the matching scan order can be burden when cascading other fast motion estimation algorithm such as SEA.

In this paper, we propose a fast motion estimation algorithm to reduce computational amount of the PDE algorithm with almost same predicted images compared with the conventional PDE algorithm for motion estimation. We reduced only unnecessary computations which doesn't affect predicted images from the motion vector. To do that, we do block matching by initial partial SAD (sum of absolute difference). Unlike previous approaches, we try to find a good candidate with minimum error from initial partial errors. By doing that, we can find minimum error point faster than the conventional spiral search pattern. Our algorithm can easily control the trade-off between prediction quality and computational reduction for motion estimation with proper error criterion. Our algorithm reduces by about 50% of computations for block matching error compared with the conventional PDE algorithm without any degradation of prediction quality.

2   Proposed Algorithm

In fast lossless motion estimation, partial distortion elimination (PDE) algorithm reduces efficiently only unnecessary computation by stopping remaining calculation in a matching block. In PDE algorithm, if an intermediate sum of matching error is larger than the minimum value of matching error at that time, the remaining computation for matching error of the block is abandoned.

$$APSAD_k = \sum_{i=1}^{k} \sum_{j=1}^{N} |f_t(i,j) - f_{t-1}(i+x, j+y)| \tag{1}$$

where $k = 1, 2, \cdots, N$

$$APSAD_k \leq SAD_{min} \tag{2}$$

where $k = 1, 2, \cdots, N$

$$APSAD_k \leq \left(alpha * \frac{k}{N} \right) * SAD_{min} \tag{3}$$

where $k = 1, 2, \cdots, N$

Eq. (1) and (2) show the partial sum of absolute difference (SAD) matching criterion of each row in the conventional PDE. Eq. (1), $f_t(i,j)$ denotes the pixel value for matching block at position (i,j) at time t, $f_{t-1}(i+x, j+y)$ denotes the pixel value of matching block at position $(i+x, j+y)$ at time $t-1$. Alpha is some weighting value.

As described previously, conventional fast PDE algorithms focused on getting large matching error in a matching block via adaptive matching. In this paper, we try to get minimum candidate with some clues. Unlike conventional PDE methods, we get initial 1/N block matching error, and then find a minimum error and its position from initial matching errors.

Here, N is the size of a matching block. When we get initial 1/N matching error, we use not a sub-block, but sub-sampling pattern to get more even error distribution. After that, we calculate a SAD for the minimum error point of the previous step. We eliminate impossible candidates according to Eq. (3), in which, we don't use PDE condition to eliminate impossible candidates because PDE condition is too loose to filter out unlikely candidate. We can summarize our algorithm as shown in Fig. 1.

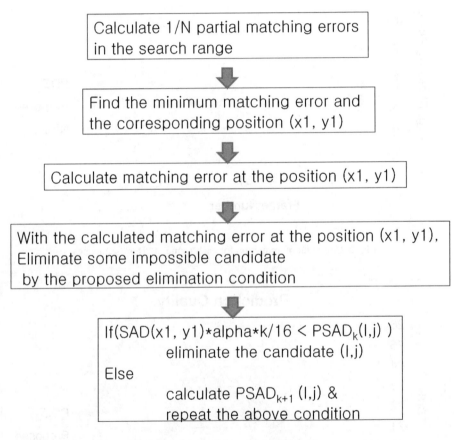

Fig. 1. Procedure of the proposed algorithm

3   Experimental Results

To compare the performance of the proposed algorithm with the conventional algorithms, we use 95 frames of "bus", "bally", "bicycle", "flower garden" and "football" video sequences. Matching block size is 16x16 pixels and the search window is ± 15 and ± 7 pixels. Image format is SD(720*480) for each sequence and only forward prediction is used. The experimental results are shown in terms of computation reduction rate with average checking rows and prediction quality with peak-signal-to-noise ratio (PSNR). We compared the proposed algorithm with the conventional full search algorithm, conventional PDE [4], and Normalized PDE algorithm [5]. We didn't compare the results of complexity based PDE algorithm [3] with ours because it can be cascaded to ours.

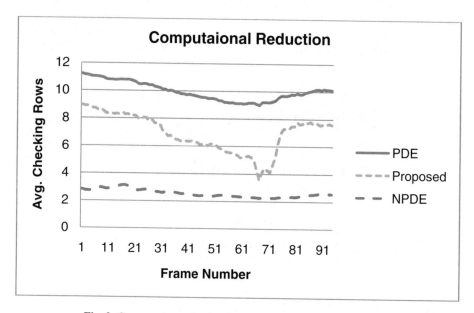

Fig. 2. Computation reduction for each frame of "Bus" sequences

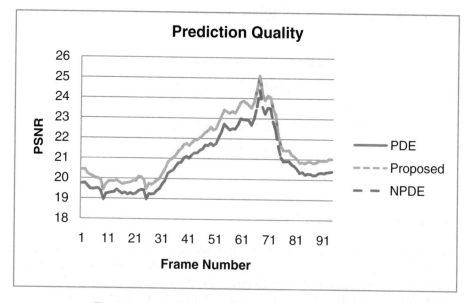

Fig. 3. Prediction quality for each frame of "Bus" sequences

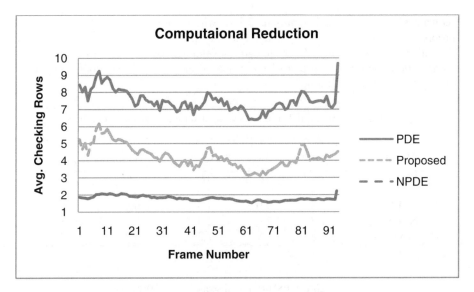

Fig. 4. Computation reduction for each frame of "Football" sequences

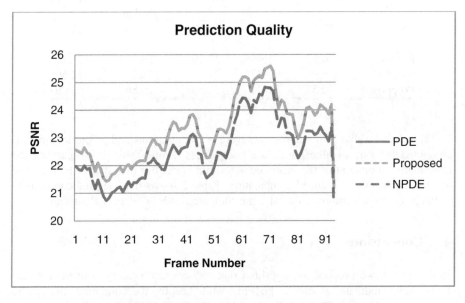

Fig. 5. Prediction quality for each frame of "Footbal" sequences

Fig. 2 and Fig 4 show the reduced average computation computed for various algorithms in "Bus" and "Football" sequences. The horizontal axis represents the number of frames and the vertical axis represents reduced computation. As you can see, the proposed method has smaller computations than the conventional PDE. Fig. 3 and Fig. 5 show prediction quality with PSNR measure in "Bus" and "Football" sequences. The vertical axis represents PSNR with the unit of dB. As you can see, the proposed method has the same prediction quality as that of than the conventional PDE and full search algorithms. Even though conventional NPDE algorithm has smaller computational amount than that of the proposed algorithm, it has poorer prediction quality than ours for all frames.

Table 1. Average number of rows computed for all sequences of 30Hz

Algorithms	bally	bus	bicycle	flower garden	foot-ball
FS	16	16	16	16	16
PDE[4]	7.5	9.9	9.5	6.6	7.5
NPDE[5]	1.7	2.5	2.3	1.6	1.8
Propposed	5.6	6.9	7.3	3.2	4.2

Table 2. PSNR   for all sequences of 30Hz

Algorithms	bally	bus	bicycle	Flower garden	football
FS	30.57	21.45	22.59	26.86	23.29
PDE[4]	30.57	21.45	22.59	26.86	23.29
NPDE[5]	29.99	20.82	21.90	26.25	22.56
Propposed	30.57	21.44	22.59	26.86	23.29

Table 1 and Table 2 summarize the average computations and prediction qualities computed for various algorithms in all sequences. From Table 1, we can see that computational amount of the proposed algorithm is smaller than that of PDE and FS algorithm. Despite of reduced computation, Table 2 shows that the prediction quality of the proposed algorithm is almost same compared with full search algorithm.

4   Conclusions

In this paper, we propose an algorithm that reduces unnecessary computations and finds likely candidates as soon as possible, while keeping the same prediction quality as that of the full search. In the proposed algorithm, we set initial matching error with the result of first 1/16 partial error in the search range. By checking initial SAD from first 1/16 partial error, we can increase the probability of finding minimum error points. We can decrease computational amount by 50% for motion estimation with almost same prediction quality. Our algorithm will be useful in real-time video coding applications using MPEG-2/4 AVC standards.

Acknowledgement

This work was supported from LEADER by RIS Project by KIAT.

References

[1] Dufaus, F., Moscheni, F.: Motion estimation techniques for digital TV: A review and a new contribution. Proceedings IEEE 83, 858–876 (1995)

[2] Kim, J.N.: A study on fast block matching algorithm of motion estimation for video compression, Ph.D. Thesis of GIST (2001)

[3] Kim, J.N., Byun, S.C., Kim, Y.H., Ahn, B.H.: Fast full search motion estima-tion algorithm using early detection of impossible candidate vectors. IEEE Trans. Signal Processing 50, 2355–2365 (2002)

[4] http://iphome.hhi.de/suehring/tml/download/old_jm/

[5] Cheung, C.K., Po, L.M.: Normalized partial distortion search algorithm for block motion estimation. IEEE Trans. Circuits Syst. Video Technol. 10, 417–422 (2000)

A Survey on Automatic Speaker Recognition Systems

Zia Saquib, Nirmala Salam, Rekha P. Nair, Nipun Pandey, and Akanksha Joshi

CDAC-Mumbai, Gulmohar Cross Road No.9, Juhu, Mumbai-400049
{saquib,nirmala,rekhap,nipun,akanksha}@cdacmumbai.in

Abstract. Human listeners are capable of identifying a speaker, over the telephone or an entryway out of sight, by listening to the voice of the speaker. Achieving this intrinsic human specific capability is a major challenge for Voice Biometrics. Like human listeners, voice biometrics uses the features of a person's voice to ascertain the speaker's identity. The best-known commercialized forms of voice Biometrics is Speaker Recognition System (SRS). Speaker recognition is the computing task of validating a user's claimed identity using characteristics extracted from their voices. This literature survey paper gives brief introduction on SRS, and then discusses general architecture of SRS, biometric standards relevant to voice/speech, typical applications of SRS, and current research in Speaker Recognition Systems. We have also surveyed various approaches for SRS.

Keywords: SRS, Speaker Recognition Systems, Voice Biometrics, Speech.

1 Introduction

1.1 Brief Overview of Speaker Recognition

Voice biometrics specifically was first developed in 1970, and although it has become a sophisticated security tool only in the past few years, it has been seen as a technology with great potential for much longer. The most significant difference between voice biometrics and other biometrics is that voice biometrics is the only commercial biometrics that process acoustic information. Most other biometrics is image-based. Another important difference is that most commercial voice biometrics systems are designed for use with virtually any standard telephone or on public telephone networks. The ability to work with standard telephone equipment makes it possible to support broad-based deployments of voice biometrics applications in a variety of settings. In contrast, most other biometrics requires proprietary hardware, such as the vendor's fingerprint sensor or iris-scanning equipment. By definition, voice biometrics is always linked to a particular speaker. The best-known commercialized forms of voice biometrics are Speaker Recognition. Speaker recognition is the computing task of validating a user's claimed identity using characteristics extracted from their voices.

A speaker's voice is extremely difficult to forge for biometrics comparison purposes, since a myriad of qualities are measured ranging from dialect and speaking style to pitch, spectral magnitudes, and format frequencies. The vibration of a user's vocal chords and the patterns created by the physical components resulting in human

T.-h. Kim et al. (Eds.): SIP/MulGraB 2010, CCIS 123, pp. 134–145, 2010.

Table 1. Typical Applications of Speaker Recognition Systems

Areas	Specific Applications
Authentication	Remote Identification & Verification, Mobile Banking, ATM Transaction, Access Control
Information Security	Personal Device Logon, Desktop Logon, Application Security, Database Security, Medical Records, Security Control for Confidential Information
Law Enforcement	Forensic Investigation, Surveillance Applications
Interactive Voice Response	Banking over a telephone network, Information and Reservation Services, Telephone Shopping, Voice Dialing, Voice Mail

speech are as distinctive as fingerprints. Voice Recognition captures the unique characteristics, such as speed and tone and pitch , dialect etc associated with an individual's voice and creates a non-replicable voiceprint which is also known as a speaker model or template. This voiceprint which is derived through mathematical modeling of multiple voice features is nearly impossible to replicate. A voiceprint is a secure method for authenticating an individual's identity that unlike passwords or tokens cannot be stolen, duplicated or forgotten.

1.2 Voice Production Mechanism

The origin of differences in voice of different speakers lays in the construction of their articulatory organs, such as the length of the vocal tract, characteristics of the vocal chord and the differences in their speaking habits. An adult vocal tract is approximately 17 cm long and is considered as part of the speech production organs above the vocal folds (earlier called as the vocal chords). As shown in Figure 1, the speech production organs includes the laryngeal pharynx (below the epiglottis), oral pharynx (behind the tongue, between the epiglottis and vellum), oral cavity (forward of the velum and bounded by the lips, tongue, and palate), nasal pharynx (above the velum, rear end of nasal cavity) and the nasal cavity (above the palate and extending from the pharynx to the nostrils). The larynx comprises of the vocal folds, the top of the cricoids cartilage, the arytenoids cartilages and the thyroid cartilage. The area between the vocal folds is called the glottis. The resonance of the vocal tract alters the spectrum of the acoustic as it passes through the vocal tract. Vocal tract resonances are called formants. Therefore the vocal tract shape can be estimated from the spectral shape (e.g., formant location and spectral tilt) of the voice signal. Speaker recognition systems use features generally derived only from the vocal tract. The excitation source of the human vocal also contains speaker specific information. The excitation is generated by the airflow from the lungs, which thereafter passes through the trachea

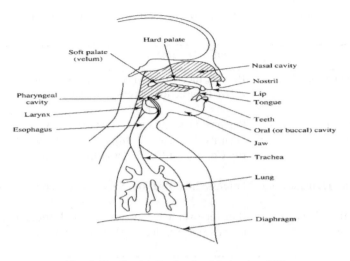

Fig. 1. The Speech Production Mechanism [29]

and then through the vocal folds. The excitation is classified as phonation, whispering, frication, compression, vibration or a combination of these. Phonation excitation is caused when airflow is modulated by the vocal folds. When the vocal folds are closed, pressure builds up underneath them until they blow apart. The folds are drawn back together again by their tension, elasticity and the Bernoulli Effect. The oscillation of vocal folds causes pulsed stream excitation of the vocal tract. The frequency of oscillation is called the fundamental frequency and it depends upon the length, mass and the tension of the vocal folds. The fundamental frequency therefore is another distinguishing characteristic for a given speaker.

1.3 How the Technology Works

The underlying premise for speaker recognition is that each person's voice differs in pitch, tone, and volume enough to make it uniquely distinguishable. Several factors contribute to this uniqueness: size and shape of the mouth, throat, nose, and teeth, which are called the articulators and the size, shape, and tension of the vocal cords. The chance that all of these are exactly the same in any two people is low. The manner of vocalizing further distinguishes a person's speech: how the muscles are used in the lips, tongue and jaw. Speech is produced by air passing from the lungs through the throat and vocal cords, then through the articulators. Different positions of the articulators create different sounds. This produces a vocal pattern that is used in the analysis.

A visual representation of the voice can be made to help the analysis. This is called a spectrogram also known as voiceprint, voice gram, spectral waterfall, and sonogram. A spectrogram displays the time, frequency of vibration of the vocal cords (pitch), and amplitude (volume). Pitch is higher for females than for males.

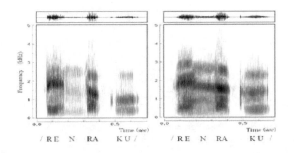

Fig. 2. These voiceprints are a visual representation of two different speakers saying "RENRAKU" [1]

1.4 Methodology

Each speaker recognition system has two phases: Enrollment and verification. During enrollment, the speaker's voice is recorded and typically a number of features are extracted to form a voice print, template, or model. In the verification phase, a speech sample or "utterance" is compared against a previously created voice print

Speaker recognition systems fall into two categories: Text-Dependent and Text-Independent.In a text-dependent system, text is same during enrollment and verification phase. In Text-independent systems the text during enrollment and test is different. In fact, the enrollment may happen without the user's knowledge, as in the case for many forensic applications.

2   General Speaker Recognition System Architecture

There are two major commercialized applications of speaker recognition technologies and methodologies: Speaker Identification and Speaker Verification.

2.1   SIS (Speaker Identification System)

Speaker Identification can be thought of as the task of finding who is talking from a set of known voices of speakers. It is the process of determining who has provided a given utterance based on the information contained in speech waves. Speaker identification is a 1: N match where the voice is compared against N templates.

2.2   SVS (Speaker Verification System)

Speaker Verification on the other hand is the process of accepting or rejecting the speaker claiming to be the actual one. Speaker verification is a 1:1 match where one speaker's voice is matched to one template.

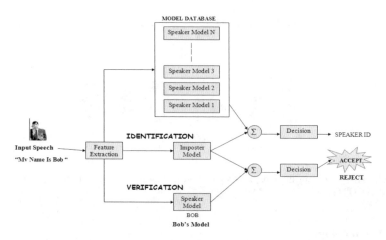

Fig. 3. General SIS and SVS Architecture [2]

3 Voice Biometric Standards

Standards play an important role in the development and sustainability of technology, and work in the international and national standards arena will facilitate the improvement of biometrics. The major standards work in the area of speaker recognition involves:

- Speaker Verification Application Program Interface(SVAPI)
- Biometric Application Program Interface (BioAPI)
- Media Resource Control Protocol (MRCP)
- Voice Extensible Markup Language (VoiceXML)
- Voice Browser (W3C)

Of these, BioAPI has been cited as the one truly organic standard stemming from the BioAPI Consortium, founded by over 120 companies and organizations with a common interest in promoting the growth of the biometrics market.

4 Speech Database

Table 2. Publicly available speech databases

Database	Website
TIMIT	http://www.ldc.upenn.edu/Catalog/CatalogEntry.jsp?catalogId=LDC93S1
NIST	http://itl.nist.gov/iad/mig/tests/sre
NOIZEUS	http://www.utdallas.edu/~loizou/speech/noizeus/
NTIMIT	http://www.ldc.upenn.edu/Catalog/CatalogEntry.jsp?catalogId=LDC93S2
YOHO	http://www.ldc.upenn.edu/Catalog/CatalogEntry.jsp?catalogId=LDC9z4S16

5   Leading Vendors of Speaker Recognition Systems

Table 3. List of Vendors

S.No	Vendor	Website
1.	Persay (NY, USA)	www.persay.com
2.	Agnito (Spain)	www.agnitio.es
3.	TAB Systems Inc. (Slovenia, Europe)	www.tab-systems.com
4.	DAON (Washington DC)	www.daon.com
5.	Smartmatic (USA)	www.smartmatic.com
6.	Speech Technology Center (Russia)	www.speechpro.com
7.	Loquendo (Italy)	www.loquendo.com/en/
8.	SeMarket (Barcelona)	www.semarket.com
9.	Recognition Technologies Ltd.(NY)	www.speakeridentification.com

6   Current Research in Speaker Recognition Systems

Indian Institute of Technology, Guwahati: Study of Source Features for Speech Synthesis and Speaker Recognition & Development of Person Authentication System based on Speaker Verification in Uncontrolled Environment.

Indian Institute of Technology, Kharaghpur: Development of speaker verification software for single to three registered user(s) & Development of Speaker Recognition Software for Telephone Speech.

Speech Technology and Research Laboratory, SRI International, CA: Speaker Recognition and Talk Printing.

Speech and Speaker Modeling Group, University Of Texas at Dallas: Dialect / Accent Classification & In-Set Speaker Recognition & Speaker Normalization.

The Centre for Speech Technology Research, University of Edinburgh, United Kingdom: Voice transformation.

Human Language Technology Group, Lincoln Laboratory, Massachusetts Institute of Technology: Forensic Speaker Recognition Project.

7   Issues Pertaining to SRS

Speaker recognition and verification has been an area of research for more than four decades and thus have many challenges that are needed to overcome.

(a) Hackers might attempt to gain unauthorized access to a voice authenticated system by playing back a pre-recorded voice sample from an authorized user..
(b) A major issue facing all biometric technologies that store data is maintaining the privacy of that data. As soon as a user registers with a voice biometric system, that

voiceprint is stored somewhere just like an address or a phone number. What if companies decide to sell voiceprints like addresses?

(c) If the data is encrypted in storage and in transport, there is always the possibility of cracking the encryption and stealing the data.

(d) Designing long-range features (which by definition occur less frequently than very short-range features) that provide robust additional information even for short (e.g., 30 seconds) training and test spurts of speech.

(e) Develop methods for feature selection and model combination at the feature level, that can cope with large numbers of interrelated features, odd feature space distributions, inherent missing features (such as pitch when a person is not voicing), and heterogeneous feature types.

8 SRS Modules

8.1 Preprocessing

The captured voice may contain unwanted background noise, unvoiced sound, and there can be a device mismatch, environmental mismatch between training and testing voice data which subsequently leads to degradation in the performance of Speaker Recognition System. The process of removal of this unwanted noise, dividing sounds into voiced and unvoiced sounds and channel compensation etc for the enhancement of speech/voice is called pre-processing.

Speech Enhancement (Denoising): Numerous schemes have been proposed and implemented that perform speech enhancement under various constraints/assumptions and deal with different issues and applications.

Table 4. Various approaches for Speech Enhancement

S.No.	Approach	Characteristic
1.	Chowdhary et al. [2008][3]	Improvement over the conventional power spectral subtraction method.
2.	Jun et al. [2009][4]	Based on fast noise estimation.
3.	Hansen et al. [2006][5]	A Generalized MMSE estimator (GMMSE) is formulated after study of different methods of MMSE family.
4.	Hasan et al. [2010][6]	Considers the constructive and destructive interference of noise in the speech signal.
5.	Lev-Ari et al. [2003][7]	This algorithm is an extension to signal subspace approach for speech enhancement to colored-noise processes.

Channel Compensation: Channel effects, are major causes of errors in speaker recognition and verification systems. The main measures to improving channel robustness of speaker recognition system are channel compensation and channel robust features.

Table 5. Various approaches for Channel Compensation

S.No	Approach	Characteristics
1.	Wu et al. [2006][8]	Utilizes channel-dependent UBMs as a priori knowledge of channels for speaker model synthesis.
2.	Han et al. [2010] [9]	Applies MAP channel compensation, pitch dependent feature and Speaker model.
3.	Calvo et al. [2007][10]	This paper examines the application of Shifted Delta Cepstral (SDC) features in biometric speaker verification and evaluates its robustness to channel/handset mismatch due by telephone handset variability.
4.	Zhang et al. [2008][30]	GMM super vectors generated by stacking means of speaker models can be seen as combination of two parts: universal background model (UBM) super vector and maximum a posteriori (MAP) adaptation part
5.	Neville et al. [2005][31]	Blind equalization techniques with QPSK modulation

9.2 Feature Extraction

The purpose of this module is to convert the speech waveform to some type of parametric representation (at a considerably lower information rate).The heart of any speaker recognition system is to extract speaker dependent features from the speech.

They are basically categorized into two types: low level and high level features.

Low Level Features: Low level features are short range features.

Table 6. Various approaches for extraction of low level features

S.No	Approach	Characteristic
1.	Prahallad et al. [2007][11]	Auto-associative neural network (AANN) and formant features
2.	Chakroborty et al. [2009][12]	Gaussian filter based MFCC and IMFCC scaled filter bank is proposed in this paper
3.	Revathi et al. [2009][13]	Perceptual Features & Iterative clustering approach for both speech and speaker recognition
4.	Huang et al. [2008][14]	Fusion of pitch and MFCC GMM Supervectors Systems on score level
5.	Deshpande et al. [2010][15]	AWP for Speaker Identification and multiresolution capabilities of wavelet packet transform are used to derive the new features.

High Level Features: Higher level features are long range features of voice that have attracted attention in automatic speaker recognition in recent years.

Table 7. Various approaches for extraction of high level features

S.No	Approach	Characteristic
1.	Campbell et al. [2007][16]	SVM and new kernel based upon linearizing a log likelihood ratio scoring system
2.	Baker et al. [2005][17]	Presegmentation of utterances at word level using ASR system, HMM & GMM used
3.	Mary et al. [2008][18]	Syllable-like unit is chosen as the basic unit for representing the prosodic characteristics
4.	Dehak et al. [2007][19]	Prosodic feature is extracted using a basis consisting of Legendre polynomials and is modeled using joint factor analysis.
5.	Campbell et al. [2004][32]	Proposes the use of support vector machines and term frequency analysis of phone sequences to model a given speaker.

9.3 Modeling

Speaker Model Generation: The feature vectors of speech are used to create a speaker's model/template. The recognition decision depends upon the computed distance between the reference template and the template devised from the input utterance.

Table 8. Various approaches for Speaker Model Generation

S.No	Approach	Characteristic
1.	Aronowitz et al. [2005][20]	Used GMM simulation & Compression algorithm
2.	Zamalloayz et al. [2008][21]	GA(Genetic Algorithm) & Comparison with LDA and PCA
3.	Aronowitz et al. [2007][22]	Based on approximating GMM likelihood scoring using ACE ,GMM compression algorithm
4.	Apsingekar et al. [2009][23]	GMM-based speaker models are clustered using a simple k-means algorithm

9.4 Matching /Decision Logic

Score Normalization: Score normalization has become a crucial step in the biometric fusion process. It makes uniform input data that comes from different sources or processes, hence reduces the biased information created by the differences between the different pre-processors.

Table 9. Various approaches for Score Normalization

S.No	Approach	Characteristic
1.	Puente et al. [2010][24]	New normalization algorithm DLin is proposed.
2.	Guo et al. [2008][25]	An unsupervised score normalization is proposed.
3.	Castro et al. [2007][26]	Score normalization technique based on test-normalization method (Tnorm) is presented
4.	Zajic et al. [2007] [27]	Unconstraint cohort extrapolated normalization, is introduced.
5.	Sturim et al. [2005][28]	A new method of speaker Adaptive-Tnorm that offers advantages over the standard Tnorm by adjusting the speaker set to the target model is presented.

10 Conclusions

In this paper, we have presented an extensive survey of automatic speaker recognition systems. We have categorized the modules in speaker recognition and discussed different approaches for each module. In addition to this, we have presented a study of the various typical applications of Speaker Recognition Systems, list of vendors worldwide and the current research being carried out in the field of speaker recognition. We have also discussed issues and challenges pertaining to the Speaker Recognition Systems.

References

1. SANS Information Security Reading Room, http://www.sans.org
2. Biometrics.gov,
   http://www.biometrics.gov/Documents/SpeakerRec.pdf
3. Chowdhury, M.F.A., Alam, M.J., Alam, M.F.A., O'Shaughnessy, D.: Perceptually weighted multi-band spectral subtraction speech enhancement technique. In: International Conference on Electrical and Computer Engineering, ICECE 2008, pp. 395–399 (2008)
4. Jun, L., He, Z.: Spectral Subtraction Speech Enhancement Technology Based on Fast Noise Estimation. In: ICIECS (2009)
5. Hansen, J.H.L., Radhakrishnan, V., Arehart, K.H.: Speech Enhancement Based on Generalized Minimum Mean Square Error Estimators and Masking Properties of the Auditory System. IEEE Transactions on Audio, Speech, and Language Processing 14(6), 2049–2063 (2006)
6. Hasan, T., Hasan, M.K.: MMSE estimator for speech enhancement considering the constructive and destructive interference of noise. Signal Processing, IET 4(1), 1–11 (2010)
7. Lev-Ari, H., Ephraim, Y.: Extension of the signal subspace speech enhancement approach to colored noise. IEEE Signal Processing Letters 10(4), 104–106 (2003)

8. Wu, W., Zheng, T.F., Xu, M.: Cohort-Based Speaker Model Synthesis for Channel Robust Speaker Recognition. In: ICASSP (2006)
9. Han, J., Gao, R.: Text-independent Speaker Identification Based on MAP Channel Compensation and Pitch-dependent Features. In: IJECSE (2010)
10. Calvo, J.R., Fernandez, R., Hernandez, G.: Channel / Handset Mismatch Evaluation in a Biometric Speaker Verification Using Shifted Delta Cepstral Features. In: Rueda, L., Mery, D., Kittler, J. (eds.) CIARP 2007. LNCS, vol. 4756, pp. 96–105. Springer, Heidelberg (2007)
11. Prahallad, K., Varanasi, S., Veluru, R., Bharat Krishna, M., Roy, D.S.: Significance of Formants from Difference Spectrum for Speaker Identification. In: INTERSPEECH 2006 (2006)
12. Chakroborty, S., Saha, G.: Improved Text-Independent Speaker Identification using Fused MFCC & IMFCC Feature Sets based on Gaussian Filter. In: IJSP (2009)
13. Revathi, A., Ganapathy, R., Venkataramani, Y.: Text Independent Speaker Recognition and Speaker Independent Speech Recognition Using Iterative Clustering Approach. In: IJCSIT, vol. 1(2) (2009)
14. Huang, W., Chao, J., Zhang, Y.: Combination of Pitch and MFCC GMM Supervectors for Speaker. In: ICALIP (2008)
15. Deshpande, M.S., Holambe, R.S.: Speaker Identification Using Admissible Wavelet Packet Based Decomposition. International Journal of Signal Processing 6, 1 (2010)
16. Campbell, W.M., Campbell, J.P., Gleason, T.P., Reynolds, D.A., Shen, W.: Speaker Verification using Support Vector Machines and High-Level Feature. IEEE Transactions on Audio, Speech, And Language Processing 15(7) (2007)
17. Baker, B., Vogt, R., Sridharan, S.: Gaussian Mixture Modeling of Broad Phonetic and Syllabic Events for Text-Independent Speaker Verification. In: Euro speech (2005)
18. Mary, L., Yegnanarayana, B.: Extraction and representation of prosodic features for language and speaker recognition. ELSEVIER Speech Communication 50, 782–796 (2008)
19. Dehak, N., Dumouchel, P., Kenny, P.: Modeling Prosodic Features with Joint Factor Analysis for Speaker Verification. IEEE Transactions on Audio, Speech and Language Processing 15(7), 2095–2103 (2007)
20. Aronowitz, H., Burshtein, D.: Efficient Speaker Identification and Retrieval. In: Proc. Interspeech 2005, pp. 2433–2436 (2005)
21. Zamalloayz, M., Rodriguez-Fuentesy, L.J., Penagarikanoy, M., Bordely, G., Uribez, J.P.: Feature Dimensionality Reduction Through Genetic Algorithms For Faster Speaker Recognition. In: EUSIPCO 2008 16th European Signal Processing Conference (2008)
22. Aronowitz, H., Burshtein, D.: Efficient Speaker Recognition Using Approximated Cross Entropy (ACE). IEEE Transactions on Audio, Speech, and Language Processing 15(7) (2007)
23. Apsingekar, V.R., De Leon, P.L.: Speaker Model Clustering for Efficient Speaker Identification in Large Population Applications. IEEE Transactions on Audio, Speech, and Language Processing 17(4) (2009)
24. Puente, L., Poza, M., Ruiz, B., García-Crespo, A.: Score Normalization for Multimodal Recognition Systems. In: JIAS (2010)
25. Guo, W., Dai, L., Wang, R.: Double Gauss Based Unsupervised Score Normalization in Speaker Verification. In: ISCSLP 2008, pp. 165–168 (2008)
26. Castro, D.R., Fierrez-Aguilar, J., Gonzalez-Rodriguez, J., Ortega-Garcia, J.: Speaker Verification using Speaker and Test Dependent Fast Score Normalization. Pattern Recognition Letters 28, 90–98 (2007)

27. Zajíc, Z., Vaněk, J., Machlica, L., Padrta, A.: A Cohort Method for Score Normalization in Speaker Verification System, Acceleration of On-line Cohort Methods. In: SPECOM (2007)
28. Sturim, D.E., Reynolds, D.A.: Speaker Adaptive Cohort Selection for Tnorm in Text-independent Speaker Verification. In: Proceedings of ICASSP (2005)
29. Gupta, C.S.: Significance of Source Feature for Speaker Recognition. In: A M.S Thesis IIIT Madras (2003)
30. He, L., Zhang, W., Shan, Y., Liu, J.: Channel Compensation Technology in Differential GSV–SVM Speaker Verification System. In: APCCAS (2008)
31. Neville, K., Jusak, J., Hussain, Z.M., Lech, M.: Performance of a Text-Independent Remote Speaker Recognition Algorithm over Communication Channels with Blind Equali sation. In: Proceedings of TENCON (2005)
32. Campbell, W.M., Campbell, J.P., Reynolds, D.A., Jones, D.A., Leek, T.R.: Phonetic Speaker Recognition with Support Vector Machines. In: Proc. NIPS (2004)

FITVQSPC: Fast and Improved Transformed Vector Quantization Using Static Pattern Clustering

R. Krishnamoorthy and R. Punidha

Computer Vision Lab, Department of CSE,
Anna University of Technology Tiruchirappalli,
Tiruchirappalli – 620 024, Tamilnadu, India
rkrish26@hotmail.com, r_punidha@yahoo.co.in

Abstract. In this paper, a new fast transformed vector quantization (TVQ) algorithm with static pattern clustering is proposed for coding of color images. To speed up the design process of VQ with better compression ratio, the features of transform coding and VQ are combined in this work. The transformed training set is obtained using integer based orthogonal polynomials transform with reduced computational complexity. The proposed method generates a single codebook for all the three color components, because of the inter-correlation property of the proposed transformation. The proposed algorithm reduces codebook construction time by clustering the static input patterns and eliminates redundant computations at the successive iterations of codebook optimization phase of LBG codebook training process. The computation time is also reduced by choosing only non homogeneous vectors for codebook initialization from the large volume of transformed input training set. The experiments are conducted and compared with existing technique.

Keywords: Vector Quantization, Orthogonal Polynomials Transform, Static Pattern Clustering.

1 Introduction

In the current scenario, Vector Quantization (VQ) is found to be an efficient technique for data compression, and is used in numerous applications including image encoding and image recognition [1]. Various type of VQ techniques such as Address VQ [2], Classified VQ[3], [4], Side-match VQ[5], [6], [7], Finite state VQ[8], Adaptive VQ[9], [10] have been reported for various purposes. Vector Quantization in frequency domain is reported by combining both transform coding and VQ (TVQ) that takes advantage over VQ in spatial domain. The transformed vector components in high frequency regions are the low energy components and can be discarded. Consequently the dimension of transformed vectors and the complexity of VQ are both reduced. This yields a reduced codebook size which means higher compression ratio than that of a VQ alone. Vector quantization based image compression [11], [12], [13], [14], [15], [16] exhibits good compression performance at low bit rates while computational complexity has been always a tough problem for its time

T.-h. Kim et al. (Eds.): SIP/MulGraB 2010, CCIS 123, pp. 146–155, 2010.
© Springer-Verlag Berlin Heidelberg 2010

consuming encoding system. In general, researches on VQ codebook generation can be divided into two categories: one on improving the quality of codebook and the other on reducing the computation time of the codebook generation. Linde–Buzo-Gray (LBG)[17] also known as Generalized Lloyd Algorithm (GLA) is one of the best-known algorithm for codebook generation due to its simplicity and relatively good fidelity. This algorithm has finite sequence of steps, and consists of two major phases: initialization of the codebook and optimization. The codebook initialization is an important task, since a bad choice of the initial code words generally leads to a final quantizer with high Mean Quantization Error (MQE). There are two methods of codebook initialization: Random initialization and by splitting. In random initialization, the initial code words are randomly chosen and are generally inside the convex hull of the input data set. In the case of splitting technique, the initialization requires that the number of code words is a power of two. The procedure starts from only one codeword, that, recursively splits it in two distinct code words until desired number of code words is generated. The codebook optimization phase starts from an initial codebook and after some iteration, generates a final codebook with a distortion corresponding to a local minimum.

Since the LBG codebook generation algorithm is time consuming, various techniques are proposed for fast and efficient codebook generation of vector quantization. Shen et. al.[18] proposed a vector quantization method based on GLA, in which new code words are inserted in input vector space region where the distortion error is highest until desired number of code words is achieved. In the LBG codebook construction, various improvements [19], [20] have been adopted to minimize the computing time and to generate better codebook for representing the input vector. In [21], the computation time of GLA or GLA based algorithm is reduced by compressing the input vectors that are static for successive iterations, but it suffers from edge degradation due to random initialization of initial codebook generation. Jerome Yeh et. al.[22] extended the GLA to G^2LA by utilizing the measurement of grey relational analysis(GRA) and the new selection of initial code vectors was adopted to choose distinct vectors so as to have a better representative codebook.

To summarize, the LBG codebook generation suffers from the serious drawback that, it requires extensive computations during the training process for generating a codebook and its performance is also depend on the initial starting conditions. In order to overcome these drawbacks, in this paper we propose a technique for the construction of better codebook with reduced computation time of LBG or LBG based algorithm. In this work, a new transformed vector quantization is designed so as to obtain higher compression ratio because of the reduced dimension in codebook design. The proposed transform is based on a set of orthogonal polynomials and is configured to take not only the interactions in the individual color planes but also takes into account interactions between the color planes, and so there is no need to design codebook for all the three color components separately. The proposed work constructs initial codebook from non homogeneous transformed training vectors to get better quality codebook with random initialization and reduces the codebook construction time by clustering the static patterns of input training vectors in successive iterations of codebook optimization phase of LBG algorithm.

This paper is organized as follows: The proposed orthogonal polynomials based transform coding of color images is presented in section 2. The basis operator of the proposed transform coding that serves for reconstruction is given in section 3. The proposed Fast Improved Transformed Vector Quantization using Static Pattern Clustering (FITVQSPC) is given in section 4. The experimental results are presented in section 5, and conclusion is presented in section 6.

2 Orthogonal Polynomials Based Transform Coding

In order to devise a color image transform coding technique, we first analyze the image formation system. As per the classical definition, the color image formation can be described as

$$I(x, y, z) = \int \int \int f(\xi, \eta, \gamma) \, d\mu(\xi) d\mu(\eta) d\mu(\gamma) \tag{1}$$

where the object function f (ξ, η, γ) is expressed in terms of derivatives of the image function I relative to its spatial and color coordinates, which is very useful for color image compression. The representation of object function f in terms of I is considered to be an ill-posed problem and hence, it is desirable that the differentiation of image function I must undergo smoothing process. The smoothing operation can be either of the following two types of linear transformation:

1. The integral convolution transformation

$$f(x) = \int \delta(x-t) I(t) dt \tag{2}$$

2. The ordinary linear transformation

$$f_i = \sum_{k=0}^{n-1} u_{ik} I_k, (i=0,...,n-1) \tag{3}$$

The linear transformation defined in equation (2) by the matrix $\left|\delta(x_i - t_i)\right|$ is a smoothing operation provided it is totally positive, whereas the linear transformation defined in equation (3) by the matrix $U = \left|u_{ik}\right|$ is a smoothing operation provided it is totally positive. The matrix U is totally positive provided all the minors of all orders of its determinant $\left|u_{ik}\right|$ are non-negative. The point-spread function M(s, x) is considered to be a real valued function defined for (s, x) ∈ S × X where S and X are ordered subsets of real values. In the case where S consists of a finite set {0, 1, ..., n-1}, the function M(s, x) reduces to a sequence of functions

M(i, x) = u$_i$ (x) , i = 0, 1, ..., n-1

Consequently the linear three dimensional image transformation can be shown as

$$\beta'(s, \varsigma, \eta) = \int_{x \in X} \int_{y \in Y} \int_{z \in Z} M(\varsigma, x) M(\eta, y) M(s, z) I(x, y, z) \, dx dy dz \tag{4}$$

The point-spread function M(s, t) (M(i, t) = $u_i(t)$) is a row, column and color smoothing operation provided the set of functions $\{u_0(t), ..., u_{n-1}(t)\}$ is a T-system over a closed interval [a, b]. Considering each of S, X, Y and Z to be finite set of values {0, 1, ..., n-1}, the matrix notation of equation (4) is

$$\left.\left|\beta_{ijk}^{'}\right|\right|_{i,j,k=0}^{n-1} = (|M|\otimes|M|\otimes|M|)'|I|$$

(5)

where the point spread operator $|M|$ is

$$|M| = \begin{vmatrix} u_0(x_0) & u_1(x_0) & \cdots & u_{n-1}(x_0) \\ u_0(x_1) & u_1(x_1) & \cdots & u_{n-1}(x_1) \\ & & \vdots & \\ u_0(x_{n-1}) & u_1(x_{n-1}) & \cdots & u_{n-1}(x_{n-1}) \end{vmatrix}$$

(6)

$\otimes$ is the outer product and $|\beta'_{ijk}|$ be the n^3 matrices arranged in the dictionary sequence. $|I|$ is the color image and $|\beta'_{ijk}|$ be the coefficients of transformation.
We consider the set of orthogonal polynomials $u_0(x)$, $u_1(x)$, ..., $u_{n-1}(x)$ of degrees 0, 1, 2, ..., n-1, respectively. The generating formula for the polynomials is as follows.

$$u_{i+1}(x)=(x-\mu)u_i(x)-b_i(n)u_{i-1}(x) \quad \text{for i}\geq 1,$$
$$u_1(x) = x - \mu, \text{ and } u_0(x) = 1,$$

(7)

where

$$b_i(n)= \frac{\langle u_i, u_i \rangle}{\langle u_{i-1}, u_{i-1} \rangle} = \frac{\sum_{t=1}^{n} u_i^2(x)}{\sum_{t=1}^{n} u_{i-1}^2(x)}$$

$$\mu = \frac{1}{n}\sum_{t=1}^{n} x$$

and

Considering the range of values of t to be $x_i = i$, i = 1, 2, 3, ...,,n, we get

$$b_i(n)=\frac{i^2(n^2 -i^2)}{4(4i^2-1)},$$

$$\mu=\frac{1}{n}\sum_{t=1}^{n} t =\frac{n+1}{2}$$

The point-spread operators $|M|$ of different size can be constructed from equation (6) for n ≥ 2 and $t_i = i$. For the convenience of point-spread operations, the elements of $|M|$ are scaled to make them integers as represented in section 3 and hence the proposed coding involves only integer arithmetic.

3   The Orthogonal Polynomials Basis

In case of R-G-B color space, the elements of the finite set Z for convenience can be labeled as {1,2,3}. For the sake of computational simplicity, the finite Cartesian coordinate set S, X and Y are also labeled in the identical manner. The point spread operator in equation (5) that defines the linear orthogonal transformation of color images can be obtained as $|M| \otimes |M| \otimes |M|$ in which $|M|$ can be computed and scaled from equation (7) as follows.

$$|M| = \begin{vmatrix} u_0(x_0) & u_1(x_0) & u_2(x_0) \\ u_0(x_1) & u_1(x_1) & u_2(x_1) \\ u_0(x_2) & u_1(x_2) & u_2(x_2) \end{vmatrix} \quad = \quad \begin{vmatrix} 1 & -1 & 1 \\ 1 & 0 & -2 \\ 1 & 1 & 1 \end{vmatrix} \tag{8}$$

The set of 27 three dimensional polynomial basis operators O_{ijk} $(0 \le i, j, k \le n\text{-}1)$ can be computed as

$$O_{ijk} = \hat{u}_i \otimes \hat{u}_j \otimes \hat{u}_k$$

where $\hat{u}_i$ is the $(i + 1)^{st}$ column vector of $|M|$. The operator O_{ijk} is arranged in the dictionary sequence in such a manner that it becomes the $(i \times 3^2 + j \times 3 + k) + 1^{st}$ column vector of the point-spread operator $|M| \otimes |M| \otimes |M|$ in equation (5). Having described the Orthogonal Polynomials based transform coding and its basis, we present the proposed vector quantization in the next section.

4   Proposed Codebook Generation Technique

The proposed VQ encoding algorithm, called Fast and Improved Transformed Vector Quantization using Static Pattern Clustering (FITVQSPC) optimizes LBG codebook generation technique, by reducing computation time of successive iterations for constructing the final codebook and improving the quality of initial codebook design. The proposed VQ algorithm uses features that are extracted with orthogonal polynomials based transform coding as training input vectors for codebook construction. The advantage of combining both the transformation and VQ is that, when a linear transform is applied to the vector signal, the information is compacted into a subset of the vector components. In the frequency domain, the high energy components are concentrated in the low frequency region. This means that the transformed vector components in the high frequency regions have very little information. These low energy components might be discarded entirely. Let us consider a k-dimensional transformed input vector Y, containing the orthogonal polynomial coefficients β_{ijk}. We map it into a d-dimensional vector X $(d < k)$, by discarding the low frequency components. This d-dimensional VQ is used to quantize the truncated vector, forming the quantized approximation $\hat{X}$. Since the k-dimensional training vectors are reduced to d-dimension, the computational time of the proposed VQ algorithm is considerably reduced.

From the d-dimensional large set of input transformed training vectors X, the non homogeneous vectors $\overline{X}$ are chosen using similarity test on first and third feature of input vector, which improves the quality of initial codebook, because no homogeneous vectors are involved. The proposed algorithm, firstly, from the vector pool selects the first block as the mother block, then strikes the high similar blocks with the mother from the vector pool. Here the mother block is selected as the one of the non homogeneous input training vector. Next, we pick up another mother block at the vector pool and perform similarity test on all undecided blocks again, and delete the high relational blocks from the vector pool as well as the current mother block is to be the other non homogeneous training vector. Repeating the above procedure for all the transformed training vectors, the smaller set of non homogeneous training vectors is obtained.

The proposed algorithm uses non homogeneous input training vectors $\overline{X}$ for creating better initial codebook to start with random initialization by sampling, since it does not need any calculation. Next, is the codebook optimization phase that starts from an initial codebook and, after some iteration generates a final codebook with a distortion corresponding to a local minimum. The characteristics of convergence of the LBG and LBG based algorithms make it very unlikely to move the input vectors from one codeword to another at the later iterations of the evolution process and those vectors are called "static" patterns. To reduce the computation time required for generating the codebook, these static patterns are clustered to prevent it from being involved in the computations at the later iterations of the evolutionary process. The mean (μ) and standard deviation (σ) are used to find the static patterns. For instance if the distance(d_{ij}) between the input vector x_i and codeword c_j is smaller than $\mu+\sigma$ ($\left\| x_i - c_j \right\| \leq (\mu + \sigma)$), then the input vector is "static" and near the codeword. These static patterns are clustered for further iterations of codebook optimization, so that the computation time is greatly reduced.

Algorithm for the Proposed Vector Quantization:

Input: Color image having three components R,G, and B each of size ROW $\times$ COL. [ ] denotes the matrix   and the suffix denotes the elements of the matrix. Let $|\mathcal{M}| = |M| \otimes |M| \otimes |M|$ be 3-D polynomial operator of size (27 $\times$ 27), and [I] be the 3 $\times$3 color image region extracted from each of the R, G, B components and arranged in dictionary sequence.
Output: Encoded color image with the proposed vector quantization technique.
Steps:

1. Partition the input image into non-overlapping blocks of size (m $\times$ m) (typically m = 3)   and obtain input block of pixel values I_j { j = 1, 2...k; $k = m \times m \times 3$ } from three individual R, G, B color components.
2. Obtain the total number of input training vectors V, with each vector representing the input block I from the image of size (ROW $\times$ COL) { where V=(ROW $\times$ COL $\times$3) / k }.

3. Compute the Orthogonal Polynomials transformed co-efficients $[\beta'] = [\mathcal{M}]^t [I]$ for all the training vectors V as described in section 2.

4. Discard the high frequency coefficients based on energy preserving property of the proposed transform coding to form the training vectors X, where $X = \{X(j), j = 1,2,\ldots,d;\quad d < k\}$.

5. Find the non homogeneous vectors $\overline{X}$ from large set of training vectors X, using similarity test on first and third features of each training vector X.

6. Construct the Initial codebook C with N code words using random sampling from the non homogeneous vectors $\overline{X}$, and each codeword consists of d-dimensional code vectors represented as c_{ij} (i=1 to N, j=1,2,..,d, where $d < k$).

7. Let iter=1 and and static pattern cluster CL=NULL.

8. Partition the input training vector $\{X - CL\}$ into set of disjoint blocks B_j, j=1,2,…,m using minimum distance rule.

9. Determine the centroid $\overline{C}_j$ of B_j, j=1, 2, …,m.

10. For each x_i, $x_i \in (X - CL)$ and $x_i \in B_j$, if $(\|x_i - c_j\| \le (\mu + \sigma))$, then $CL = CL \bigcup \{x_i\}$.

11. If the average distortion satisfies the stopping criterion, go to step 13.

12. Otherwise iter=iter+1 and go to step 8.

13. Output the final codebook C* with N code words.

14. For each input vector, find the closest codeword from the codebook C* using the distance measure.

15. Encode the closest codeword index using entropy encoding for further compression.

5 Experimental Results

The proposed algorithm for color image compression has been experimented with various test images. Sample test image viz. lena and pepper which are of size (128×128) with pixel values in the range of 0-255 are shown in Fig. 1(a) and (b). The images are partitioned into non-overlapping blocks of size (3×3) in the three R, G, B color space and the proposed transformation is applied as described in section 2. Since this transform provides interaction among the three color planes, the transformed co-efficients (β_{ijk}) are made available in one dimensional dictionary sequence (k-dimensional). After discarding half of the high frequency transformed coefficients, the vector dimension is reduced to 13 (d-dimensional). The similarity test is conducted on first and third feature of transformed training vectors and non homogeneous vectors are chosen as codebook training vectors. The codebook is initialized from these non homogeneous vectors by random initialization. The mean(μ) and standard deviation(σ) are calculated based on the first feature. The static input patterns ($\|x_i - c_j\| \le (\mu + \sigma)$) are clustered for successive iterations to reduce the computation time of the codebook optimization phase.

The proposed vector quantization is experimented with Intel Core(2)-Quad, 2.3 GHZ speed processor system for various codebook sizes. The computation time of

(a)   **(b)**

Fig. 1. Original test images

Table 1. Comparison of computing time (seconds) for various codebooks generated for lena and pepper images

Code Book size	Computation Time (seconds)			
	Lena		Pepper	
	FITVQSPC	LBG	FITVQSPC	LBG
2048	06.346	11.148	06.778	11.886
1024	03.441	06.351	03.643	06.779
512	02.284	04.146	02.785	03.946
256	01.200	02.265	01.669	02.021
128	00.482	01.372	00.649	01.436

codebook construction for various codebook (CB) sizes are presented in table 1 for lena and pepper images. For the codebook of size 2048, the proposed algorithm takes 06.346 seconds(s) and 06.778s for lena and pepper images. To construct the codebook of size 512, the FITVQSPC algorithm takes 02.284s and 02.785s for lena and pepper images respectively. Similarly, for the codebook of size 128 the proposed algorithm takes 0.482s and 0.649s for the same test images. The performance of proposed vector quantization is measured with Peak-Signal-to-Noise-Ratio (PSNR). For the codebook of size 2048, the proposed algorithm gives a PSNR value of 41.05 dB and 40.10 dB for lena and pepper images respectively. Similarly, for the codebook of size 512, the proposed algorithm gives 37.23 dB and 36.12 dB for lena and pepper images. The reconstructed images corresponding to the images shown in fig. 1 with the proposed vector quantization are presented in fig. 2(a) and (b) for the codebook of size 512. Similarly, for the codebook of size 128, the proposed algorithm achieves 33.32dB and 32.25dB for the same test images. The PSNR values obtained with the proposed algorithm for various codebook sizes are presented in table 2.

In order to measure the performance of the proposed vector quantization algorithm, we conduct experiments with LBG codebook generation algorithm. For the codebook of size 2048, the LBG algorithm took 11.148s and 11.886s for lena and pepper images respectively with the same computer system. Similarly for codebook of size 512, the LBG algorithm takes 04.146s for lena image and 03.946s for pepper image. When the codebook size is 128, the LBG technique took 1.372s and 1.436s for lena and pepper images. The PSNR values obtained using LBG algorithm is also incorporated into the

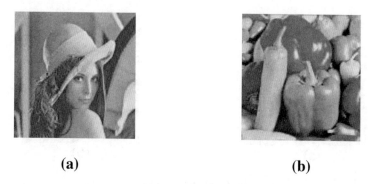

(a)   **(b)**

Fig. 2. Reconstructed images with the proposed VQ scheme

Table 2. PSNR values (dB) for various codebook sizes using proposed vector quantization for lena and pepper images

CB size	PSNR (dB)			
	Lena		Pepper	
	FITVQSPC	LBG	FITVQSPC	LBG
2048	41.05	41.53	40.10	40.58
1024	39.11	39.46	38.23	38.86
512	37.23	37.65	36.12	36.54
256	34.98	35.38	36.12	36.54
128	33.32	33.74	32.25	32.74

same table 2 for various codebook sizes. From the experimental results, it is evident that the proposed FITVQSPC codebook generation algorithm greatly reduces the codebook construction time, when compared with LBG algorithm.

6 Conclusion

In this paper a new fast transformed vector quantization encoding algorithm with static pattern clustering is presented for color image compression. This scheme uses orthogonal polynomials based transformation to extract the features of the training image. The proposed vector quantization gives better compression ratio and reduces codebook construction time because of the reduced dimension due to energy compaction property of the proposed transform. Due to the inter-correlation property of the proposed transform, a single codebook is generated for all the three color components of training image. The codebook is initialized using non homogeneous training vectors and static patterns are clustered in the codebook optimization phase for constructing the final codebook.

References

1. Gersho, A., Gray, R.M.: Vector Quantization and Signal Compression. Kluwer Academic Publishers, Dordrecht (1991)
2. Nasrabad, N.M., Feng, Y.: Image Compression using Address Vector Quantization. IEEE Transactions on Communications 38(12), 2166–2173 (1990)
3. Ramamoorthy, B., Gersho, A.: Classified Vector Quantization of Images. IEEE Transactions on Communications COM-34(11), 1105–1115 (1986)
4. Tseng, H.-w., Chang, C.: A Very Low Bit Rate Image Compression using Transformed Classified Vector Quantization. Informatica 29, 335–341 (2005)
5. Yang, S.B.: Smooth Side-Match Weighted Vector Quantizer with Variable Block size for Image Coding. IEEE Proceedings on Vision, Image and Signal Processing 152(6), 763–770 (2005)
6. Chang, R.F., Chen, W.-M.: Adaptive Edge based Side Match Finite State Classified Vector Quantization with Quadtree Map. IEEE Transactions on Image Processing 5(2), 378–383 (1996)
7. Kannan, N., Krishnamoorthi, R.: Edge Oriented Image Coding based on Orthogonal Polynomials. In: International Conference on Emerging Trends in Computing, ICETiC 2009, pp. 77–82 (2009)
8. Foster, J., Gray, R.M., Dunham, O.M.: Finite State Vector Quantization for Waveform Coding. IEEE Transactions on Information Theory 31(3), 348–355 (1985)
9. Goldberg, M.: Image Compression using Adaptive Vector Quantization. IEEE Transaction on Communications COM-34(2), 180–187 (1986)
10. Shen, G., Zeng, B., Liou, M.-L.: Adaptive Vector Quantization with Codebook Updating based on Locality and History. IEEE Transactions on Image Processing 12(3), 283–295 (2003)
11. Shelley, P., Li, X., Han, B.: A Hybrid Quantization Scheme for Image Compression. Image and Vision Computing 22(3), 203–213 (2004)
12. Makwana, M.V., Nandurbarkar, A.B., Joshi, S.M.: Image Compression using Tree Structured Vector Quantization with Compact Codebook. In: IEEE International Conference on Computational Intelligence and Multimedia Applications, pp. 102–107 (2007)
13. Li, R., Kim, J.: Image Compression using Fast Transformed Vector Quantization. In: IEEE Proceedings of Applied Imagery Pattern Recognition Workshop, pp. 141–145 (2000)
14. Al-Otum, H., Shahab, W., Smadi, M.: Color Image Compression using a Modified Angular Vector Quantization. Journal of Electrical Engineering 57(4), 226–234 (2006)
15. Wu, F., Sun, X.: Image Compression by Visual Pattern Vector Quantization. In: Data Compression Conference, pp. 123–131 (2008)
16. Hsieh, C.-H., Shao, W.-Y., Jing, M.-H.: Image Compression based on Multistage Vector Quantization. Visual Communication and Image Representation 11(4), 374–384 (2000)
17. Linde, Y., Buzo, A., Gray, R.M.: An Algorithm for Vector Quantizer Design. IEEE Transactions on Communications COM-28(1), 84–94 (1980)
18. Shen, F., Hasegawa, O.: An Adaptive Incremental LBG for Vector Quantization. Neural Networks 19(5), 694–704 (2006)
19. Patane, G., Russo, M.: The Enhanced LBG algorithm. Neural Networks, 1219–1237 (2001)
20. Tsai, C.-W., Lee, C.-Y.: A Fast VQ Codebook Generation Algorithm via Pattern Reduction. Pattern Recognition Letters 30(7), 653–660 (2009)
21. Gang, L., Jing, L., Quan, W.: A Robust Lin-Buzo-Gray Algorithm in Data Vector Quantization. IEEE International forum on Information Technology and Applications 1, 464–467 (2009)
22. Yeh, J., Hsu, Y.-T.: A G^2LA Vector Quantization for Image Data Coding. Expert Systems with Applications 36(3), 5660–5665 (2009)

New Data Hiding Scheme Using Method of Embedding Two Bits Data into Two DCT Coefficients

Makoto Fujimura[1], Tetsuya Takano[1], Satoshi Baba[1], and Hideo Kuroda[2]

[1] Dept. of Computer and Information Science
Faculty of Engineering, Nagasaki University,
1-14 Bunkyo-machi Nagasaki, Japan
makoto@cis.nagasaki-u.ac.jp
[2] Computing and Fundamental Dep., FPT University
Hanoi, Vietnam
kuroda@fpt.edu.vn

Abstract. Data hiding technologies are applicable in various application technologies. Generally, more the hiding volume of data is the better. One of the current methods embeds the data of one bit per two coefficients on the frequency domain that orthogonal converts the image. This paper proposes the method of embedding two bits per two DCT coefficients aiming to increase the embedding volume of data on the current method. First bit of embedding data is represented by amplitude relation between two DCT coefficients. Second bit is represented by mean of two DCT coefficients. These two use embedding strength parameters P and Q, respectively. These parameters were decided by preliminary experiment. Experimental results show that the proposed scheme achieved good image quality and robustness against JPEG data compression.

Keywords: Data hiding, Image watermarking, DCT.

1 Introduction

Recently, some data hiding technologies are used for various applications because from the point of technical view any data are available to be hidden data. For examples, the water marking hides the personal IDs for protecting copyright[1]. The steganography hides the confidential information for confidential communication like communication data of spies[2]. The objective estimation of image quality hides the marker signals for the automatic objective estimation of image quality[3]. The image compression scheme using data hiding technique hides a part of coded data extracted in the coding process which is hidden into the other parts of coded data of own image,[4]-[6].

From the point of view of targeted cover image, there are the embedding method in the still picture[7] and that in video stream[8][9].

On the other hand, there are two types of technologies from the viewpoint of the embedding technology. One is a method using pixel space, for example the data is embedded into the lower bit of pixel values[10]. The other is a method using

T.-h. Kim et al. (Eds.): SIP/MulGraB 2010, CCIS 123, pp. 156–164, 2010.

frequency space (in the frequency domain), like wavelet conversion coefficients, and DCT coefficients in JPEG and JPEG2000 etc.[11]-[13]

In general, the hiding technology can be applied to any application, and more hided volume of data is, the preferable.

In this paper, it proposes new technology that can hide two bits per two coefficients as a method of improving the method which hide only one bit per two coefficient by [14].

This paper is organized as follows. Section 2 presents our proposed data hiding scheme. Section 3 explains the experimental results. Section 4 describes discussion. Section 5 concludes the paper.

2   Proposed Data Hiding Scheme

2.1   Using a Pair of DCT Coefficients

An original cover image is transformed to DCT coefficients by discrete cosine transform.

Fig. 1 shows a block of DCT whose size is 8 x 8. Upper left, the DC element and the remainder are AC elements. The numbers show the order when the AC element is scanned zigzag. In the proposed method a pair of the adjacent coefficients in scanning order of the block is used for embedding. This is because the ratio to original coefficient value of the changed value is few even if the coefficient is changed for the embedding because the probability of the adjacent coefficient to which the value is close is higher.

In the proposed method, embedding is carried out into comparatively low frequency coefficient except coefficient of which value is 0.

After embedding, the DCT coefficients embedded information is transformed to the image data by inverse discrete cosine transform.

DC	1	5	6	14	15	27	28
2	4	7	13	16	26	29	42
3	8	12	17	25	30	41	43
9	11	18	24	31	40	44	53
10	19	23	32	39	45	52	54
20	22	33	38	46	51	55	60
21	34	37	47	50	56	59	61
35	36	48	49	57	58	62	63

Fig. 1. Zigzag scan of DCT coefficients of 8x8 block

2.2 Embedding Process

Fig. 2 shows the overview of an embedding process. In the proposed scheme, two bits of hiding information (W_1, W_2) are embedded into two DCT coefficients (a_i, b_i) from transformed original image by DCT. First bit W_1 is represented by magnitude relation between two DCT coefficients. Second bit W_2 is represented by mean of two DCT coefficients. The embedding procedure is as follows.

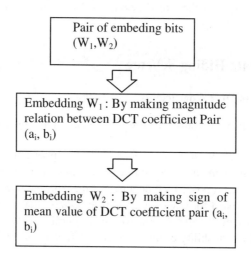

Fig. 2. Overview of an embedding process

2.2.1   Embedding Data W_1 by Making Magnitude Relation between DCT Coefficient Pair

As shown in Fig. 2, a pair of DCT coefficients (a_i, b_i) is used for embedding data W_1. If W_1 is "0", then a DCT coefficient is modified greater than the other one of pair of two DCT coefficients. If "1", then modified one smaller than the other one. For these two DCT coefficients a strength parameter Q is used for making robustness.
As explained by Equation 1, if W_1 is "1" it is forced $a<b$ by using Q.

$$\begin{cases} a'_i = \dfrac{1}{2}(a_i + b_i - Q) \\ b'_i = \dfrac{1}{2}(a_i + b_i + Q) \end{cases} \quad \text{for } W_1 = 1 \qquad (1)$$

where

(a_i', b_i') is a pair of DCT coefficients after modification.

If W_1 is "0", it is forced a>b as shown by Equation (2).

$$\begin{cases} a'_i = \dfrac{1}{2}(a_i + b_i + Q) \\[2mm] b'_i = \dfrac{1}{2}(a_i + b_i - Q) \end{cases} \quad \text{for } W_1 = 0 \tag{2}$$

2.2.2 Embedding Data W_2 by Making Sign of Mean Value of DCT Coefficient Pair (a_i, b_i)

As shown in Fig. 2, a pair of DCT coefficients (a_i', b_i') is used for embedding data W_2, too, where (a_i', b_i') are the DCT coefficients embedded W_1 as mentioned above in section 2.2.1.

If W_2 is "0", then mean value of two DCT coefficients is changed to minus quantity by modifying the DCT coefficients as shown at Equation (3). If W_2 is "1", then the mean value is changed to plus quantity (Equation (3)). The changed mean value is controlled the robustness by strength parameter P. After embedding W_2, (a_i'', b_i'') are gotten as final DCT coefficients(Equation (4)).

$$M = \pm 2P - \dfrac{a'_i + b'_i}{2} \quad \begin{cases} +; W_2 = 1 \\ -; W_2 = 0 \end{cases} \tag{3}$$

$$\begin{cases} a''_i = a'_i + M \\ b''_i = b'_i + M \end{cases} \tag{4}$$

2.3 Extracting Process

Fig. 3 shows the overview of an extracting process.

First data W_1 is extracted by checking a pair of DCT coefficient (a_i'', b_i''). If $a_i'' < b_i''$, then W_1 is "1". If $a_i'' > b_i''$, then W_1 is "0".

Next data W_2 is extracted by checking the mean value of a pair of DCT coefficient (a_i'', b_i''). As explained by Equations (5) and (6), if M > P, then W_2 is "1". If M < -P, then W_2 is "0".

$$M = \dfrac{a''_i + b''_i}{2} > P \; : \; W_2 \text{ is "1"} \tag{5}$$

$$M = \dfrac{a''_i + b''_i}{2} < -P \; : \; W_2 \text{ is "0"} \tag{6}$$

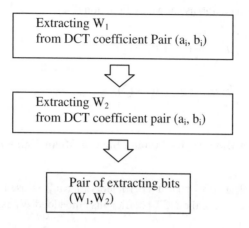

Fig. 3. Overview of a extracting information

3 Experimental Results

3.1 Image Quality by Embedding Data

Experiment about the basic characteristic of new proposed method is carried out. The image quality is one of the most important characteristics for embedding method. Lenna as the representative of a smooth image, and the mandrill as the representative of a complex image were chosen as test images. And, the data only of 100 bits was embedded under the upper part of the image for confirmation of the image quality of the area. Investigating the influence to image quality, P and Q are used as parameters to measure PSNR of embedded cover images. Fig. 4 shows the PSNR of proposed method. Good PSNR is obtained at the vicinity of P=20 and Q=10.

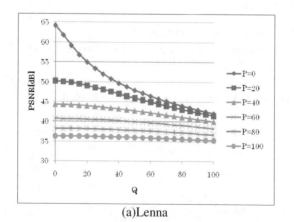

(a)Lenna

Fig. 4. PSNRof proposed method

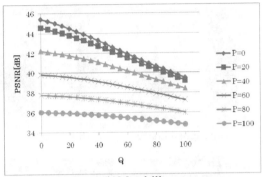

(b)Mandrill

Fig. 4. *(Cont.)*

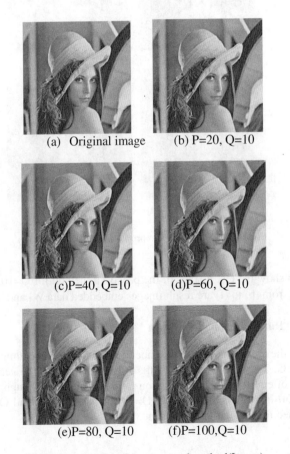

(a)   Original image   (b) P=20, Q=10

(c)P=40, Q=10   (d)P=60, Q=10

(e)P=80, Q=10   (f)P=100,Q=10

Fig. 5. Test images of the proposed method(Lenna)

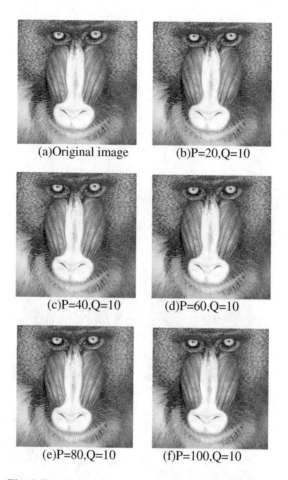

(a)Original image   (b)P=20,Q=10

(c)P=40,Q=10   (d)P=60,Q=10

(e)P=80,Q=10   (f)P=100,Q=10

Fig. 6. Test images of the proposed mothod(Mandrill)

Fig. 5 and 6 show the test images of the proposed method for Q=10. (a) is an original image, and from (b) to (f) are result images embedded data W_1 and W_2.

3.2 Detection Ratio of Embedded Data

Fig. 7 shows the detection ratio of embedded data over encoding and decoding process by JPEG. The horizontal axis is the JPEG quality, and the vertical axis is the detection ratio of embedded data. Detection ratio 1 is obtained though it encodes with JPEG quality 30 in combination (P=20, Q=10) of parameter P and Q chosen by the above-mentioned image quality experiment.

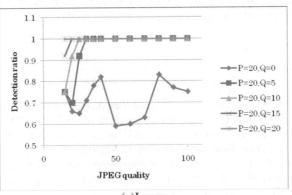

(a)Lenna

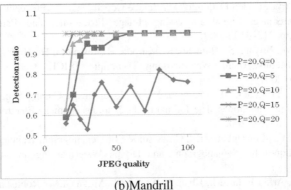

(b)Mandrill

Fig. 7. Detection ratio of embedded data

4 Discussion

Fig. 5(Lenna) and Fig. 6(Mandrill) where the outcome of an experiment concerning the image quality is shown are compared. Quality degradation cannot be detected with Mandrill that is a complex image on the sight even at the time of (P=100, Q=10) though the picture quality degradation is slightly detected in Lenna at (P=20, Q=10). In the actual experiment the data was embedded into only under the upper part of the image (It was a smooth area in Lenna) for confirmation of influence to image quality.

As a result, a complex area is chosen even if it is Lenna, and it has been understood that the proposed method is applicable if it embeds the data into there.

5 Conclusion

We proposed a new data hiding scheme using method of embedding two bits data into two DCT coefficients. Experimental results show that the proposed scheme achieved good image quality and robustness against JPEG data compression, especially for the mandrill as the representative of a complex image. In the future we are going to improve the proposed scheme as adaptive embedding method by extracting a complex area.

References

1. Iwata, M., Miyake, K., Shiozaki, A.: Digital Watermarking Method to Embed Indes Data into JPEG Images. IEICE, Trans. Fundamentals E85-A(10), 2267–2271 (2002)
2. Iwata, M., Miyake, K., Shiozaki, A.: Digital Steganography Utilizing Features of JPEG Images. IEICE, Trans. Fundamentals E87-A(4), 929–936 (2004)
3. Sugimoto, O., Kawada, R., Koike, A., Matsumoto, S.: Automatic Objective Picture Quality Measurement Method Using Invisible Marker Signal. IEICE Trans. Information and Systems J88-D-II(6), 1012–1023 (2005)
4. Kuroda, H., Fujimura, M., Hamano, K.: Considerations of Image Compression Scheme Hiding a Part of Coded Data into Own Image Coded Data. In: Proceedings of the International Conference, Signal Processing, Image Processing, and Pattern Recognition (SI 2009), pp. 312–319 (2009)
5. Fujimura, M., Miyata, S., Hamano, K., Kuroda, H., Imamura, H.: High Efficiency Coding of JPEG Method Using Watermarking Technique. IEICE, Trans. ISS, D J92-D(9), 1672–1676 (2009)
6. Kuroda, H., Fujimura, M., Imamura, H.: A Novel Image Coding Scheme using Digital Watermarking Techniques. The Journal of the Institute of Image Electronics Engineers of Japan 37(1), 63–68 (2008)
7. Johnson, N.F., Katezenbeisser, S.C.: A survey of steganographi techniques. In: Information Techniques for Steganography and Digital Watermarking, pp. 43–75 (December 1999)
8. Ishtiaq, M., Arfan Jaffar, M., Khan, M.A., Jan, Z., Mirza, A.M.: Robust and Imperceptible Watermarking of Video Stream for Low Power Devices. In: Proceedings of the International Conference, Signal Processing, Image Processing, and Pattern Recognition (SI 2009), pp. 177–184 (2009)
9. Dittman, J., Stabenau, M., Steinmetz, R.: Robust MPEG video Watermarking Technologies. In: ACM Multi,edia, Bristol, UK, pp. 71–80 (August 1998)
10. Henmandez, J.R., Amado, M., Perez-Gonzelez, F.: DCT-Domain watermarking techniques for still images: Detector performance analysis and a new structure. IEEE Transactions on image processing 9(1), 55–68 (2000)
11. Liu, Z., Karam, L.J., Watson, A.B.: JPEG2000 Encoding With Perceptual Distortion Control. IEEE Transactions on Image Processing 15(7), 1763–1778 (2006)
12. Yen, E., Tsai, K.S.: HDWT-based grayscale watermarking for copyright protection. An International Journal Source Expart Systems with Applications 35(1-2), 301–306 (2008)
13. Hassanin, A.E.: A Copyright Protection using Watermarking Algorithm. Informatica 17(2), 187–198 (2006)
14. Morita, K., Kawakami, K., Yamada, A., Fujimura, M., Imamura, H., Kuroda, H.: An Adaptive Data Hiding Method Using Best Combination Neighboring Parts of WTC for Division into Flat/Non-Flat Pairs. In: Proc. of The Pacific Rim Workshop on Digital Steganography, pp. 68–74 (2003)

Design and Implementation of a Video-Zoom Driven Digital Audio-Zoom System for Portable Digital Imaging Devices

Nam In Park[1], Seon Man Kim[1], Hong Kook Kim[1],
Ji Woon Kim[2], Myeong Bo Kim[2], and Su Won Yun[2]

[1] School of Information and Communications
Gwangju Institute of Science and Technology, Gwangju 500-712, Korea
{naminpark,kobem30002,hongkook}@gist.ac.kr
[2] Camcoder Business Team, Digital Media Business
Samsung Electronics, Suwon-si, Gyenggi-do 443-742, Korea
{jiwoon.kim,kmbo.kim,ysw1926.yun}@samsung.com

Abstract. In this paper, we propose a video-zoom driven audio-zoom algorithm in order to provide audio zooming effects in accordance with the degree of video-zoom. The proposed algorithm is designed based on a super-directive beamformer operating with a 4-channel microphone system, in conjunction with a soft masking process that considers the phase differences between microphones. Thus, the audio-zoom processed signal is obtained by multiplying an audio gain derived from a video-zoom level by the masked signal. After all, a real-time audio-zoom system is implemented on an ARM-CORETEX-A8 having a clock speed of 600 MHz after different levels of optimization are performed such as algorithmic level, C-code, and memory optimizations. To evaluate the complexity of the proposed real-time audio-zoom system, test data whose length is 21.3 seconds long is sampled at 48 kHz. As a result, it is shown from the experiments that the processing time for the proposed audio-zoom system occupies 14.6% or less of the ARM clock cycles. It is also shown from the experimental results performed in a semi-anechoic chamber that the signal with the front direction can be amplified by approximately 10 dB compared to the other directions.

Keywords: Audio-zoom, sound focusing, beamforming, digital imaging devices.

1 Introduction

Most portable digital imaging devices produced today have video zoom capabilities, typically achieved by either optical or digital zoom [1]. In terms of use, video zoom aims to bring people clarity when viewing objects as the objects move in to and out of the video viewfinder. However, video zoom becomes more realistic, and thus it enhances user's satisfaction of movement if audio content could also be zoomed in conjunction with video zoom. That is, since video zoom only reflects the images of objects, audio signals are recorded regardless of the video-zoom level. Therefore, the degree of video zoom should be in accordance with audio contents by amplifying the

T.-h. Kim et al. (Eds.): SIP/MulGraB 2010, CCIS 123, pp. 165–174, 2010.

audio signal as objects come closer or by decreasing the audio contents as objects move further away, which is referred to here as *video-zoom driven audio zoom.*

In this paper, we propose a video-zoom driven audio-zoom algorithm and implement it on a portable digital imaging device having an ARM-CORETEX-A8 processor. In particular, the proposed audio-zoom algorithm is designed based on super-directive beamforming [2], which operates with a 4-channel microphone system. Then, to realize the proposed audio-zoom algorithm in real time while recording with a portable digital imaging device, several different levels of optimization are performed such as algorithmic level, C-code level, and memory configuration optimization. The performance of the proposed audio-zoom algorithm is subsequently evaluated by comparing the sound pressure level of the original audio signal with that of the processed audio signal for each direction (front, both sides, and back) in a semi-anechoic chamber, which has minimal reverberation.

The remainder of this paper is organized as follows. Following this introduction, Section 2 presents the proposed video-zoom driven audio-zoom algorithm. After that, Section 3 describes implementation issues of a system employing the proposed video-zoom driven audio-zoom algorithm on a portable digital imaging device that has limited resources. Section 4 then demonstrates the performance of the implemented video-zoom driven audio-zoom system, and the paper is concluded in Section 5.

2   Proposed Video-Zoom Driven Audio-Zoom Algorithm

The proposed video-zoom driven audio-zoom algorithm is designed using a 4-channel microphone system. The proposed algorithm consists of three parts: an audio focus control using a super-directive beamformer, a soft masking process based on the direction of arrival (DOA), and an audio gain control based on the video-zoom level. First, the input signals from the 4-channel microphone system are transformed into the frequency domain using a fast Fourier transform (FFT), followed by applying a fixed super-directive beamformer [2][3], as shown in Fig. 1. The signals processed by the beamformer are then mixed with the original signals according to a mixing ratio parameter, r, which is determined by the video-zoom level. In other words, as the

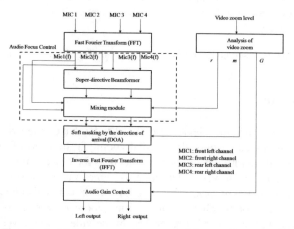

Fig. 1. Flowchart of the proposed video-zoom driven audio-zoom algorithm

video-zoom level increases, r decreases. Next, soft masking is processed to the mixed signals, where a masking threshold parameter, m, which is determined according to the video-zoom level. That is, the mixed signals are multiplied by frequency-dependent mask coefficients that are obtained by the DOA estimation. Then, the masked signals are transformed back into the time-domain signals using an inverse FFT (IFFT), and an audio gain, G, corresponding to the video-zoom level, is used to amplify the time-domain signals. Finally, a sigmoid function is applied to prevent audio signal from being clipped when the signal is amplified.

2.1 Audio Focusing Control Using a Beamformer

The proposed video-zoom driven audio-zoom algorithm adjusts the audio zooming effects according to the degree of video zoom, in which objects appear to come closer or go further away. In order to provide more realistic audio-zoom effects, we design a super-directive beamformer as [2]

$$W = \frac{\Gamma_{vv}^{-1}d}{d^H \Gamma_{vv}^{-1}d} \tag{1}$$

where d represents a delay incurred by the distance between microphones, Γ_{vv} is the power spectral density matrix of each microphone input signal [3], and W is the optimum weight of the super-directive beamformer. In Eq. (1), the power spectral density matrix for a fixed super-directive beamformer is given by [2]

$$\Gamma_{v_n v_m}(m) = \text{sinc}\left(\frac{\omega \cdot l_{mn}}{c}\right) \tag{2}$$

where ω is the angular frequency, c is the speed of sound, and l_{mn} is the distance between the m-th and n-th microphones. The weighting vector described in Eq. (1) can be obtained by the minimum variance distortionless response (MVDR) algorithm [2].

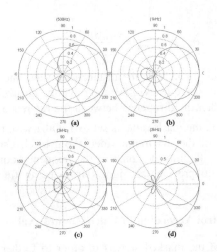

Fig. 2. Super-directive beam patterns at (a) 0.5 kHz, (b) 1 kHz, (c) 2 kHz, and (d) 3 kHz when the video-zoom is activated with the maximum level

Fig. 2 shows the beam patterns by varying the frequency from 0.5 kHz to 3 kHz when video zoom is activated with the maximum level. As shown in the figure, the front direction is seen to be more concentrated than the other directions. The original signals are then mixed with the signals that are obtained after applying the fixed super-directive beamformer. Table 1 summarizes the mixing ratios according to the different audio-zoom levels. As shown in the table, the audio-zoom level is divided into five levels according to the video-zoom level. As the video-zoom level increases, the audio-zoom level also increases and thus the sound from the objects should be expected to come louder.

Table 1. Mixing ratio and audio-zoom level according to the video-zoom level

Video-zoom level	1-20	21-40	41-60	61-80	81-100
Audio-zoom level	1	2	3	4	5
Mixing ratio	1	0.75	0.5	0.25	0

2.2 DOA-Based Soft Masking

In order to further suppress the signals coming from the directions except for the front direction, we apply a soft mask to the mixed signals obtained in Section 2.1. The mask value for each frequency bin is estimated by comparing the normalized phase difference between the microphone distances and the normalized phase difference of the reference signal [4][5]. That is, the phase difference is expressed by

$$\phi_{m,n}(i,k) = \theta_m(i,k) - \theta_n(i,k) \tag{3}$$

where $\theta_n(i,k)$ and $\phi_{m,n}(i,k)$ are the phase of the n-th microphone and the phase difference between the m-th and n-th microphone for the i-th frame number and k-th frequency bin. Note that $\phi_{m,n}(i,k)$ can also be represented by the normalized phase difference as

$$\phi_{m,n}(i,k)_{norm} = \frac{\phi_{m,n}(i,k) \cdot c}{2\pi \cdot d \cdot f} \tag{4}$$

where c is the speed of sound, d is the distance between microphones, and f is the corresponding continuous frequency. Assuming that the normalized phase difference, $\phi_{m,n}(i,k)_{norm}$, is less than the difference between the reference signal that is coming from the front direction, the mask value is set to 1. Otherwise the mask value is then set to 0.2 in the case of the maximum video-zoom level. Consequently, the mask values are multiplied to the mixed signals in the frequency domain, and the masked signals are smoothed to prevent the signals from being fluctuated over adjacent frames [6].

2.3 Audio Gain Control According to Video-Zoom Level

We control the gain of the masked signals in order to further emphasize when the signals are coming from the front direction. In particular, the proposed audio-zoom algorithm is designed to amplify the input signals up to 12 dB. This is because human ear perceives the loudness difference if the amplified audio signals are 10 dB larger than the original input signals [7]. Fig. 3 shows a curve of the audio gain according to the video-zoom level. As already shown in Table 1, there is a relationship between the video-zoom level and the audio-zoom level. Here, the maximum audio-zoom level is bounded to 4 in terms of gain control. As shown in Fig. 3, if the audio-zoom level approaches an audio-zoom level of 2 or 4, the audio gain is set to 1.75 or 3.25, respectively. Therefore, the output, $S_G(n)$ in Eq. (5), is defined as

$$S_G(n) = G \cdot S_m(n) \tag{5}$$

where $S_m(n)$ is the masked signal and G is the audio gain.

However, the amplified audio signals can be overflowed after multiplying the audio gain, thus an audio level control (ALC) is applied on the basis of a sigmoid function, such as

$$S_S(n) = \begin{cases} S_G(n), & if \ |S_G(n)| < 16384 \\ \dfrac{32767}{1 + \exp(-0.0001 \times (S_G(n) - sgn(n) \times 16384))}, & otherwise \end{cases} \tag{6}$$

where $sgn(x)$ is a function whose value is set to 1 if x is greater than or equal to 0, otherwise it is set to -1. In other words, $S_S(n)$ is the amplified audio signal after applying a sigmoid function. The magnitude of the amplified signal, $S_G(n)$, is first compared with a predefined value, 16384. If it is indeed greater than 16384, $S_S(n)$ is obtained by applying the sigmoid function as defined in Eq. (6).

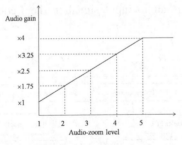

Fig. 3. Audio gain control according to the audio-zoom level

3   Implementation of Proposed Video-Zoom Driven Audio-Zoom Algorithm on a Portable Digital Imaging Device

In this section, we first discuss implementation issues for achieving real-time operation of the proposed algorithm on a portable digital imaging device equipped with limited resources; an ARM processor with a 600 MHz CPU and 111 MB of memory. Note that since the implemented system should be executed in the recording mode of the device, the processing time has to be occupied less than 20% of system resources, considering the processing time of a video codec. Thus, we need to optimize the proposed algorithm using several techniques for the computational complexity reduction.

3.1 C-Code Level Optimization

As a first step for the optimization, we first perform fixed-point arithmetic programming since fixed-point operations are much faster than their corresponding floating-point operations, although the ARM processor supports both fixed-point and floating-point arithmetic. In other words, most 64-bit or 32-bit floating variables (*double, float*) are converted into 32-bit or 16-bit integer variables (*int, short*) [9]. Next, we carry out c-code level optimization, including inline functions, intrinsic functions, redundant code removal, reordering operations, loop unrolling, and so on. In particular, basic operations are redefined using relatively fast ARM instruction sets [9][10].

3.2 Algorithmic Level Optimization

In the case of algorithmic level optimization, the computational complexity is reduced by around 50% by efficiently dealing with the super-directive beamformer. That is, two FFTs for of 2-channel input signals is merged into one complex FFT, which reduces the complexity by almost half [11], as shown in Fig 4. Here, since the pre-processing and post-processing are composed of adders and shift operations, the structure of the proposed audio-zoom algorithm has a lower complexity than performing four FFTs.

Next, the DOA estimation requires an arctangent function for each frequency bin in order to estimate the direction of the audio source. As mentioned above, since a floating-point operation is highly complex, it needs to be converted into a fixed-point operation. Fig. 5 shows a flowchart for a fixed-point implementation of the arctangent function. Here, $X(e^{j\omega})$ is audio input signal in the frequency domain, θ_{out} is the

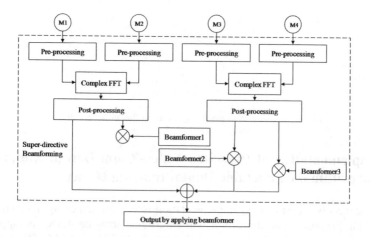

Fig. 4. Illustration of algorithmic level optimization for a super-directive beamformer

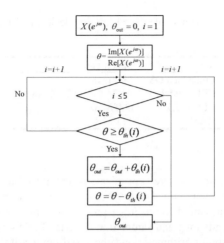

Fig. 5. Flowchart of a fixed-point implementation of arctangent function

Table 2. Arctangent parameter settings for fixed-point implementation

i	1	2	3	4	5
$\theta_{th}(i)$	6434	3798	2007	1019	511
Degree	45°	26.565°	14.036°	7.125°	3.576°

result of arctangent, and i is the number of iteration. First of all, θ_{out} and i are initially set at 0 and 1, respectively; Table 2 represents a threshold, $\theta_{th}(i)$, and the angle in degree corresponding to each threshold. The table is used to check whether or not the phase of the input signal, θ, that is obtained from $X(e^{j\omega})$ is greater than $\theta_{th}(i)$. For the i-th iteration, $\theta_{th}(i)$ is added to θ_{out} if θ is greater than $\theta_{th}(i)$. Simultaneously, the input signal phase, θ, is adjusted into $\theta - \theta_{th}(i)$. In this work, we allow five iterations for obtaining θ_{out} [12].

4   Performance Evaluation

In this section, we evaluated the performance of the proposed video-zoom driven audio-zoom algorithm implemented on a portable digital imaging device having an ARM processor. The performance of the proposed algorithm was measured in terms of processing time and the relative sound pressure level (SPL) of the audio signals with the front direction over the other directions. Table 3 describes the specifications of the portable digital imaging device used for the implementation. Note that Montavista was used as the embedded operating system (OS) with a CPU clock of 600 MHz and RAM size of 111 MB. The OS also had an eight-stage pipeline, a branch-prediction with return stack, and a vector floating-point coprocessor.

Table 3. Specifications of the portable digital imaging device used for the implementation of the proposed algorithm

Item	Specification
Embedded OS	Montavista
CPU model	ARM-CORETEX-A8
CPU clock	600 MHz
RAM	111 MB

4.1 Processing Time

As described in Section 3, several optimization techniques to reduce the computational complexity had been applied to overcome the resource limitations of the ARM processor. We then measured the average processing time for a series of test data, which were 21.3 seconds long and sampled at 48 kHz. Table 4 shows the comparison between the floating-point arithmetic and the fixed-point arithmetic implementation of the proposed algorithm. As shown in the table, the processing time yielded a 65.7% processing time reduction for the fixed-point arithmetic programming, compared to the floating-point arithmetic implementation.

Table 5 compares the average processing times before and after performing the optimization techniques to the fixed-point arithmetic programming. It was shown from the table that the proposed video-zoom driven audio-zoom algorithm occupied less than 15% of ARM clock cycles after the optimization process.

Table 4. Complexity comparison between floating-point and fixed-point implementation for the proposed video-zoom driven audio-zoom algorithm

Item	Average processing time (sec)	Percentage of CPU occupation (%)
Floating-point arithmetic programming	30.2	141.8
Fixed-point arithmetic programming	16.2	76.1

Table 5. Complexity comparison of the proposed video-zoom driven audio-zoom algorithm before and after performing the optimization process to the fixed-point arithmetic programming

Item	Average processing time (sec)	Percentage of CPU occupation (%)
Before optimization	16.2	76.1
After optimization	3.1	14.6

4.2 Sound Pressure Level According to Audio Source Directions

The performance of the implemented audio-zoom algorithm was evaluated in a semi-anechoic chamber in order to minimize the effects of reverberation, as shown in Fig. 6. A loudspeaker was used as the audio source, in which the white noise was adjusted to 60 dB by using a noise level meter. We then measured the sound pressure level (SPL) of the processed audio signal in each direction (front, both side, back). As a result, with the audio-zoom off, the SPL in all directions was equivalent, as shown

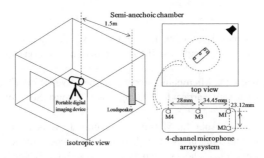

Fig. 6. Experimental environment for measuring the sound pressure levels in a semi-anechoic chamber

Table 6. Sound pressure levels (SPLs) measured in dB depending on the DOA of the audio source

	Direction of arrival of audio source (white noise)			
	0°	90°	180°	-90°
Audio-zoom off	0	0	0	0
Audio-zoom on (minimum video-zoom)	5.77	5.31	0	-0.98
Audio-zoom on (maximum video-zoom)	13.87	5.73	0	1.32

in Table 6. However, with the audio-zoom on, the front audio signal was amplified by approximately 5 dB and 10 dB at minimum and maximum video-zoom, respectively.

5   Conclusion

In this paper, we proposed a beamforming-based video-zoom driven audio-zoom algorithm as a means of focusing a front audio signal on a portable digital imaging device. The proposed algorithm was comprised of three parts: an audio-focus control based on super-directive beamforming, a soft masking process using the DOA estimation, and an audio gain control based on the video-zoom level. In order to realize the proposed algorithm on a portable digital imaging device with limited resources in real time, several optimization techniques to reduce the computational complexity and required memory size were applied. The performance of the implemented algorithm was then measured in terms of the processing time and the sound pressure level (SPL) between the front and the other directions at the maximum video-zoom level in a semi-anechoic chamber. As a result, it was shown from the experiments that average processing time was reduced to around 14.6% of the entire CPU clock cycles. Moreover, the front signal was amplified by approximately 10 dB compared to the other directions.

References

1. Matsumoto, M., Naono, H., Saitoh, H., Fujimura, K., Yasuno, Y.: Stereo zoom microphone for consumer video cameras. IEEE Transactions on Consumer Electronic 35(4), 759–766 (1989)
2. Brandstein, M., Ward, D.: Microphone Arrays. Springer, New York (2001)
3. Kates, J.M.: Super directive arrays for hearing aids. J. Acoust. Soc. Am. 94(4), 1930–1933 (1993)
4. Alik, M., Okamoto, M., Aoki, S., Matsui, H., Sakurai, T., Kaneda, Y.: Sound source segregation based on estimating incident angle of each frequency component of input signals acquired by multiple microphones. Acoustical Science and Technology 22(2), 149–157 (2001)
5. Wang, D.L., Brown, G.J.: Computational Auditory Scene Analysis: Algorithms and Applications. IEEE Press, Wiley-Interscience (2006)
6. Jeong, S.Y., Jeong, J.H., Oh, K.C.: Dominant speech enhancement based on SNR-adaptive soft mask filtering. In: Proceedings of ICASSP, pp. 1317–1320 (2009)
7. Hansen, C.H.: Noise control: from concept to application. Taylor & Francis, Inc., Abington (June 2005)
8. Chaili, M., Raghotham, D., Dominic, P.: Smooth PCM clipping for audio. In: Proceedings of AES 34th International Conference, pp. 1–4 (2008)
9. ARM Limited.: ARM1176JZF-S Technical Reference Manual (2006)
10. Texas Instruments: TMS320C6000 CPU and Instruction Set Reference Guide (2000)
11. Oppenheim, A.V., Schafer, R.W., Buck, J.R.: Discrete-time Signal Processing, 2nd edn. Prentice-Hall, Englewood Cliffs (1999)
12. http://www.mathworks.com/

An Item-Based Collaborative Filtering Algorithm Utilizing the Average Rating for Items

Lei Ren[1,2], Junzhong Gu[1], and Weiwei Xia[1]

[1] Department of Computer Science & Technology,
East China Normal University
[2] Department of Computer Science & Technology,
Shanghai Normal University, Shanghai, China
{lren,jzgu,wwxia}@ica.stc.sh.cn

Abstract. Collaborative filtering is one of the most promising implementation of recommender system. It can predict the target user's personalized rating for unvisited items based on his historical observed preference. The majority of various collaborative filtering algorithms emphasizes the personalized factor of recommendation separately, but ignores the user's general opinion about items. The unbalance between personalization and generalization hinders the performance improvement for existing collaborative filtering algorithms. This paper proposes a refined item-based collaborative filtering algorithm utilizing the average rating for items. The proposed algorithm balances personalization and generalization factor in collaborative filtering to improve the overall performance. The experimental result shows an improvement in accuracy in contrast with classic collaborative filtering algorithm.

Keywords: Recommender system, user-based collaborative filtering, item-based collaborative filtering.

1 Introduction

With the rapid development of online information systems in recent years, there has been too much information for users to select according to their individual information need, which is called information overload. As a solution for information overload, recommender systems have been proposed in academia and industry, and are becoming a core component in the adaptive information system. Recommender system can produce individualized recommendations as output or has the effect of guiding the user in a personalized way to interesting or useful objects in a large space of possible options [1]. Now recommender system has been applied in various domains, such as E-Commerce, digital libraries, E-Learning etc. It has been a successful case that Amazon.com and other mainstream E-commerce websites have deployed recommender system to help recommending books, CDs and other products [2].

As an intelligent component in integrated systems, the work mechanism of recommender system can be depicted as three main steps:

T.-h. Kim et al. (Eds.): SIP/MulGraB 2010, CCIS 123, pp. 175–183, 2010.
© Springer-Verlag Berlin Heidelberg 2010

1. At the layer of HCI, the user agent records the user's actions or ratings for visited items and profiles all users' preferences as a user-item rating matrix.
2. The predictor estimates the target user's ratings for unvisited items based on his historical preference stored in the user-item rating matrix.
3. Finally, a personalized unvisited item list including the top-η items sorted by corresponding estimated ratings of the target user are recommended to him.

Collaborative filtering is the most promising approach employed in recommender systems. One of the advantages of collaborative filtering is its independence of contents in contrast to content-based filtering, in other words, it can deal with all kinds of items including those unanalyzable ones such as videos and audios. The other distinct advantage of collaborative filtering is that the target user's implicit interest can be induced from the preference of his like-minded neighbors, thus the corresponding potential interesting items will be introduce to him. This feature is valuable for the commercial target marketing.

In a recommender system deploying collaborative filtering, all users' observed ratings are integrated and represented as the user-item rating matrix. According to the ways of using the user-item rating matrix in horizontal direction or vertical direction, collaborative filtering is usually categorized as user-based collaborative filtering and item-based one.

User-based collaborative filtering takes each row of the user-item matrix as a user's preference vector. For a given target user, it searches his like-minded neighbors based on the similarity between his preference vector and other users'. The prediction for unobserved ratings of the target user's unvisited items is achieved through a weighted sum of his neighbors' observed ratings. But in user-based collaborative filtering, the problem of matrix sparsity and computational complexity are issues limiting its performance.

As an alternative for user-based collaborative filtering, item-based collaborative filtering takes each column of the user-item matrix as the item's profile vector based on which the similarity of different items is computed. The similarity between two given items shows their distance in rating space with respect to the users rating both ones. As for each of the target user's unvisited items, the k-nearest items are taken as its neighbors from the target user's visited items based on their similarity to it. The ratings for the target user's unvisited items are predicted through a weighted sum of the observed ratings for its item neighbors [3]. In contrast to user-based one, item-based collaborative filtering doesn't suffer the sparsity problem seriously, therefore its recommendation quality is more accurate [3]. Furthermore, item-based collaborative filtering has better scalability for the stability of the item similarity [2]. As a result, item-based collaborative filtering has been becoming the most popular and promising approach in the domain of recommender systems.

Even with the popularity of item-based collaborative filtering, there is still space for accuracy improvement for it. Most of the existing research has concentrated on the aspect of personalization in recommendation, but the user's general opinion about items has not been considered sufficiently. As a result of this, the accuracy of item-based collaborative filtering without concerning about generalization can hardly be improved. In this paper, we focus on the leverage of personalization and generalization in collaborative filtering, and propose a refined item-based collaborative filtering algorithm utilizing the average rating for items namely ICFAR, which incorporates the classic item-based collaborative filtering with the utilization of all users' global general opinions about items. In addition to its scalability equivalent to the classic one, ICFAR achieves the accuracy improvement as the experiment demonstrates.

The rest of this paper is organized as follows: we survey the related work in section 2. Section 3 describes the proposed ICFAR algorithm. Then, ICFAR is evaluated experimentally in section 4. Finally, the conclusion is made in section 5.

2 Related Work

Collaborative filtering algorithm emerged in 2000s, the classic collaborative filtering algorithmic framework was defined in [4], which depicted a user-based collaborative filtering algorithm and experimentally analyzed the performance of itself and its various variants. For the sake of poor scalability in the user-based one, item-based collaborative filtering was proposed alternatively in [3], it was proved experimentally more effective in performance. Item-based collaborative filtering has been approved by lots of researchers, and the majority of collaborative filtering algorithms are following its filtering mechanism. Although item-based collaborative filtering has been deployed in some fields such as E-commerce, E-learning, digital libraries and so on, its recommendation quality can hardy satisfy users. One of reasons for its poor accuracy is that personalization is emphasized excessively. Collaborative filtering aims at providing personalized information, but the user's historical rating data are often noisy in reality, consequently the filtering achieved based on these data will perhaps overfit the noise. So collaborative can't meet what the user need individually indeed.

On the other side, empirically, the average rating for a given item reflects users' general opinions which are users' global views. Specifically in recommender systems, generalization means the user's global average rating for items. Actually in daily life, other people's general view on a given object is an important reference for our personal judgments. For instance, if someone intends to see the film of "Titanic", he will usually first check its general rating on some movie rating websites, e.g. filmratings.com or IMDB.com etc. Then he will ask his like-minded friends' review about it, and at last he can make his final decision based on both opinions. In other words, generalization is also an important factor in recommendation except for personalization, and therefore both factors should be balanced.

Few researchers have noticed the aspect of generalization in recommender systems or made full use of it. In existing research incorporating clustering with recommender systems [5] [6], the average rating of users in a given cluster represents its local opinion, which is not the global one. The clustering aims at improving the scalability, but helps little in the improvement of accuracy. On the contrary, [7] proposes an E-learning recommender system which predicts the unobserved ratings for unvisited items based on their similarity to good learner's average ratings. In essence, the proposed approach is not a real personalized recommender system, because the target user's personal preference is not considered in prediction, it can hardly ensure the accuracy.

3 The Proposed Algorithm

The proposed ICFAR is a variant of the classic item-based collaborative filtering algorithm, it can be viewed as a weighted sum of the personalized collaborative filtering and the item's non-personalized average rating, and its working flow can be divided into five steps:

Step 1. The similarity of different items is measured based on users' historical ratings for them, and stored in a item-item similarity matrix;

Step 2. For the target item, the k-nearest items rated by the target user are selected as its neighbors according to their similarity;

Step 3. The average rating of every item is computed offline based the its observed rating;

Step 4. For each of the target user's unvisited items, its personalized rating is achieved through a weighted sum of the observed ratings for its neighbor items rated by the target user, and the personalized rating is combine equally with its global average rating as its final estimated rating;

Step 5. For the target user, an item list containing his unvisited items with the corresponding top-η estimated ratings is recommended to him.

In a recommender system, the set $U = \{u_1, u_2, \cdots, u_i, \cdots, u_m\}$ denotes the user set sized m, and the set $T = \{t_1, t_2, \cdots, t_j, \cdots, t_n\}$ denotes the item set sized n. All users' historical observed ratings are represented and stored in the user-item rating matrix as Table 1 illustrating.

Table 1. The user-item rating matrix

	The Matrix	...	The Matrix Reloaded	Titanic
user_1	5	...	4	4
user_2	3	...	5	3
⋮	⋮	⋮	⋮	⋮
user_n	1	...	2	0

Generally, the matrix M ($m \times n$) denotes the user-item rating matrix with m users and n items, in which the element $r_{ij} \in M$ ($1 \le i \le m$, $1 \le j \le n$) indicates that the user u_i has given a rating r_{ij} for the item. The rating r_{ij} is usually defined as a discrete value in a given range such as 1 to 5, or continuous value is optional. The value of 1 means "dislike the item", on the contrary, 5 means "like the item". The value of 0 in table 1 means the user has not rated the item which will be predicted by the algorithm. Specifically, this paper set the rating range as 1 to 5.

In the first step, the similarity between different items can be measured in the manner of vector angle cosine similarity or Pearson correlation coefficient. Both metrics have the equivalent performance, but the cosine similarity is slightly more accurate than the other one as the evaluation section shows. So the cosine similarity is employed in ICFAR, the similarity between the item t_i and item t_j is defined as $sim(i, j)$ given by

$$sim(i, j) = \frac{\sum\limits_{u \in U_{ij}} r_{ui} \cdot r_{uj}}{\sqrt{\sum\limits_{u \in U_{ij}} r_{ui}^2} \sqrt{\sum\limits_{u \in U_{ij}} r_{uj}^2}} \tag{1}$$

Where the user set U_{ij} comprises those users who have rated both item t_i and item t_j. According to the above formula the range of $sim(i, j)$ is in the interval from 0 to 1, the value approaching 0 means these two items are dissimilar, the value approaching 1 means they are similar.

In item-based collaborative filtering, only the target user's observed ratings can be used to predict the ratings for his unvisited items, thus the similarity between his unvisited items is useless. In addition, not whole the observed ratings will engage in the prediction for the reason of efficiency. Therefore in the second step, a part of the visited item set will be selected as the target item's neighbors according to their similarity to the target one. The neighborhood size is an important factor for performance, and it varies with the status of the target system, i.e. the parameter k is system-dependent which can be optimized empirically. Although most of the existing collaborative filtering algorithms set k as an absolute threshold, we argues that different users have not rated the same amount of items, thus an identical absolute threshold of k is not reasonable for all items. So we set k as a relative percentage of the amount of the target user's visited items, and the value of k in this paper is set at the level of 35, i.e. 35% as the evaluation section discusses.

As discussed in section 2, for a given item, users' general opinion about it is represented as its average rating, and the average rating can be computed periodically offline or incrementally online. In the view of a long-term duration, the general opinion about a given item is stable, i.e. a few users' ratings for it can't change its average rating drastically. In addition, the online computation consumes more resources than the offline one. So all items' average ratings are computed periodically offline in ICFAR and the corresponding computed results are stored in the item average rating vector which will be utilized in the next step.

The prediction is a core procedure of collaborative filtering algorithm. ICFAR is a variant of the classic item-based collaborative filtering. The average rating of the target item is employed in the rating prediction. For the target user u_t, his estimated rating r_{ta} for the unvisited item t_a is predicted as

$$r_{ta} = \frac{1}{2} \times \frac{\sum\limits_{j \in T_a} sim(a, j) \times r_{tj}}{|sim(a, j)|} + \frac{1}{2} \times \overline{r_a} \tag{2}$$

Where the neighbors of the target item t_a comprises the item set T_a obtained in the second step. $\overline{r_a}$ denotes the average rating of the item t_a, $\overline{r_a}$ is obtained in the third step. The similarity $sim(a, j)$ computed in the first step is the similarity between t_a and its neighbor t_j, and is regarded as the weight of different neighbors to differentiate their importance in prediction. On the whole, the prediction formula utilizes the target user's personalized rating and other users' average rating for the target item equally, i.e. both types of the rating contribute a half of the ultimate estimated rating, and this reflects both personalization and generalization in recommendation.

In the final step, the unvisited items with the top-η estimated ratings comprise a personalized recommendation set for the target user, and it will be recommended to him on demand. The parameter η is predefined by the target user as he needs.

4 Evaluation

In order to compare ICFAR with the classic collaborative filtering algorithm, some experiments have been conducted to evaluate the accuracy performance of ICFAR. In those experiments we select the known MovieLens [8] dataset as the test data. The MovieLens dataset has been adopted in most of the research in recommender systems, and is becoming the reference dataset. The dataset are composed of 1,000,209 anonymous discrete ratings ranging from 1 to 5 for approximately 3900 movies made by 6040 users, each of which has at least 20 ratings, and each item has at least one rating.

For the metric selection, we employ mean absolute error (abr. MAE) to measure accuracy. MAE is a statistical metric which describes the deviation of the estimated rating from the actual rating. For instance, if the actual ratings of N users are $R = \{r_1, r_2, \cdots, r_N\}$ and their estimated ratings are $R' = \{r'_1, r'_2, \cdots, r'_N\}$, MAE denoted as ε_N is formalized as

$$\varepsilon_N = \frac{|R - R'|}{N} = \frac{\sum_{i=1}^{N} |r_i - r'_i|}{N} \tag{3}$$

The lower MAE, the more accurately the algorithm can predict unobserved ratings for users.

In the following experiments, the classic item-based collaborative filtering algorithm [3] (abr. ICF), non-personalized item average algorithm [4] (abr. NPIA) and non-personalized user average algorithm [4] (abr. NPUA) are taken as the benchmark for comparison with ICFAR. We use the 10-fold cross-validation as the test and partitioning method in the experiment, i.e. the dataset is partitioned randomly into ten subsets each of which contains 10 percent of every user's observed ratings respectively, and the experiment was repeated ten times with one of the subsets as the test set and the other nine subsets as the training set, finally the results of ten repeated experiments are averaged as the final statistical result.

The similarity computation is an important process to differentiate the importance of different neighbors of the target item, so an experiment compares the accuracy of ICFAR employing cosine similarity (abr. COSIM) and Pearson correlation coefficient (abr. PCC) respectively, with the neighborhood size varying from 5% to 100% by step of 5%. The result is illustrated as Figure 1.

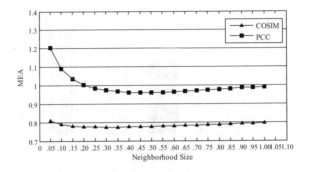

Fig. 1. The comparison between PCC-similarity and COSIM-similarity

Figure 1 demonstrates that the cosine-based ICFAR get lower MAE, i.e. more accurate, than the PCC-based one at any level of neighborhood size. So the following experiments will employ the cosine similarity.

As discussed in section 4, the size of neighborhood is a factor impacting the performance. The second experiment repeats the ICFAR with the neighborhood size varying from 5% to 100% by step of 5%. Figure 2 shows the effect of varying neighborhood size for ICFAR and ICF.

The curve in Figure 2 demonstrates that MAE of ICFAR and ICF both decreases with the neighborhood size increasing. It reaches the minimum around the level of

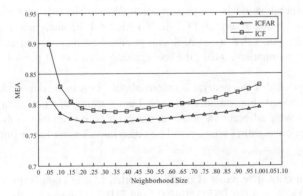

Fig. 2. The impact of neighborhood size

0.35 (35%) and begins to level off. This means that excessive neighbors for the target user cannot help improving accuracy, but deteriorates the recommendation quality with the growth of computational cost. Therefore we set the parameter k of neighborhood size as 35, i.e. 35%, in ICFAR and the following experiment.

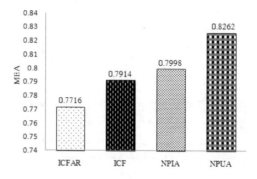

Fig. 3. The accuracy comparison of ICFAR, ICF, NPIA and NPUA

In the term of accuracy, we compare ICFAR with ICF, NPIA and NPUA in the same condition in which we use the cosine similarity and set the neighborhood parameter k as 35. As the figure shows, ICFAR outperforms both ICF and NPIA which have been combined in the proposed algorithm. In addition, NPUA get the worst accuracy. So our proposed ICFAR is proven effective.

5   Conclusions and Future Work

In this paper we propose a refined item-based collaborative filtering algorithm ICFAR in which the item-based personalized rating and the item general average rating are combined to produce the final rating. ICFAR maintains the balance between personalization and generalization in recommendation. We empirically prove that the cosine similarity is more effective than PCC in ICFAR, and the analytical result shows the relative neighborhood size set as 35% can get the best recommendation quality. The result of the comparison with the benchmark algorithms proves ICFAR more effective.

We also notice that the general opinions about items in recommender systems are diverse, what means the variance of each item's rating are not identical. Thus the average rating with different variance should contribute unequal weight in prediction, and the dynamic weighting scheme will be taken into account in our future work.

Acknowledgments. The paper is partly supported by the grant from STCSM project (No. 09510703000 and No. 08511500303) and SHNU project (No. A-3101-10-033).

References

1. Burke, R.: Hybrid Recommender Systems: Survey and Experiments. User Modeling and User-Adapted Interaction 12(4), 331–370 (2007)
2. Linden, G., Smith, B., York, J.: Amazon.com Recommendations: Item-to-item Collaborative Filtering. Internet Computing 7(1), 76–80 (2003)
3. Sarwar, B., Karypis, G., Konstan, J., Reidl, J.: Item-based collaborative filtering recommendation algorithms. In: 10th ACM International Conference on World Wide Web, pp. 285–295. ACM Press, New York (2001)
4. Herlocker, J.L., Konstan, J.A., Borchers, A.I., Riedl, J.: An algorithmic framework for performing collaborative filtering. In: 22nd ACM SIGIR International Conference, pp. 230–237. ACM Press, New York (1999)
5. Kim, T., Park, S., Yang, S.: Improving Prediction Quality in Collaborative Filtering based on Clustering. In: 2008 IEEE/WIC/ACM International Conference on Web Intelligence and Intelligent Agent Technology, vol. 01, pp. 704–710. IEEE Press, New York (2008)
6. Sarwar, B., Karypis, G., Konstan, J., Reidl, J.: Recommender Systems for Large-Scale E-Commerce- Scalable Neighborhood Formation Using Clustering. In: 5th International Conference on Computer and Information Technology (2002),
   http://www.grouplens.org/
7. Ghauth, K.I.B., Abdullah, N.A.: Building an E-Learning Recommender System using Vector Space Model and Good Learners Average Rating. In: 9th IEEE International Conference on Advanced Learning Technologies, pp. 194–196 (2009)
8. GroupLens Group, http://www.grouplens.org/node/73

Validation of Classifiers for Facial Photograph Sorting Performance

Daryl H. Hepting[1], Dominik Ślęzak[2,3], and Richard Spring[1]

[1] Department of Computer Science, University of Regina
3737 Wascana Parkway, Regina, SK, S4S 0A2 Canada
{dhh,spring1r}@cs.uregina.ca
[2] Institute of Mathematics, University of Warsaw
Banacha 2, 02-097 Warsaw, Poland
[3] Infobright Inc.
Krzywickiego 34 pok. 219, 02-078 Warsaw, Poland
slezak@infobright.com

Abstract. We analyze how people make similarity judgements amongst a variety of facial photos. Our investigation is based on the experiment conducted for 25 participants who were presented with a randomly ordered set of 356 photos (drawn equally from Caucasian and First Nations races), which they sorted based on their individual assessment. The number of piles made by the participants was not restricted. After sorting was complete, each participant was asked to label each of his or her piles with description of the pile's content.

Race, along with labels such as "Ears" or "Lips", may be treated as qualities that are a part of the judgement process used by participants. After choosing as many photos as possible, with the stipulation that half have the quality (called QP – for Quality Present) and half do not (QM – Quality Missing), we analyze the composition of each pile made by each participant. A pile is rated as QP (QM), if it contains significantly more of QP (QM) photos. Otherwise, it is QU (Quality Undecided). For the group of 25 participants, we form binary decision classes related to the QU percentages. For example, for the quality "Ears", a participant can drop into the class "Uses-Ears", if his or her QU is not more than, e.g., 60%, or the class "Uses-Not-Ears" otherwise.

In our previous research, given the above decision classes, we constructed classifiers based on binary attributes corresponding to the judgements made by each participant about a specific pairs of photos; whether a participant rated them as similar (placed in same pile) or dissimilar (placed in different piles). Our goal was to find pairs indicating the quality occurrences in participants' judgement processes. We employed rough set based classifiers in order to provide maximally clear interpretation of dependencies found in data. In this paper, we further validate those results by exploring alternative decision class compositions. For example, we find that manipulation of the QU threshold (fixed as 60% in our previous analyzes, as mentioned above) provides a useful parameter. This leads to an interaction between building a classifier and defining its training data – two stages that are usually independent from each other.

T.-h. Kim et al. (Eds.): SIP/MulGraB 2010, CCIS 123, pp. 184–195, 2010.

1   Introduction

Our work is motivated by the goal of improving the accuracy of eyewitness identification [2]. There are two problems when a witness may attempt to identify a suspect: they may be asked to describe the face, which could lead to verbal overshadowing [12], or they may be shown a large amount of facial images, which creates the risk of incorrect source attribution [4].

We hypothesize that if we can understand how a person makes judgements about the similarity of faces, we can personalize the presentation of facial images for that person in order to reduce the total number of facial images seen before an accurate identification is made. Could a concise classifier lead to an effective screening test that would allow this personalization to be applied? This paper continues this stream of investigation.

Our inquiry is indirectly related to binary classification – all photos are either QP or QM, as described in the Abstract. However, we are not interested in the participant's accuracy in correctly classifying each photo, because we realize that our division of photos into QP and QM is subjective. Rather, we are interested in how participants' classification of photos can help to identify a strategy that is being used in the sorting. For example, we may find that some participants are making use of a quality in their sorting. However, it is not sufficient to say that other participants are not using that quality in their sorting. Further difficulties arise, e.g., when participants place photos together not because they are similar but because they are not sufficiently different.

The idea of sorting photos is not new. There are also some approaches to reduce the number of photos viewed before finding the target (see e.g. [11]). However, when comparing to the others, our methodology requires less information (for example, we do not ask about any photo similarity ratings) and leads to a far simpler data layout (see the next section for details). Consequently, when building classification models and considering attribute selection criteria, we attempt to apply very straightforward techniques, which do not make any assumptions about the training data (cf. [10]).

In this paper, we look further into the quality "Ears". Of those trials conducted, the classifiers based on this quality were most accurate. However, we seek to determine if it is possible to improve the accuracy of the classifiers by making modifications to the decision classes. As we modify the original decision classes, does that change the strategy or the meaning?

The rest of this paper is organized as follows. Section 2 presents our basic approach described in previous papers. Section 3 presents some refinements, as a means to improve our understanding of what is being recognized. Section 4 presents conclusions and opportunities for future work.

2   Data Preparation

For the "Ears" quality, we found that 44% of all the photos were QP. We then randomly selected an equal number of QM photos. The unused QM photos

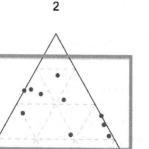

Fig. 1. Plot of participants with respect to the quality "Ears". Each point corresponds to one of 25 participants. A point in the center of the triangle represents an equal mix of photos classified as QP (Vertex 1), QM (Vertex 3), and QU (Vertex 2). The lower and upper rectangles contain those participants in the "Uses-Ears" (C_{E_0}) and "Uses-Not-Ears" (C_{N_0}) decision classes [6], respectively.

(12% of the original 356 photos) were removed from consideration, which meant that the existing piles were analyzed in terms of only the selected QP and QM photos (88% of the original amount). For each participant, the percentages of QP, QM, and QU photos were calculated according to the procedure outlined in the Abstract. These are displayed in the trilinear plot in Figure 1. As reported by Hepting et al. [6], the decision classes "Uses-Ears" (C_E) and "Uses-Not-Ears" (C_N) were formed based on a threshold of 60% for QU.

Once the decision classes are set in this fashion, we may employ various attribute reduction methodologies in order to find pairs which indicate occurrences of the quality in the participants' judgement processes. With the full set of 356 photos, there are 63,190 pairs possible. In order to reduce that, we seek to find pairs which participants in one decision class predominantly rate as similar and which participants in the other decision class predominantly rate as dissimilar. Consider two decision classes, C_E and C_N. We compute the normalized distance (d) for each pair, for both decision classes (d_{C_E}, d_{C_N}). A distance of 0 means that all participants in a decision class have rated the pair as similar, whereas a distance of 1 means that all participants in a decision class have rated the pairs as dissimilar. We group the pairs on the basis of 2 parameters: $d = min(d_{C_E}, d_{C_N})$ and $\Delta = |d_{C_E} - d_{C_N}|$.

This way, we can compute the so-called pair table, where the numbers of attributes (pairs of photos) are aggregated for particular (d, Δ) levels. Valid combinations of parameters are those such that $d + \Delta \leq 1$. We present the information in increments of 0.1, beginning with $d = 0$ to $d \leq 0.4$, and $\Delta = 1.0$ to $\Delta \geq 0.6$. Comparing to [5,6], we add the line for $d = 0$, as it is important on its own and should be highlighted separately. On the other hand, the line for $d = 0.5$ is removed because those pairs are not able to contribute substantially to the discrimination between decision classes.

Pairs found at $d = 0, \Delta = 1$ divide the participants exactly into the two decision classes and herein they are referred to as "splitting pairs". We refer to Tables 7-14 as the examples of the pair tables obtained for various settings considered in Section 3.

3 Validation and Refinement

In our previous work, we used the pairs in each cell of the pairs table as input to one of rough-set-based classifier construction systems [1,10]. We observed that there appears to be a relationship between the pairs table and the accuracy of classifiers. Table 1 summarizes the pair tables' properties and the observed accuracy of classifiers obtained for particular decision classes. This relationship surely requires further investigation. Here, we use it as heuristics replacing more thorough classifier training processes. Intuitively, decision classes that more clearly separate the participants will have more differing pairs of photos. Table 7 shows this information for the decision classes illustrated in Figure 1.

We denote by C_{E_0} and C_{N_0} the original "Uses-Ears" and "Uses-Not-Ears" decision classes, respectively. We will use notation C_{E_i}/C_{N_i}, $i = 1, 2, ...$, for their refinements studied in the next subsections.

Table 1. Comparison of the total number of pair table entries with the measured average accuracy of classifiers presented by Hepting et al. [5]. This is the heuristic which we adopt for use in the rest of this paper.

Quality	Pair Table Count	Average Accuracy
Ears	1782	99.2, σ 1.69
Lips	303	94.4, σ 3.86
Race	237	92.38, σ 4.38

As it can be seen in Table 1, the classifiers built for the "Ears" quality in [5,6] were very accurate. Yet, none of the 63,190 pairs split the participants exactly into the decision classes C_{E_0}/C_{N_0}.

Our further experimental efforts begin by trying to improve the overall model by not only optimizing the classifier but also by tuning the definitions of the decision classes. This way, we hope to converge on the most significant pairs of photos.

3.1   Small Perturbations

We seek to make small adjustments to the composition of the decision classes for "Ears". We notice that some pairs are very close to splitting C_{E_0}/C_{N_0}, according to Tables 2 and 3. We begin with a preference for an approximately equal split of similarity votes to dissimilarity votes. Following this idea, we identify pairs 152-153 (Figure 2) and 4833a-9948a (Figure 3) with $12/25$ similarity votes each as the best candidates with which to create new decision classes. It seems reasonable to choose modifications which make small changes, but with high discriminability. We try new decision class pairs based on 152-153 (C_{E_1}/C_{N_1}) and 4833a-9948a (C_{E_2}/C_{N_2}), first separately and then in combination (C_{E_3}/C_{N_3}). The resulting pair tables are shown in Tables 8, 9, and 10, respectively.

Table 2. Pairs with highest similarity rating in C_{E_0}, across both decision classes

Pair	$C_{E_0} (n = 15)$	$C_{N_0} (n = 10)$	Total Similar
1907a-3600a	12	3	15/25
1182a-1391a	14	3	17/25
3905a-9948a	12	5	17/25
050-105	13	6	19/25
089-137	12	7	19/25
4426a-9700a	12	7	19/25

Table 3. Pairs with highest similarity rating in C_{N_0}, across both decision classes

Pair	$C_{N_0} (n = 10)$	$C_{E_0} (n = 15)$	Total Similar
152-153	10	2	12/25
4833a-9948a	10	2	12/25
023-114	10	4	14/25
105-136	10	5	15/25
1306a-2038a	10	5	15/25
2660a-7669a	10	6	16/25
5774a-9717a	10	6	16/25
060-098	10	7	17/25
1969a-4158a	10	7	17/25
2094a-3722a	10	7	17/25
3669a-3943a	10	7	17/25

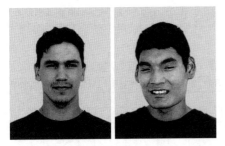

Fig. 2. Photos 152 (left) / 153 (right)

Fig. 3. 4833a (left) / 9948a (right)

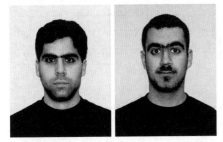

Fig. 4. 1182a (left) / 1391a (right)

Fig. 5. 1907a (left) / 3600a (right)

Considering the original decision classes for the quality "Ears" (C_{E_0}/C_{N_0}), all of the most similar pairs have $d = 0$ for C_{N_0} but $\Delta \neq 1$ for C_{E_0} (Table 3). Therefore, another reasonable strategy may be to move participants from C_E to C_N, so that $d = 0$ for C_E. Based on Table 2, we have chosen pairs 1182a-1391a (Figure 4) and 1907a-3600a (Figure 5) for this purpose. The results of this experiment are shown in Tables 11 and 12. The new decision class pairs are denoted as C_{E_4}/C_{N_4} and C_{E_5}/C_{N_5}, respectively.

Table 6 compares the heuristic (sum of pair table entries) for each of the decision class pairs. It is somewhat surprising that none of the perturbed decision class pairs that have been contemplated in this subsection would seem to be more accurate than the original, at least according to the applied heuristic. It changes, however, for the decision classes described in further subsections.

3.2 Intersection of Reduct Pair Similarities

Hepting et al. [6] presented a list of popular pairs in so called rough-set-based reducts – minimal subsets of attributes (pairs of photos) determining decisions to a satisfactory degree [10] – for the quality "Ears" (with decision classes C_{E_0}/C_{N_0}). For each of these popular pairs, we examined which participants rated them as similar. The results are presented in Table 4. We notice that there are 4 participants who rate all 10 of the popular pairs as similar. We constructed the

Table 4. The most popular pairs which appeared in reducts of classifiers with full accuracy and coverage, as well as information which participants rated those pairs as similar (indicated below with an 'S')

Pairs	$C_{N_0}(n=10)$										$C_{E_0}(n=15)$														
	7	13	14	17	19	20	21	22	24	25	1	2	3	4	5	6	8	9	10	11	12	15	16	18	23
038-068	S		S	S	S	S	S	S	S	S									S						
038-095	S	S		S	S	S	S	S	S	S	S			S											
061-065	S		S	S	S	S	S	S	S	S				S				S							
095-106	S	S	S	S	S	S	S		S	S															
146-172	S	S	S	S		S	S	S	S	S															
111-121	S	S	S	S	S	S		S	S	S						S									
037-176	S	S	S	S	S	S		S	S	S															S
033-121	S	S	S	S	S	S		S	S	S					S										
058-157	S	S	S		S	S	S	S	S	S							S								
4833a-9948a	S	S	S	S	S	S	S	S	S	S								S							S
	10	8	9	9	9	10	7	9	10	10	1	0	0	2	1	1	1	2	1	0	0	0	0	0	2

Table 5. Discrimination values calculated for all participants. Membership in decision classes is also displayed. An 'X' indicates membership in C_N.

Discrim	57	59	62	63	66	71	74	75	76	77	79	81	82	85	85	85	87	89	89	90	92	94	94	95	96
Particip	25	9	20	24	7	14	18	13	17	21	22	10	23	4	11	16	15	8	12	19	2	1	6	5	3
C_{E_0/N_0}	X		X	X	X	X		X	X	X	X										X				
C_{E_6/N_6}	X		X	X	X																				

Table 6. Comparison of the heuristic value for each of 8 decision class pairs. Further details can be found in Tables 7-14.

Decision Classes	Pair Table Count
C_{E_0}/C_{N_0}	1794
C_{E_1}/C_{N_1}	278
C_{E_2}/C_{N_2}	1261
C_{E_3}/C_{N_3}	678
C_{E_4}/C_{N_4}	491
C_{E_5}/C_{N_5}	80
C_{E_6}/C_{N_6}	30736
C_{E_7}/C_{N_7}	26166

decision class pair C_{E_6}/C_{N_6} to capture this division. Table 13 shows that much better accuracy of classification might be expected for this decision class pair. This remains to be verified in further experiments, by applying both rough-set-based and not-rough-set-based classification models [1,9].

Table 7. Pair table for C_{E_0}/C_{N_0}. (The $d + \Delta > 1$ is always greyed out). Note that there are no splitting pairs.

d	Gap (Δ)				
	1.0	≥ 0.9	≥ 0.8	≥ 0.7	≥ 0.6
0.0	-	-	2	3	7
≤ 0.1		2	9	31	59
≤ 0.2			22	77	250
≤ 0.3				159	498
≤ 0.4					675

Table 8. Pair table for C_{E_1}/C_{N_1}. 152-153 is the only splitting pair. (The upper left corner, $d = 0, \Delta = 1$).

d	Gap (Δ)				
	1.0	≥ 0.9	≥ 0.8	≥ 0.7	≥ 0.6
0.0	1	1	1	2	3
≤ 0.1		2	3	4	8
≤ 0.2			4	8	39
≤ 0.3				17	70
≤ 0.4					115

Table 9. Pair table for C_{E_2}/C_{N_2}. 4833a-9948a is the only splitting pair.

d	Gap (Δ)				
	1.0	≥ 0.9	≥ 0.8	≥ 0.7	≥ 0.6
0.0	1	1	1	3	8
≤ 0.1		2	3	15	52
≤ 0.2			9	50	194
≤ 0.3				90	355
≤ 0.4					477

Table 10. Pair table for C_{E_3}/C_{N_3}. There are no splitting pairs.

d	Gap (Δ)				
	1.0	≥ 0.9	≥ 0.8	≥ 0.7	≥ 0.6
0.0	-	1	2	2	2
≤ 0.1		1	2	3	4
≤ 0.2			6	10	28
≤ 0.3				40	231
≤ 0.4					346

Table 11. Pair table for C_{E_4}/C_{N_4}. Although there is $d = 0$ for C_{E_4}, we have $\Delta \neq 1$ for any pair in C_{N_4}.

d	Gap (Δ)				
	1.0	≥ 0.9	≥ 0.8	≥ 0.7	≥ 0.6
0.0	-	-	-	1	1
≤ 0.1		-	1	4	13
≤ 0.2			4	13	55
≤ 0.3				31	135
≤ 0.4					233

Table 12. Pair table for C_{E_5}/C_{N_5}. Although there is $d = 0$ for C_{E_5}, we have $\Delta \neq 1$ for any pair in C_{N_5}.

d	Gap (Δ)				
	1.0	≥ 0.9	≥ 0.8	≥ 0.7	≥ 0.6
0.0	-	-	-	1	2
≤ 0.1		-	-	1	2
≤ 0.2			-	1	3
≤ 0.3				1	10
≤ 0.4					59

Table 13. Pair table for C_{E_6}/C_{N_6}. While C_{N_6} includes only 4 participants, there are 26 splitting pairs.

d	Gap (Δ)				
	1.0	≥ 0.9	≥ 0.8	≥ 0.7	≥ 0.6
0.0	26	322	1073	1904	2436
≤ 0.1		322	1073	1904	2436
≤ 0.2			1073	1904	2437
≤ 0.3				2641	5592
≤ 0.4					5593

Table 14. Pair table for C_{E_7}/C_{N_7}, where C_{N_7} comprises 5 participants with the lowest scores in Table 5.

d	Gap (Δ)				
	1.0	≥ 0.9	≥ 0.8	≥ 0.7	≥ 0.6
0.0	9	211	649	1119	1417
≤ 0.1		211	649	1119	1417
≤ 0.2			742	2219	4585
≤ 0.3				2219	4585
≤ 0.4					4965

Table 15. Comparison of the heuristic value for different thresholds. Those marked with an asterisk (*) have not changed from the line above.

QU%	C_E	C_N	Heuristic	Splitting Pairs	Participants added to C_E
10	1	24	59058	259	12
20	3	22	420	0	3, 15
*30	3	22	420	0	
40	5	20	347	0	2, 6
50	11	14	1972	0	4, 10, 1, 8, 5, 16
60	15	10	1794	0	18, 9, 23, 11
*70	15	10	1794	0	
80	20	5	16187	4	22, 19, 21, 13, 7
*90	20	5	16187	4	

3.3 Discrimination

Going further, to pursue the undecidedness of participants, we compute a measure of discrimination that each participant applies to the set of photos, computed as 100, less the fraction of all possible pairs made by the participant. Consider the two extremes: if a participant made 356 piles (1 photo in each), there would be no pairs and so discrimination would be 100. If the participant made only 1 pile containing all photos, the participant's sorting could make all possible pairs so discrimination would be 0. In Table 5, we present the discrimination values calculated for all participants along with those participants who are members of 2 different decision class pairs – C_{E_0}/C_{N_0} and C_{E_6}/C_{N_6}.

From Table 5, participant 9 seems to be a good candidate to add to C_N, even though Table 4 does not agree. This configuration, as described, is evaluated as the decision class pair C_{E_7}/C_{N_7} (Table 14).

3.4 QU Threshold

As a final set of manipulations, let us return to the threshold for QU. It was set originally to be 60%. Here we consider other possible values, which correspond to the horizontal lines in Figure 1. Note that there are no participants found between 20% and 30%, between 60% and 70%, or above 90%. Table 15 summarizes various details about the decision classes at various thresholds for QU. Even though some of the heuristic values seem low, they all exceed values computed for highly accurate classifiers (Table 1). Because very few of these binary decision classes have splitting pairs, the application of the rough set attribute reduction method will be especially important, and the top photos for each of the binary decision classes will likely change. With that caveat, we present the top 5 photos for C_{E_6}, C_{N_6}, as well as for selected values of the QU threshold (Figures 6 to 10). C_{E_6} contains 4 participants, 3 of which overlap with the C_E for QU = 80%. The photos were determined by separating all pairs in the respective pair table and then computing the frequency for each individual photo. Imperfect as it may be, it is possible to observe some trends moving through Figures 6 to 10.

Fig. 6. Top 5 photos extracted from the pairs table for C_{E_6}, C_{N_6} (Table 13)

Fig. 7. Top 5 photos extracted from pairs table for QU% = 80

Fig. 8. Top 5 photos extracted from pairs table for QU% = 50

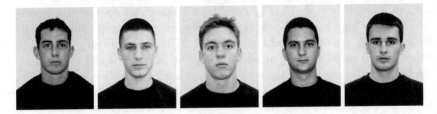

Fig. 9. Top 5 photos extracted from pairs table for QU% = 20

Fig. 10. Top 5 photos extracted from pairs table for QU% = 10

4   Conclusions and Future Work

In this paper, we investigated whether our results reported in [5,6] can be further improved by means of more careful data analysis, as well as more careful data preparation. We employed a simple attribute subset selection and classifier construction procedure based on the theory of rough sets [1,10], in order to understand better the dependencies hidden behind particular qualities and, consequently, to use the obtained knowledge to seek for guidance how to optimize, e.g., the eyewitness identification process [2,11].

We obtained a mechanism for selecting those pairs that may help to classify people according to the qualities they use while making facial similarity judgements. Yet, the applied rough set algorithms did not lead to unique pairs of photos that would be sufficient to discriminate particular qualities. Such a pair might be context-dependent, which makes it difficult to discover. The sorting behavior might also be described by multiple pairs.

In our previous research, the process of labeling the experiment's participants according to their use of particular qualities when comparing faces was somewhat arbitrary, with some parameters to be decided. Therefore, we were motivated to investigate various modifications to our initial labeling of the training data and this paper presents a number of strategies attempted.

We verified that our basic method works well, but there is more work to be done. There appears to be no simple answer about how to re-interpret the data that would lead to convincing solutions. At the same time, the relative success of our straightforward parameterization based on QU seems to indicate that the participants did a good job in assessing similarities.

In order to systematically explore all potential decision class pairs in the neighborhood of C_{E_0}/C_{N_0}, a numerical experiment to assess all binary decision classes over the 25 participants will be pursued. From the resulting complete data set, we will be able to fully consider all variations on the strategies proposed herein. Computation of the pair tables is much less demanding than k-fold validation with more advanced approaches [1,9]. With more study, we could gain further insight into the relationship between the pair tables and the accuracy of classifiers that are built for the associated decisions.

It may be quite valuable to examine the accuracy of participants in classifying photos as QP (QM) in order to gain some further information about how each photo is viewed and what impact that may have on a person's classification. Looking at Figure 1, it is clear that it may be useful to explore thresholds of QP% and QM% instead of QU%. We may also employ a non-binary decision for which several levels of the "undecided" threshold are defined [3], as an alternative to many experiments with different binary decisions.

This quantitative analysis may also benefit from a qualitative approach, using a repertory grid [7] to elicit the basis for similarity between photos. Finally, one may combine our efforts with various aspects of image processing, in order to achieve a better understanding of photos [8,13,14].

Acknowledgements. The first and third authors acknowledge support by the Natural Sciences and Engineering Research Council (NSERC) of Canada and the University of Regina. The second author was supported by the grants N N516 368334 and N N516 077837 from the Ministry of Science and Higher Education of the Republic of Poland.

References

1. Bazan, J.G., Szczuka, M.: The rough set exploration system. In: Peters, J.F., Skowron, A. (eds.) Transactions on Rough Sets III. LNCS, vol. 3400, pp. 37–56. Springer, Heidelberg (2005)
2. Bergman, P., Berman-Barrett, S.J.: The Criminal Law Handbook: Know Your Rights, Survive the System, 11th edn., Nolo (2009)
3. Błaszczyński, J., Greco, S., Słowiński, R.: Multi-criteria classification – a new scheme for application of dominance-based decision rules. European Journal of Operational Research 181(3), 1030–1044 (2007)
4. Dysart, J., Lindsay, R., Hammond, R., Dupuis, P.: Mug shot exposure prior to lineup identification: Interference, transference, and commitment effects. Journal of Applied Psychology 86(6), 1280–1284 (2002)
5. Hepting, D., Maciag, T., Spring, R., Arbuthnott, K., Ślęzak, D.: A rough sets approach for personalized support of face recognition. In: Sakai, H., Chakraborty, M.K., Hassanien, A.E., Ślęzak, D., Zhu, W. (eds.) RSFDGrC 2009. LNCS, vol. 5908, pp. 201–208. Springer, Heidelberg (2009)
6. Hepting, D., Spring, R., Maciag, T., Arbuthnott, K., Ślęzak, D.: Classification of facial photograph sorting performance based on verbal descriptions. In: Szczuka, M. (ed.) RSCTC 2010. LNCS, vol. 6086, pp. 570–579. Springer, Heidelberg (2010)
7. Kelly, G.: The Psychology of Personal Constructs. Norton (1955)
8. Kimchi, R., Amishav, R.: Faces as perceptual wholes: The interplay between component and configural properties in face processing. Visual Cognition 18(7), 1034–1062 (2010)
9. Kuncheva, L.: Combining Pattern Classifiers: Methods and Algorithms. Wiley, Chichester (2004)
10. Pawlak, Z.: Rough set approach to knowledge-based decision support. European Journal of Operational Research 99, 48–57 (1997)
11. Pryke, S., Lindsay, R., Pozzulo, J.: Sorting mug shots: methodological issues. Applied Cognitive Psychology 14(1), 81–96 (2000)
12. Schooler, J.W., Ohlsson, S., Brooks, K.: Thoughts beyond words: when language overshadows insight. Journal of Experimental Psychology: General 122, 166–183 (1993)
13. Schwaninger, A., Lobmaier, J.S., Collishaw, S.M.: Role of featural and configural information in familiar and unfamiliar face recognition. In: Bülthoff, H.H., Lee, S.-W., Poggio, T.A., Wallraven, C. (eds.) BMCV 2002. LNCS, vol. 2525, pp. 643–650. Springer, Heidelberg (2002)
14. Ślęzak, D., Sosnowski, L.: SQL-based compound object comparators – a case study of images stored in ICE. In: Proc. of ASEA, Springer CCIS, vol. 117, pp. 304–317 (2010)

Integrated Environment for Ubiquitous Healthcare and Mobile IPv6 Networks

Giovanni Cagalaban and Seoksoo Kim[*]

Department of Multimedia, Hannam University,
Ojeong-dong, Daedeok-gu,
Daejeon 306-791, Korea
gcagalaban@yahoo.com, sskim0123@naver.com

Abstract. The development of Internet technologies based on the IPv6 protocol will allow real-time monitoring of people with health deficiencies and improve the independence of elderly people. This paper proposed a ubiquitous health-care system for the personalized healthcare services with the support of mobile IPv6 networks. Specifically, this paper discusses the integration of ubiquitous healthcare and wireless networks and its functional requirements. This allow an integrated environment where heterogeneous devices such a mobile devices and body sensors can continuously monitor patient status and communicate re-motely with healthcare servers, physicians, and family members to effectively deliver healthcare services.

Keywords: Ubiquitous healthcare, Mobile IPv6, Network mobility.

1 Introduction

Traditionally, healthcare services for a person's well-being have been provided by medical institutions such as hospitals. With the advancement of medical technology for healthcare diagnosis and treatment, ubiquitous healthcare delivering services anytime and anywhere emerges as a new computing paradigm [1]. The advent of ubiquitous healthcare technologies enables the removal of spatial and time barriers in information services. These enabling technologies range from biosensors, context-aware sensors, handheld computing terminals, to wireless communication technologies, and informa-tion processing technologies such as database managements, knowledge processing and decision support, workflow management, data exchange, security and privacy protection.

The emergence of mobile computing technologies has facilitated the use of com-puter-based devices in different and changing settings. Mobile technologies are bene-ficial to deliver ubiquitous healthcare services available everywhere, at anytime, without disruption of services. The Internet Engineering Task Force (IETF) has de-fined some IP-layer protocols that enable terminal mobility in IPv6 [2] networks without stopping their ongoing sessions. The mobility support in IPV6 allows an

---

[*] Correspomding author.

T.-h. Kim et al. (Eds.): SIP/MulGraB 2010, CCIS 123, pp. 196–203, 2010.
© Springer-Verlag Berlin Heidelberg 2010

entire network, referred to as a mobile network, to migrate in the Internet topology. It is envisioned that anything will soon be connected to the Internet, particularly wireless body area networks (WBAN), personal area networks (PANs), and networks of sensors deployed in vehicles to provide Internet access and connectivity.

Mobile telecommunication technologies could highly help elderly people check their health conditions, arrange healthcare schedules, and administer diagnosis of their diseases. For instance, mobile devices could keep them in touch and even to improve their independence by delivering personalized healthcare services in home regardless of one's location. It also helps physicians and personal healthcare consultants to care about their patients and clients, such as remote access to patients' medical records and provide remote medical recommendations. However, most ubiquitous healthcare systems are not supported by mobile IPv6. These researches do not provide the integration of heterogeneous sensors, devices and distributed resources.

This paper proposes an integrated environment for delivering ubiquitous healthcare services and mobile IPv6 networks. Here, we consider augmentation of mobile devices with awareness of their environment and situation as context, thus, the integrated system uses context information generated by location, health and home environment information collected from heterogeneous devices. Specifically, we look at sensor integration in mobile devices at abstraction of sensor data to more general characterizations. We consider device situation and the use of context information to improve user interaction and to support new types of application.

The rest of this article is organized as follows. Section 2 discusses related works. In section 3, we present the system requirements of the ubiquitous healthcare system. Section 4 follows with the ubiquitous healthcare system with network mobility support. Section 5 presents the implementation of the prototypical testbed network and lastly, section 6 concludes with summary and future work.

2 Related Works

Numerous researchers have applied ubiquitous healthcare technologies into the healthcare sector. Many ubiquitous healthcare services will be available soon and will provide significant benefits to our everyday life. These researches include a mobile healthcare project called MobiHealth [3]. The MobiHealth system allows patients to be fully mobile while undergoing health monitoring. Another healthcare project, CodeBlue [4], has a scalable software infrastructure for wireless medical devices designed to provide routing, naming, discovery, and security for wireless medical sensors, personal digital assistants (PDAs), personal computers (PCs), and other services that may be used to monitor and treat patients. General issues related to using wearable and implantable sensors for distributed mobile monitoring is addressed by UbiMon [5]. This healthcare project is a network of sensors implanted on someone's body sending data to devices such as PDAs, phones and deliver to a medical facility for analysis and storage. Healthcare service applications have also been applied in Smart Medical Home [6] and Aware Home [7] projects. Ubiquitous healthcare services have been developed from the scratch without following the standardized architecture or reusing of available components.

Information technologies such as the latest Internet technology using IPv6 are incorporated with various computing devices to support people in maintaining their health, needing assistance from healthcare providers and as well as their families. Research on this area includes the E-Care Town Project [8] composed of monitoring programs which aimed at creating a town for people of all ages to feel comfortable, safe and enjoy living in. The project utilized IPv6 equipments and data communication methods are being used. However, the project currently doesn't consider IPv6 mobility support mechanisms such as mobile IP, network mobility (NEMO), fast mobile IPv6 (FMIP) and other communication protocols.

3 System Requirements

In this research, we discuss the functional requirements of a ubiquitous healthcare system supporting mobile IPv6 network. The ubiquitous healthcare services which include patient services, personal healthcare services and physician assistant services are to help people get some assistance for their healthcare, such remote monitoring physical states with automatic acquisition of various vital signs, remote assistance in case of accidents and emergencies, follow-up of personal physical exercise and nutrition status. The ubiquitous healthcare system must allow continuous communication between the patient and the healthcare servers, physicians, nurses, and family members, provide real-time monitoring, and improve the independence of elderly people. The requirements are discussed below:

- Permanent Connectivity. This requirement is needed to maintain connectivity between the patient and the server, physicians, and family members while on the move.
- Ubiquity. This is necessary in order to ensure the system is reachable at any time, any place by anyone.
- Real-Time Monitoring. This is required to perform a permanent health check and to guarantee that proper action would be taken in a timely manner and the most effective way in case of health troubles.
- Reliability: This is essential to guarantee against loss of access coverage of a particular wireless technology and against failure of equipments.
- Scalability. This requirement is significant in guaranteeing an expanding number of equipments and users.
- Multimedia Applications. The physical limitation of the elderly people requires enhanced means to communicate with their healthcare server, physicians, healthcare providers and family members. The overall system must be able to integrate different kinds of sensed data such as images, video, voice, and environmental data such as humidity, temperature, air pressure, and location.

4 Ubiquitous Healthcare System with Network Mobility Support

In this section, we discuss in detail the networking features required to meet the functional needs of the ubiquitous healthcare system.

4.1 System Architecture

The ubiquitous healthcare system supports intelligent healthcare with network mobility support is shown in Fig. 1. Here, we design the system architecture which aims to provide an efficient infrastructure support for building healthcare services in ubiquitous computing environments. A platform deploys the architecture on the heterogeneous hardware and integrates software components with the web service technologies. As a result, the flexible system is capable of attending elder persons by providing appropriate services such as interactions, healthcare activities, etc. in an independent and efficient way.

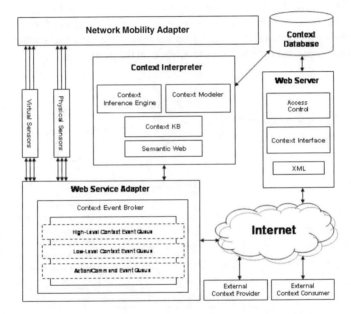

Fig. 1. System Architecture

4.2 IPv6

Internet Protocol (IP) is a protocol designed for use in interconnected systems of computer communication networks to integrate different technologies. IP appears to be the base technology of future networks to provide all kind of services and through different access technologies, both fixed and mobile. Initially, IPv4 [9] is the first major version of addressing structure of IP to be deployed widely. However, its successor, the IPv6, is being deployed actively worldwide due to a number of reasons. First, IPv6 has built-in features that effectively the new services requested by new demanding applications. These features include support for mobility, multicast, traffic reservation, security, etc. Another reason is that IPv6 offers a much larger space of addresses compared to IPv4. IPv4's address space limitation coupled with dramatic rise of devices connected to the Internet, it is imperative to have a sufficiently large number of addresses. Eventually, future devices will be connected to the Internet through IPv6.

4.3 Network Mobility Support

To provide ubiquitous healthcare services, connection to the network must be available everywhere, at anytime, without disruption of service, even when devices are on the move. IP networks were not fully designed for mobile environments. A mechanism is required to manage the mobility of the equipment within the Internet topology. Mobile IPv6 is usually sought to manage mobility of a single IPv6 device. The drawback of this approach is that all embedded devices must be designed to operate this protocol. Recently, IETF NEMO Working Group [10] put on an effort to address issues related to network mobility, i.e. entire IPv6 networks that change their point of attachment to the Internet topology. Such work would set up network of sensors and exchange data between those sensors and remote computers while on the move.

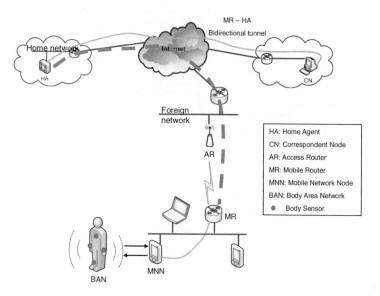

Fig. 2. Network mobility support

For the terminologies, a router that provides connectivity to the mobile network is Mobile Router (MR). Devices belonging to the mobile network that obtain connectivity through the MR are called Mobile Network Nodes (MNNs). A Mobile Node (MN) is a node that implements the mobile IP protocol and has its home network outside the mobile network and is roaming around the mobile network.

The network mobility support as shown in Fig. 2 for IPv6 [11] [12] is conceptually similar to that of terminal devices. It follows the configuration of a bidirectional tunnel between the MR and its Home Agent (HA). The HA is located in the home network of the mobile network, that is, in a location where the addressing of the mobile network is topologically correct. All the traffic addressed to the mobile network is delivered to HA then sends it to the MR through the tunnel. The MR removes the tunnel header and forwards the traffic to its destination within the mobile network. The traffic originated in the mobile network is sent by the MR towards the HA through the tunnel, the HA removes the tunnel header and forwards the packets to their destination.

4.4 Multicast Support

To support multicast traffic for mobile networks, MR can use the bidirectional tunnel (BT) between the HA and the MR's CoA located in the foreign network. An alternative approach is a remote subscription (RS) to a multicast group within the foreign network. However, the BT approach may prove inefficient in terms of non-optimal triangular routing, breach of the multicast nature of the flow, and limited scalability if we consider the multicast traffic to and from mobile networks. The main drawback of applying RS for multicast services emerging or terminating within mobile networks is the required frequent re-construction of the multicast tree. This happens especially if the traffic source is moving fast which results in high latency and network traffic overhead.

MR forwards the request or the traffic to the HA utilizing the Multicast Listener Discovery protocol [13] upon the subscription of a node within the mobile network to a multicast group or transmission of multicast traffic. Subsequently, the HA forwards the corresponding data traffic or group control messages back to the MR. In the case of reduced mobility of the sub-network detected by means of low handover rate, MR initiates routing of multicast traffic via the remote access point.

4.5 Seamless Mobility Features

The abrupt degradation of the network connectivity is unpredictable so a smooth handover from one access point to another is necessary. This mechanism could be provided by protocols such as FMIPv6, HMIPv6, and others. On one hand, vertical handover between distinct access technologies will necessarily result into changes of network conditions. Applications must then be able to dynamically adapt to the available bandwidth as in the case of transmitting video over a GSM link. If it is not possible, the application has to switch to a less bandwidth-consuming mode such as refreshing a picture every ten seconds.

5 Implementation of Ubiquitous Healthcare System

For our ubiquitous healthcare system, the goal is to demonstrate and validate the underlying IPv6 technologies. The system architecture addresses three different issues: the provision of network mobility support, by implementing part of the NEMO Basic Support protocol, the provision of Multicast support to mobile networks, and the route optimization of unicast traffic due to the use of the bidirectional MR-HA tunnel introduced by the NEMO Basic Support protocol is clearly inefficient, in terms of delay, packet overhead and reliability.

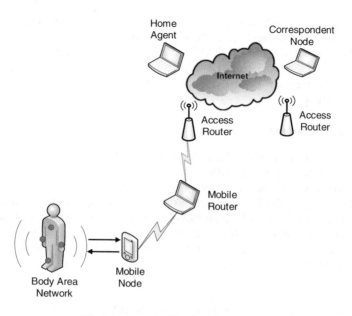

Fig. 3. Prototypical Testbed Network

The configuration of the prototypical testbed network is shown in Fig. 3 which consists of the following devices: a laptop that acts as Home Agent, two hardware routers that act as Access Routers, a laptop that acts as Mobile Router, two other laptops and a PDA, acting as Mobile Network Node, a laptop acting as Correspondent Node, and a collection of body sensors. In this system, we rely on NEMO Basic Support to manage the mobility of the mobile network and provide permanent monitoring. We use several wireless technologies to connect to the Internet. Applications sit on other CPUs in the mobile network; a protocol is being designed to exchange preference settings and network quality between demanding applications and the mobile router. The mobile router connecting the mobile network to the Internet will thus be able to choose the most effective available link, according to the need of the application. Security between the mobile router and the home agent is provided by IPSec, while authentication and confidentiality will be ensured by proper AAA and encryption mechanisms.

5   Conclusion

The success of cellular communications networks shows the interest of users in mobility. These networks are evolving to provide not only the traditional voice service but also data services. Mobile devices are augmented with the environment and everyday objects to improve user interaction and to support new types of applications. This paper demonstrates how evolving Internet technologies based on IPv6 and network mobility will improve ubiquitous healthcare monitoring, mobility and diagnosis of patient health status. In the system, we defined the essential functional requirements of the system based on IPv6 and NEMO protocols. For future study, more in-depth performance evaluation of the system will be performed.

Acknowledgements

This paper has been supported by the 2010 Hannam University Research Fund.

References

1. Weiser, M.: Some Computer Science Problems in Ubiquitous Computing. Communications of the ACM (1993)
2. Johnson, D., Perkins, C., Arkko, J.: Mobility Support in IPv6; RFC 3775, Internet Engineering Task Force (2004)
3. MobiHealth Project, http://www.mobihealth.org
4. CodeBlue: Wireless Sensor Networks for Medical Care,
   http://www.eecs.harvard.edu/~mdw/proj/codeblue/
5. UbiMon, http://www.ubimon.org
6. Smart Medical Home,
   http://www.futurehealth.rochester.edu/
   smart_home/Smart_home.html
7. Aware Home Research Initiative,
   http://www.cc.gatech.edu/fce/house/house.html
8. E-Care Town Fujisawa Project Consortium,
   http://www.e-careproject.jp/english/index.html
9. IPv4, http://en.wikipedia.org/wiki/IPv4
10. IETF Network Mobility (NEMO) Working Group,
    http://www.ietf.org/html.charters/nemo-charter.html
11. Devarapalli, V., Wakikawa, R., Petrescu, A., Thubert, P.: Network Mobility (NEMO) Basic Support Protocol; RFC 3963, Internet Engineering Task Force (January 2005)
12. De la Oliva, A., Bernardos, C.J., Calderón, M.: Practical evaluation of a network mobility solution. In: EUNICE 2005: Networked Applications - 11th Open European Summer School (July 2005)
13. Vida, R., Costa, L.: Multicast Listener Discovery Version 2 (MLDv2) for IPv6, RFC 3810, Internet Engineering Task Force (June 2004)

Fusion Render Cloud System for 3D Contents Using a Super Computer

E-Jung Choi and Seoksoo Kim[*]

Department of Multimedia, Hannam University, 133 Ojeong-dong, Daedeok-gu,
Daejeon-city, Korea
ejchoi@hnu.kr, sskim0123@naver.com

Abstract. This study develops a SOHO RenderFarm system suitable for a lab
environment through data collection and professional education, implements a
user environment which is the same as a super computer, analyzes rendering
problems that may arise from use of a super computer and then designs a
FRC(Fusion Render Cloud) system. Also, clients can access the SOHO RenderFarm system through networks, and the FRC system completed in a test environment can be interlinked with external networks of a super computer.

Keywords: Super Computer, Animation, RenderFarm, 3D Modeling, Fusion, Render.

1 Introduction

Rendering refers to a process that animation data are converted into images, and RenderFarm[1] is a cutting edge technology that makes such images look more natural
and real by calculating illumination reflected on a character and backgrounds[2] in
each scene. The following figure shows the process of 3D image contents production.

Fig. 1. 3D video content creation process

---

[*] Corresponding author.

T.-h. Kim et al. (Eds.): SIP/MulGraB 2010, CCIS 123, pp. 204–211, 2010.
© Springer-Verlag Berlin Heidelberg 2010

In order to produce a full 3D video, a tremendous amount of rendering hours is required. In general, a computer using one CPU may need about 2~10 rendering hours for one frame of a film. Since 24 frames are played in a second, about 259,200 ~ 1,296,000 rendering hours are needed for a 90-minute film.

For example, it took 20 million rendering hours to make Shrek 3, which can be done by a general computer over 2283 years without a break. Using the RenderFarm technology, however, the film could be produced within 3 years.

Likewise, in order to solve the problem of long rendering hours, 3D CG production requires a multitude of servers integrated into a large RenderFarm system using network devices. Then, a producer can use and control the servers as if they are a single system[3].

Nevertheless, using the technology of RenderFarm requires a lot of initial expenses and continuous maintenance costs, which could be quite burdensome for small 3D CG production companies. So, these small productions outsource professional rendering services after paying fees.

In many cases, small companies spend expenses allotted for 3D contents production on rendering services because they can not afford RenderFarm management, resulting in a lower-quality production. If they can use a super computer, however, these small companies can obtain high-quality rendering at the expenses 17 times cheaper than those of outsourcing services. The following table compares functions of Digimars, a rendering service company, and those of KISTI, a super computer.

Table 1. Digimars renderfarm spec

Classification	CPU Spec	Operation speed (Tflops)
Main center	400CPUs, 1000core	7.4
2nd center	100CPUs, 400Core	2.9
3rd center	120CPUs, 240Core	1.0
Maximum operation speed	620CPUs(1,640Core)	11.3

Table 2. KISTI Super computer spec (Total 3)

Classification	CPU Spec	Operation speed (Tflops)
TACHYON (SUN B6048)	3,008CPUs	24.0
GAIA (IBM p595)	640CPUs	0.6
GAIA (IBM p6)	1,536CPUs	30.7

Likewise, although high-function H/W is available in Korea, the super computer itself is not equipped with a system to offer rendering services. Also, it is not possible that a company that needs rendering services come to use the super computer whenever necessary or access to it through networks due to an inability to solve problems(if any).

As a solution, an integrated system, combining rendering, cloud computing, and a super computer, may allow a number of clients to access the super computer through networks for rendering and, this method can solve problems related with time, costs, and the environment.

This study, therefore, develops the FRC(Fusion Render Cloud) system so as to deal with the difficulties related with 3D CG rendering using a super computer. With the FRC system, small companies may enjoy a high-efficiency working environment at lower costs compared to outsourcing of the services or establishment of a Render-Farm system.

2   Related Research

According to the 3D industry report by Thomson Reuters, an information company, 3D application areas that will lead the market in the future after 2010 are 3D TV, 3D photography, and 3D cinema [4].

Table 3. 3D technical patent TOP 3 areas(from 2003 to 2009)

Technical Classification	2003	2008	Rate of increase (2003 ~ 08)	2009 (Q1&Q2)	Placing
3D TV	612	1,034	69%	486	1
3D Photography	460	720	57%	368	2
3D Cinema	103	149	45%	61	3

In US, Pixar, DreamWorks, and LightStorm have released Up, Shrek, and Avatar while in Korea EON, Toiion, etc. have produced 3D films and animations such as Take Off and Dino Mom, drawing attention of the market.

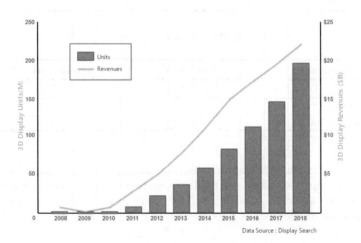

Fig. 2. 3D display technology and market report

Display Search, a research agency, expects that in 2010 3D films will become increasingly popular and this will have a positive impact on 3D contents production in various areas [5].

As the 3D entertainment market, especially the film industry, attracting consumers, the technology is expected to be further applied to the advertising, construction, and media contents industry.

Also, last year AMD and OTOY began joint server-side rendering services using the fastest computer in the world and, thus, Korea needs to make increasing efforts to explore the field.

In order to develop technologies needed to lead the market according to each application and to avoid reckless competition in the industry, Korea shall provide sustainable support to related research projects.

3 Fusion Render Cloud System

In this research, two studies are provided for 3D contents using a super computer and FRC(Fusion Render Cloud) system design, respectively.

3.1 Analysis of Rendering Problems and System Design by Establishing a Test Environment

① Establishment of a RenderFarm System for a Test Environment

This study has adopted the SOHO(Small Office/Home Office) RenderFarm system for a small-scale test environment. SOHO RenderFarm has an adequate size for an office or research, consisting of 3 server computers, 1 database(storage), and other devices.

In the RenderFarm system, computers shall be connected in parallel through clustering. Also, Beowulf technology[6] is applied to make possible computer operation close to that of a super computer, providing H/W and other utilities. And VS(Linux Virtual Server) enables the FRC system to connect web servers in parallel to offer services without cutoff and deal with technical problems.

The completed SOHO RenderFarm can be continuously used for other 3D contents rendering studies. In addition, if clustering of other computers is possible later, the function of the system can be enhanced through the expansibility. The following figure depicts the SOHO RenderFarm system used in this study.

② The Same UI Configuration as a Super Computer

The super computer No.4, operated by KISTI, uses AIX 5L 5.3 based on Unix as its O/S. AIX 5L 5.3 offers binary compatibility with previous releases and supports advanced technologies such as Virtual SCSI, Virtual Ethernet, Micro-Partitioning, and Simultaneous Multi-Threading(SMT) along with flexible and reliable tools perfectly operated under various IT environments.

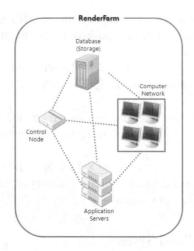

Fig. 3. SOHO renderfarm structure

③ FRC(Fusion Render Cloud) System Design Optimized for a Super Computer

Based on the AIX 5L 5.3 O/S of the super computer, the system provides 3D contents production tools and a rendering engine to implement RenderFarm functions. It is expected that a number of problems will arise during the process but solutions suggested and applied to the super computer in the near future can successfully deal with the problems. The final FRC system implemented in the test environment is as follows.

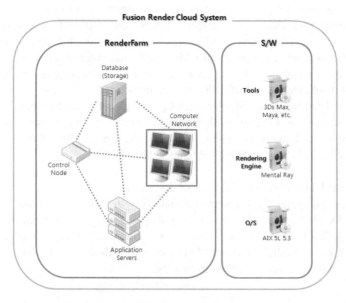

Fig. 4. FRC systems consist of experiments in an environment

3.2 Fusion Render Cloud System Development and Debug

① Analysis of Requirements by Servers and Clients

The FRC server is set up in the test environment and the system allows client computers to access through networking. The FRC server must be able to perform clients' commands and clients without difficulties request the server to carry out rendering. Therefore, in the development stage, the requirements of the server and clients shall be analyzed for convenient use and management in the future [8].

② Client UI Configuration Optimized for Rendering

Basically, rendering needs a file manager that can upload to the FRC server scene files completed by 3D contents production tools. Also, it should be possible to change all the options related with rendering and confirm the information. Lastly, the system shall display the condition of rendering processes in real time along with the progress and elapsed time. A user can download a completed file to his computer after rendering is finished.

③ 1:1 FRC System Prototype and Debug

Since a prototype is developed in a test environment, 1:1 rendering should be done first. Despite the 1:1 condition, debugs may occur frequently according to the development circumstances. Therefore, the debug information is collected to solve the same problems in the future after application to the super computer.

Due to the requirements and UI configuration analyzed and reflected in the system in advance, a client can easily perform rendering work using the completed prototype.

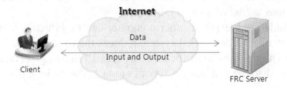

Fig. 5. Structure of the FRC system prototype

④ System design for 1:N Networking

In the near future, the system shall enable a number of clients to access and carry out rendering. Thus, this study ensures that the system has 1:N networking for future expansibility.

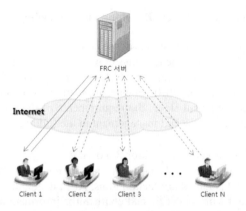

Fig. 6. Considering scale of the FRC system

4 Conclustion

This study develops a SOHO RenderFarm system suitable for a lab environment through data collection and professional education, implements a user environment which is the same as a super computer, analyzes rendering problems that may arise from use of a super computer and then designs a FRC(Fusion Render Cloud) system. This system is expected to contribute to the 3D contents industry as follows.

First, in terms of technology, the system applies cloud computing to 3D contents rendering and uses a super computer to allow a number of general clients to perform high-speed/quality rendering. And a RenderFarm system in a test environment can implement the same as that of a super computer.

In economic/industrial terms, a super computer can be effectively utilized for CG rendering. As a result, the developed system reduces the time needed for film production and cuts down expenses used for outsourcing of rendering services. In addition, companies having difficulty in using a super computer due to long distances can use it through this system.

Acknowledgement

This paper has been supported by the 2010 Hannam University Research Fund.

References

1. KISTI Visulization Wiki, http://sv.ksc.re.kr/svwiki/RenderFarm
2. Sung, K.-J., Kim, S.-R.: The Understanding about the Work Process For 3D Animation. In: Digital Interaction Design Institute Conference 9, vol. 7(2), pp. 13–21
3. Choi, Y., Rhee, Y.: A Task Decomposition Scheme for Parallel Rendering of Continuous Images. In: Korean Insitute of Information Scientists and Engineers Conference, vol. 32(2), pp. 1042–1044 (November 2005)

4. Thomson Reuters, http://thomsonreuters.com/
5. DisplaySearch, http://www.displaysearch.com/
6. Lee, y.-h., Bark, s.-c.: Computer Performance Improvement using the LINUX Clustering. In: Korean Institute of Information Technology Conference, vol. 5(1), pp. 17–21 (March 2007)
7. Park, D.-K., Lee, Y.-W.: Performance Evaluation of Organization Methods and Scheduling Algorithms in a Linux Virtual Server. In: Korean Insitute of Information Scientists and Engineers Conference, vol. 29(1), pp. 613–615 (April 2002)
8. Kim, S., Keum, T., Kim, K., Lee, W., Jeon, C.: Implementation and Performance Analysis of Testbed Clusters for Cloud Computing. In: Korean Insitute of Information Scientists and Engineers Conference, vol. 36(1(B)), pp. 545–548 (July 2005)

On Perceptual Encryption: Variants of DCT Block Scrambling Scheme for JPEG Compressed Images

Muhammad Imran Khan[*], Varun Jeoti , and Aamir Saeed Malik

Department of Electrical and Electronics Engineering,
Universiti Teknologi PETRONAS, Bander Seri Iskander,
31750 Tronoh, Perak, Malaysia
www.signalimran.com

Abstract. In this paper, a perceptual encryption scheme based on scrambling of DCT blocks in JPEG compressed images and its variants are proposed. These schemes are suitable for the environment where perceptual degradation of the multimedia content is required. The security of these techniques is shown to be sufficient against casual attacks. These discussed schemes are kept as light as possible in order to keep the encryption overhead and cost low. Also, an investigation in the progressive degradation of image by increasing the number of the DCT blocks to be scrambled has been carried out. The quality of the degraded multimedia content is measured with objective image quality assessment (IQA) metrics such as SSIM, MS-SSIM, VIF, VIFP and UQI. These IQA metrics provide choice for the selection of control factor.

Keywords: Discrete Cosine Transform, JPEG, Multimedia Commerce, Multimedia Security, Perceptual Encryption, Scrambling.

1 Introduction

Of late, multimedia security has attracted the attention of many researchers. A huge amount of research work in securing the multimedia content has been reported during the past decade after the development of the image and video compression standards. There is an extensive use of multimedia content like images, videos and audio on the internet. Due to this increase in transferring of visual data, there is a need to make sensitive content secure against any kind of adversary. With the growth of internet, some of the commercial applications like pay-TV, pay-per-view, Video on Demand, etc. [1] have also sprouted, with unique requirements of multimedia content preview. These applications allow the degraded quality of the content to be available to be viewed freely to attract the consumer. However, if the consumer wants to watch the high quality or unencrypted version then he/she has to pay for it. Perceptual encryption addresses the problem of this required degradation. A comprehensive survey on the recently proposed perceptual encryption schemes can be found in [2].

Inspired by the idea in [3] and [4], this paper proposes a perceptual encryption scheme and its variants based on 2 levels of scrambling – first the 8×8 blocks in the

---

[*] Corresponding author.

T.-h. Kim et al. (Eds.): SIP/MulGraB 2010, CCIS 123, pp. 212–223, 2010.
© Springer-Verlag Berlin Heidelberg 2010

Region Of Interest (ROI) and second, the number of AC coefficients of DCT as stipulated in JPEG compression standard [5]. Also, an investigation is carried out to analyze the effect of encryption on compression. To scramble the selected DCT blocks and AC coefficients, a chaotic scrambler [6] is used, that is described in section 3.1. For the purpose of measuring the level of degradation introduced in the image, along with PSNR [7], objective image quality assessment (IQA) metrics such as SSIM [8], MS-SSIM [9], VIF [10], VIFP [10] and UQI [11] are used. The measured values of these IQA metrics act as a control factor and determine the level of perceptual encryption. The cryptographic security of the given scheme is dependent on the key and is discussed in section 4.8.

The rest of this paper is organized as follows: related work is presented in Section 2. In Section 3, the proposed perceptual encryption scheme and its variant have been discussed. In Section 4, an analysis has been done and results have been presented. The work is concluded in Section 5.

2 Related Work

In [3], a number of simple and efficient frequency domain encryption techniques like selective bit scrambling, block shuffling and block rotation for videos were proposed. The proposed techniques in [3] are for both, wavelet based system and DCT based system. For DCT based system, the techniques include sign bit encryption and motion vector scrambling. One of the advantages of the proposed methods is that it has an insignificant effect on compression ratio. In [12], a selective encryption of human skin in JPEG compressed images was proposed. In the proposed technique [12], the amplitude part (as shown in Fig. 1) of the selected number of AC coefficients in the DCT block is encrypted with AES cipher.

In [13], a selective encryption scheme has been proposed for uncompressed and compressed images. For uncompressed images, the image is decomposed in to bitplanes and these bitplanes are encrypted starting from the least significant bitplane to

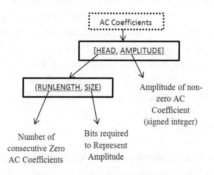

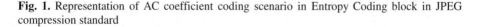

Fig. 1. Representation of AC coefficient coding scenario in Entropy Coding block in JPEG compression standard

most significant bitplane. An encryption method for compressed image was also proposed in [13], where the appended bits[1] corresponding to selected number of the AC coefficients have been encrypted. In [14], a selective encryption scheme for secure transmission of uncompressed images in mobile environment has been proposed. The image is divided in to 8 bitplanes. AES is used for the encryption of bitplanes. It is reported in [14], at least 12.5 % of the data needs to be encrypted if the encryption process starts from most significant bitplane. The metric to measure the level of distortion introduced in the images is PSNR. In [4], a notation of perceptual cryptography was presented. The alternative Huffman codewords of the AC coefficients that falls in the Zone of Encryption were encrypted. Few applications of perceptual encryption are also given in [4].

3   The Proposed Scheme

The proposed scheme consists of two levels; first the ROI, which needs to be encrypted in the image/video, is identified and encrypted. Secondly, some selected numbers of AC coefficients are scrambled in order to degrade the perceptual quality of the entire image/video. The main idea here is to identify the area which needs to be encrypted. We took advantage of the fact that usually the important information is focused in the center of the image or video. Therefore, one can degrade the content from the center. We consider several other variants of DCT block scrambling along with AC coefficients to degrade the perceptual quality of the image. DCT blocks from the center are selected. These DCT blocks are selected in such a manner that they form a square in the center. The number of selected blocks from center is increased to show progressive perceptual encryption. Besides DCT block scrambling selected number of AC coefficients are scrambled to degrade the perceptual quality of the entire image. Another possible variant can be the random selection of DCT blocks from the image and then scrambling them. Also the selected number of AC coefficients can be scrambled with it. For the purpose of scrambling, the scrambler is incorporated after the quantization block in the JPEG compression standard as shown in Fig.2. Also it is worth mentioning that the indexes of the quantized DCT blocks and AC coefficients are read in zigzag manner. If the region of interest is other than a square area then raster scan can also be used to read their indexes. This paper mainly focuses on the security against the casual listeners/observers so it provides sufficient security against casual listeners as discussed in section 4.8.

3.1   Scrambler / Encryptor

The selected indexes of DCT block and the AC coefficient are permuted to random positions. These random positions are obtained by multiplying the data that needs to be scrambled with the scrambling matrix.

---

[1] The bits that are representing the (AMPLITUDE) symbol followed by (RUNLENGTH, SIZE) symbol, see [5] JPEG image compression Standard.

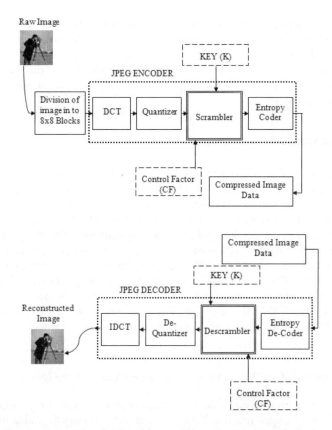

Fig. 2. JPEG Coder and Encoder with Scrambler and Descrambler embedded in it respectively

As shown below, S is the scrambling matrix, X is the input values (old position vector) and Y is the scrambled output (scrambled position vector).

$$S \cdot X = Y$$

The Scrambling matrix 'S' is obtained by using chaotic logistic map given in Eq. (1),

$$x_{n+1} = rx_n(1-x_n) \tag{1}$$

where $0 < x_n < 1$ and $3.47 < r < 4$. When the value of "r" is between 3.47 and 4, it generates random values. The scrambling matrix is generated in such a way by this logistic map that it has exactly one "1" in its each column or each row. This chaotic scrambler is described in detail in [6]. It has optimal property of spread and dispersion. Spread measures the distance between the elements that were near to each other before permutation. Dispersion measures the randomness in spread. These two properties are of great significance as the DCT block indexes and the AC coefficients needs to be randomly scrambled and there spread and dispersion should be optimal to achieve quality encryption.

4 Performance Evaluation

The performance of the abovementioned scheme and its variants is evaluated by keeping several aspects in consideration. The quality or the level of distortion introduced in the image by encryption is measured by objective image quality assessment metrics such as SSIM, MS-SSIM, VIF, VIFP and UQI for each variant separately. Another aspect, that is, the compression efficiency is also analyzed. The effect of the encryption scheme on compression ratio and the security of the scheme are also discussed in section 4.7 and 4.8 respectively. For the experiment, a set of commonly used test images are taken. The experiment is performed on the images of size 256×256. Then, the level of perceptual encryption is measured by several metrics including PSNR, SSIM, MS-SSIM, VIF, VIFP and UQI. The measured values are shown in Table 1, 2, 3 and 4 (MATLAB implementation of image quality assessment metrics can be found in [18]).

4.1 Scrambling of DCT-Blocks Selected from Center of the Image

In majority of the images and videos, the main object of focus always appears in center. So, it will be a good practice if the image or video is only degraded from the center as it will save the time required in encryption process and can result in reduction of computational cost. In this case, 8×8 quantized DCT blocks from center are selected and scrambled. For the experiment, we start by taking four 8×8 DCT blocks from center and then scrambling them. In the Fig. 3(b) shown below, the 144 (12×12) blocks are selected and scrambled using the chaotic scrambler. The number of DCT blocks selected from the center of image is then increased to 16 (4×4) and are scrambled. In a similar manner the number of the DCT blocks selected from the center is increased until the whole image is distorted. The corresponding values of the SSIM, MS-SSIM, VIF, VIFP, UQI and PSNR are computed after every step and are shown in Table 1. The corresponding graph is shown in Fig. 8(a), which will help in selecting the control factor also known as quality loss factor. As one can observe from the graph, the increase in the number of the DCT blocks from center will increase in the percentage of hidden amount of information in the image. The selection of the blocks depends upon the application and the content. One can increase the number of selected DCT blocks from the center according to one's need. As it can be seen from Table 1 that for 144 DCT blocks selected from the center of the image, on average 13.75% measured from SSIM and 22.53% measured from MS-SSIM visible information is distorted in the image.

(a)   (b)

Fig. 3. (a) 256×256 Lena Image. (SSIM = 1, MS-SSIM = 1, VIF = 1, VIFP = 1, UQI = 1), (b) 256×256 Lena Image selectively scrambled from center (PSNR = 18.83, SSIM = 0.8478, MS-SSIM = 0.7022, VIF = 0.6835, VIFP = 0.7654, UQI = 0.8429)

Table 1. The Corresponding Measured Metric Values of Scrambled DCT-Blocks from Center of Image

Metrics/ Block Size	Measured Average value for Center Selected Scrambling								
	8×8	10×10	12×12	14×14	16×16	18×18	20×20	22×22	24×24
PSNR	25.51	23.31	18.22	18.03	18.98	17.53	16.65	15.80	15.01
SSIM	0.9413	0.9078	0.8625	0.7980	0.7687	0.7061	0.6454	0.5769	0.4963
MS-SSIM	0.9068	0.8543	0.7747	0.6514	0.6377	0.5319	0.4663	0.3868	0.3118
VIF	0.8350	0.7660	0.6847	0.6017	0.5269	0.4323	0.3431	0.2540	0.1654
VIFP	0.8699	0.8088	0.7351	0.6986	0.5873	0.4977	0.4093	0.3187	0.2244
UQI	0.9330	0.8956	0.8455	0.7903	0.7349	0.6628	0.5877	0.5058	0.4062

4.2 Scrambling of DCT-Blocks Selected from Center along with 5th AC Coefficient

As in the abovementioned case, the DCT-blocks in the center are only scrambled and the information around the ROI (center of the image which is scrambled) is clearly visible which allows the availability of plaintext. To encounter this issue, one can use the AC coefficients to degrade the whole image to a certain level of quality. First few AC coefficients contain large values and are sufficient for encryption of the content [16]. So, for the given cases, we will only investigate the level of perceptual encryption one will get by encrypting first 5 AC coefficients along with the scrambled DCT-blocks from the center. In this case, the same abovementioned experiment was repeated with the 5th AC coefficient scrambled. The AC coefficients are read in zig-zag order from DCT block. Table 2 shows the measured values after every step. For 144 DCT blocks selected from the center along with 5th AC coefficient, on average 29.50% measured from SSIM and 27.09% measured from MS-SSIM visible informa-tion is distorted. The scrambling of the AC coefficients is performed in such a manner that they remain in their own orbit or plane. In other words, the AC 5 of one DCT block can only takes the place of only other AC 5 coefficients in any other DCT block.

Fig. 4. 256×256 Lena Image selectively scrambled from center along with 5th AC coefficient scrambled (PSNR = 18.27, SSIM = 0.6291, MS-SSIM = 0.6486, VIF = 0.2639, VIFP = 0.2921, UQI = 0.5360)

Table 2. The Corresponding Measured Metric Values of Scrambled DCT-Blocks from Center of Image Along With 5th AC Coefficient

Metrics/ Block Size	Measured Average value for Center Selected Scrambling								
	8×8	10×10	12×12	14×14	16×16	18×18	20×20	22×22	24×24
PSNR	23.78	22.26	20.83	19.73	18.66	17.34	16.52	15.71	14.95
SSIM	0.7684	0.7416	0.7050	0.6722	0.6318	0.5818	0.5349	0.4812	0.4189
MS-SSIM	0.8528	0.8039	0.7291	0.6754	0.6016	0.5032	0.4422	0.3680	0.2985
VIF	0.3955	0.3621	0.3217	0.2869	0.2476	0.2038	0.1632	0.1233	0.0857
VIFP	0.3640	0.3383	0.3053	0.2764	0.2442	0.2062	0.1691	0.1337	0.0966
UQI	0.5632	0.5366	0.4996	0.4679	0.4286	0.3826	0.3360	0.2875	0.2282

4.3   Scrambling of DCT-Blocks Selected from Center along with 5th, 4th and 3rd AC Coefficient

A similar experiment is repeated as in section 4.2 but with 5th, 4th and 3rd AC coefficient scrambled. The results are shown in the Table 3. It is noticed that again for 144 DCT blocks selected from the center along with 5th, 4th and 3rd AC coefficient, on average 42.60% measured from SSIM and 32.67% measured from MS-SSIM visible information is hidden.

Fig. 5. 256×256 Lena Image selectively scrambled from center along with 5th, 4th and 3rd AC coefficient scrambled (PSNR = 17.88, SSIM = 0.5167, MS-SSIM = 0.6089, VIF = 0.1472, VIFP = 0.2097, UQI = 0.4298)

Table 3. The Corresponding Measured Metric Values of Scrambled DCT-Blocks from Center of Image Along With 5th, 4th And 3rd AC Coefficient

Metrics /Block Size	Measured Average value for Center Selected Scrambling								
	8×8	10×10	12×12	14×14	16×16	18×18	20×20	22×22	24×24
PSNR	22.44	21.33	20.19	19.25	18.31	17.10	16.35	15.59	14.87
SSIM	0.6257	0.6043	0.5740	0.5481	0.5146	0.4755	0.4390	0.3971	0.3470
MS-SSIM	0.7860	0.7417	0.6733	0.6242	0.5564	0.4662	0.4114	0.3444	0.2815
VIF	0.2094	0.1937	0.1719	0.1538	0.1339	0.1118	0.0921	0.0735	0.0545
VIFP	0.2398	0.2237	0.2016	0.1829	0.1611	0.1361	0.1125	0.0902	0.0661
UQI	0.4216	0.4005	0.3701	0.3464	0.3141	0.2790	0.2442	0.2091	0.1646

4.4 Scrambling of DCT-Blocks Selected from Center along with First 5 AC Coefficients

Same experiment as in section 4.2 & 4.3 is repeated but now with all first five AC coefficients that are selected to be scrambled. Now one can see that for the case of 144 DCT blocks, on average 61.65% calculated by SSIM and 47.85% calculated by MS-SSIM is distorted. The results are shown in Table 4 and the corresponding graph is plotted in Fig. 8 (b). Another variant of block based DCT scrambling for perceptual encryption can be the random selection of the DCT blocks along with some selected number of AC coefficients.

Fig. 6. 256×256 Lena Image selectively scrambled from center along with first 5 AC coefficients scrambled (PSNR = 16.02, SSIM = 0.3035, MS-SSIM = 0.4616, VIF = 0.0576, VIFP = 0.0788, UQI = 0.2184)

Table 4. The Corresponding Measured Metric Values of Scrambled DCT-Blocks from Center of Image Along With first 5 AC Coefficients

Metrics /Block Size	Measured Average value for Center Selected Scrambling								
	8×8	10×10	12×12	14×14	16×16	18×18	20×20	22×22	24×24
PSNR	19.40	18.89	18.29	17.75	17.15	16.28	15.76	15.13	14.56
SSIM	0.4145	0.4018	0.3835	0.3679	0.3480	0.3228	0.3015	0.2766	0.2466
MS-SSIM	0.6025	0.5715	0.5215	0.4862	0.4382	0.3688	0.3339	0.2828	0.2394
VIF	0.0869	0.0820	0.0743	0.0680	0.0611	0.0532	0.0476	0.0408	0.03461
VIFP	0.0925	0.0872	0.0792	0.0729	0.0657	0.0564	0.0488	0.0405	0.0326
UQI	0.2290	0.2169	0.1995	0.1853	0.1681	0.1476	0.1304	0.1100	0.0880

4.5   Control Factor

The control factor for the above discussed variants of block based perceptual encryption scheme can be selected by plotting metrics values against the number of selected blocks as shown in Fig.8 (a, b). A number of choices have been given to select the control factor using several IQA metrics. If one wants to degrade the quality of the content to 60% (measured using VIFP) then from Fig. 8(a), 20×20 blocks from the center needs to be encrypted. Thus the control factor in this case will be 20×20 (400). The graph in Fig. 8(a) tends to show approximately linear relationship with the number of DCT blocks selected and the percentage of hidden visual information thus allowing approximately linear degradation in the image. For the rest of the variants, similar kind of graphs (design curves) can be plotted and which will help in selecting the level of perceptual encryption.

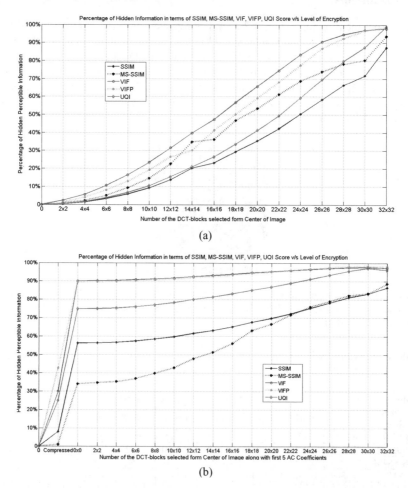

Fig. 8. Graph showing the selected DCT-blocks from the center of the image v/s the percentage of hidden information measured in terms of SSIM, MS-SSIM, VIF, VIFP, and UQI. (a) With no AC coefficient Scrambled. (b) With first 5 AC coefficient Scrambled.

4.6 Compression v/s Encryption

An analysis has been done to investigate the effect of encryption on compression for the presented perceptual encryption scheme and its variants. The corresponding graph is shown in Fig. 9. It is noticed that there is a minor effect of scrambling on compression efficiency. When the DCT blocks are scrambled, the position of the DC coefficient is also changed which will result in the change in difference between the two DC coefficients of consecutive DCT blocks (as in JPEG compression standard the difference is encoded for DC coefficients). Also, due to the scrambling of the AC coefficient, the position of the coefficients are changed that can result in different RUNLENGTHS.

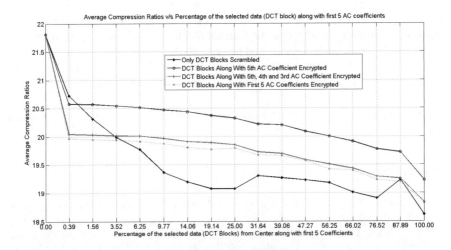

Fig. 9. Average Compression Ratios v/s Number of Selected DCT block from Center of the image along with selected number of AC coefficients

4.7 Security Analysis

The cryptographic security of the scheme is based on the scrambler used to scramble data and is dependent on the security of the key used for the scrambler. It is assumed that the key is transferred from sender to receiver through a secure channel. The analysis of the scrambler based on spread, dispersion and key sensitivity analysis is done in [6]. A multimedia encryption scheme is considered to be secure if the cost of breaking that scheme is higher than the cost of buying that unencrypted content and if the time required for breaking the encryption scheme is more than the time in which that content is considered to be useful [17]. As mentioned in [15], security is required against two types of attackers (i.e. casual listeners/ observers and professional unauthorized recipients). For casual listeners/observers, security is required only for few hours and professional unauthorized recipients (cryptanalysts) security is required for years. The variants of DCT block scrambling discussed are suitable for casual listeners/observers and may not be effective in case of professional cryptanalyst. As in our case, the focus is on the security against casual attacks. Consider a case when all the 8×8 DCT blocks of image of size 256×256 is being scrambled along with the first 5 AC coefficients. This scrambling of a coefficients and DCT blocks are independent of each other. So an exhaustive search of possible combinations of DCT blocks and AC coefficients is required to recover the image.

5 Conclusions and Future Work

In this paper, some of the variants of DCT block based scrambling and AC coefficient scrambling for perceptual encryption are proposed. The investigation is further refined by analysis of encryption on compression. It is noticed that there is a minor change in the compression ratios. So these schemes are considered to be light-weight. These schemes are designed against causal attackers and these schemes can be further refined by scrambling / encrypting the selected number of AC coefficients in the ROI to achieve a higher level of security and perceptual degradation.

In future we will implement these variants on H.264/MPEG 4 compression standard and will investigate the effectiveness of these schemes on videos.

Acknowledgments. The authors would like to gratefully acknowledge the financial support from Universiti Teknologi PETRONAS.

References

1. Macq, B.M., Quisquater, J.-J.: Cryptology for Digital TV Broadcasting. Proceedings of the IEEE 83(6), 944–957 (1995)
2. Li, S., Chen, G., Cheung, A., Bhargava, B., Lo, K.-T.: On the Design of Perceptual MPEG-Video Encryption Algorithms. IEEE Transactions on Circuits and Systems for Video Technology 17(2), 214–223 (2007)
3. Zeng, W., Lei, S.: Efficient frequency domain selective scrambling of digital video. IEEE Transactions on Multimedia 5(1), 118–129 (2003)
4. Torrubia, A., Mora, F.: Perceptual cryptography of JPEG compressed images on the JFIF bit-stream domain. In: IEEE International Conference on Consumer Electronics (ICCE), pp. 58–59 (2003)
5. Wallace, G.K.: The JPEG still picture compression standard. IEEE Transactions on Consumer Electronics 38(1), xviii–xxxiv (1992)
6. Khan, M.A.: A Novel Seed Based Random Interleaving for OFDM System and Its PHY Layer Security Implications, MSc Thesis, Department of Electrical and Electronic Engineering, Universiti Teknologi PETRONAS, Malaysia (2008)
7. Peak signal-to-noise ratio (PSNR),
   http://en.wikipedia.org/wiki/Peak_signal-to-noise_ratio
8. Wang, Z., Bovik, A.C., Sheikh, H.R., Simoncelli, E.P.: Image quality assessment: from error visibility to structural similarity. IEEE Transactions on Image Processing 13(4), 600–612 (2004)
9. Wang, Z., Simoncelli, E.P., Bovik, A.C.: Multi-scale structural similarity for image quality assessment. In: IEEE Asilomar Conference Signals, Systems and Computers, pp. 1398–1402 (2003)

10. Sheikh, H.R., Bovik, A.C.: Image information and visual quality. IEEE Transactions on Image Processing 15(2), 430–444 (2006)
11. Wang, Z., Bovik, A.C.: A universal image quality index. IEEE Signal Processing Letters 9(3), 81–84 (2002)
12. Rodrigues, J.M., Puech, W., Bors, A.G.: Selective Encryption of Human Skin in JPEG Images. In: IEEE International Conference on Image Processing, pp. 1981–1984 (2006)
13. Droogenbroeck, M.V., Benedett, R.: Techniques for a selective encryption of uncompressed and compressed images. In: Proceedings of Advanced Concepts for Intelligent Vision Systems (ACIVS), pp. 90–97 (2002)
14. Podesser, M., Schmidt, H.-P., Uhl, A.: Selective Bitplane Encryption for Secure Transmission of Image Data in Mobile Environments. In: 5th Nordic Signal Processing Symposium, on board Hurtigruten, Norway (2002)
15. Rao, Y.V.S., Mitra, A., Prasanna, S.R.M.: A Partial Image Encryption Method with Pseudo Random Sequences. In: Bagchi, A., Atluri, V. (eds.) ICISS 2006. LNCS, vol. 4332, pp. 315–325. Springer, Heidelberg (2006)
16. Meyer, J., Gadegast, F.: Security Mechanisms for Multimedia Data with the Example MPEG-1 Video, Project Description of SECMPEG, Technical University of Berlin, Germany (1995), http://www.gadegast.de/frank/doc/secmeng.pdf
17. Raju, C.N., Umadevi, G., Srinathan, K., Jawahar, C.V.: Fast and Secure Real-Time Video Encryption. In: Sixth Indian Conference on Computer Vision, Graphics & Image Processing, ICVGIP 2008, pp. 257–264 (2008)
18. Gaubatz, M.: MeTriX MuX Visual Quality Assessment Package, http://foulard.ece.cornell.edu/gaubatz/metrix_mux/

The iPad and EFL Digital Literacy*

Robert C. Meurant

Director, Institute of Traditional Studies, Seojeong University College
Yongam-Ri 681-1, Eunhyun-Myeon, Yangju-Si, Gyeonggi-Do, Seoul, Korea 482-863
Tel.: +82-10-7474-6226
rmeurant@me.com
http://web.me.com/rmeurant/INSTITUTE/HOME.html

Abstract. In future, the uses of English by non-native speakers will predominantly be online, using English language digital resources, and in computer-mediated communication with other non-native speakers of English. Thus for Korea to be competitive in the global economy, its EFL should develop L2 Digital Literacy in English. With its fast Internet connections, Korea is the most wired nation on Earth; but ICT facilities in educational institutions need reorganization. Opportunities for computer-mediated second language learning need to be increased, providing multimedia-capable, mobile web solutions that put the Internet into the hands of all students and teachers. Wi-Fi networked campuses allow any campus space to act as a wireless classroom. Every classroom should have a teacher's computer console. All students should be provided with adequate computing facilities, that are available anywhere, anytime. Ubiquitous computing has now become feasible by providing every student on enrollment with a tablet: a Wi-Fi+3G enabled Apple iPad.

Keywords: iPad, iPhone, iOS 4, Korea, EFL, ESL, L2, digital literacy, CALL.

1 Introduction

Digital resources play an increasing role in Second Language Acquisition (SLA), with more attention being given to intentional instruction in English as a Foreign Language (EFL) Digital Literacy (i.e. digital literacy in students' Second Language (L2) English). This is in part due to growing recognition of three key factors that have been identified in a number of recent papers [1, 2] as impacting contemporary SLA. Firstly, the

---

* This paper expands upon research presented to NCM 2010, the 6th International Conference on Networked Computing and Advanced Information Management, under the title *"Providing Every Student with an iPad as a Means of Helping Develop Korean EFL Digital Literacy",* published in the IEEE conference proceedings, pp.242–247; and further develops research presented to ICHIT 2010, the International Conference on Convergence and Hybrid Information Technology, under the title *"Enabling Korean EFL Digital Literacy by Implementing Student Use of the iPad",* published in the ACM conference proceedings, pp.67–73.

T.-h. Kim et al. (Eds.): SIP/MulGraB 2010, CCIS 123, pp. 224–234, 2010.
© Springer-Verlag Berlin Heidelberg 2010

predominant use of English by non-native speakers is increasingly in communication with other non-native speakers, rather than with native speakers as might be expected. Secondly, the emergence of English as a global language has meant that desired online resources and discourse are mainly in English, despite the rapidly growing Internet use of other languages. Thirdly, a critical and profoundly symbolic threshold is fast approaching whereby the majority of interpersonal communications worldwide will have become computer-mediated, rather than face-to-face. (The average Briton spends half their waking life using media and communications, in effect multi-tasking over nine hours a day using TV, mobile phones and surfing the net [3]).

These three factors indicate that the predominant use of English by non-native speakers will in future be firstly in navigating English language digital resources, in locating, editing, and contributing to online content in English; and secondly in computer-mediated communication with other non-native speakers of English [4, 5]. That both of these envisaged predominant uses of English by non-native speakers are computer-mediated has profound implications for SLA, and specifically for Korean learners of English. For Korea to be competitive in the global economy, Korean EFL needs to nurture and develop L2 Digital Literacy in English. But how best to achieve this?

Korea enjoys an enviable status as the most wired nation on the planet, with the fastest Internet connections in the world [6]. But computer facilities in Korean educational institutions are currently in need of revisualization and drastic reorganization. In particular, the opportunities for computer-mediated second language learning need to be increased, particularly for native teacher English classes. Fixed desktop computer labs need to be replaced with the provision of multimedia capable, mobile web solutions that put the Internet firmly into the hands of all students and teachers.

Existing computer labs are mainly designed for class use in learning computer applications, where students do not interact with one another, but focus attention on their individual screen, with attention also paid to the teacher, and her OHP class screen. This arrangement actively interferes with face-to-face collaboration and networking, whether structured or unstructured, as would commonly take place in the workplace, and which is usual in L2 classrooms where pair and small group activities are held, and teacher-student and student-student interactions are often demonstrated.

These existing fixed computer labs should therefore be complemented with a comprehensive ICT solution. This comprises Wi-Fi networked campuses (already being implemented) that would allow just about any campus space to act as a wireless classroom; a teacher's computer console in almost every classroom that allows synching, with high-speed Internet access, OHP, and classroom printer; and the provision of systems that allow all students adequate computing facilities, anywhere, anytime. This latter provision of almost ubiquitous computing, which previously has been impractical, has now become feasible through providing each and every student, on enrollment, with a Wi-Fi+3G enabled Apple iPad (or similar tablet computing device), to be used in tandem with students' privately owned iPhones, or other smartphones.

2 Introduction of the iPad

In January 2010, Apple announced the iPad tablet, featuring a 9.7-inch, 1024 x 768 display with 16-, 32-, and 64-GB capacities. The 13.4mm thin 0.68kg iPad is available in either Wi-Fi only, or in Wi-Fi+3G-capable models; and is designed to put the Internet into the hands of its users [7]. The iPad in Figure 1 below, designed for using fingers on a multitouch screen rather than the PC use of physical keyboard and mouse, will in my opinion likely revolutionize education. This will become particularly evident in language education, because of the integrated multimedia and telecommunications features that have particular application to language learning. The iPad is already being embraced by businesses much faster than its iPhone predecessor [8]; its acceptance as a result of the iPhone success may pave the way for a tablet invasion. I address the impact the iPad could have on Second Language Acquisition, through putting L2 English use of the Internet into the hands of Korean EFL learners.

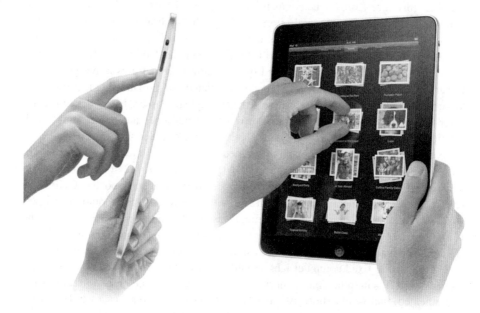

Fig. 1. The Apple iPad (courtesy www.apple.com)

In the US, Warschauer has consistently noted the increasing notion that is being given to mobile computer-mediated language learning, as American schools create one-to-one classroom environments through connecting laptops wirelessly to the Internet [9]. He argues that computers and the Internet are highly disruptive technologies that require extensive organizational restructuring and professional development for successful use [10]. Progressive universities, notably the Abilene Christian

University, have for some years provided students with free iPhones, and integrated them into their curriculum [11]. Web apps are used to turn in homework, look up campus maps, watch lecture podcasts and check class schedules and grades [12]; for classroom participation, polling software allows shy students to make choices without risking embarrassment. In 2008, Oklahoma Christian University made Apple hardware mandatory, providing MacBooks to incoming freshmen and faculty [13]. At Francis Tuttle Technology Center [14], pilot projects using iPhones and Kindle e-readers enabled administrators to save students up to 50 percent on textbook costs by buying them electronically. Computer mobility is regarded as being key and critical to the future, providing ways to get people access to learning content, no matter where they are; iPhones and iPod touches are being evaluated for nursing students to carry medical reference books electronically instead of lugging 5kg books about the clinics. In 2009, all 550 of Aoyama Gakuin University's students, and some staff, received free iPhone 3G's, which also track attendance by taking advantage of their inbuilt GPS [15].

With the release of the iPad, a number of universities and schools in America and worldwide are distributing iPads to students and faculty. The motivation includes *being a recruiting tool* at George Fox University to lure talented freshmen; *for classroom participation* at Abilene Christian to encourage faculty and students to experiment with the tablet as a learning tool; *being a shoulder saver* at Cedars School of Excellence in Scotland to reduce the amount of of paper students lug around; and *providing computing with less distraction* at Hawaii Preparatory Academy to protect young students from the unfettered mercy of the World Wide Web [16]. Educators are predicting the iPad will herald a revolution in the classroom, replacing textbooks with a mobile multimedia device to engage students in new innovative ways.

Korea enjoys a high level of broadband Internet penetration, leading the world in household broadband penetration [17]. Extensive 3G coverage and a rapidly growing provision of free Wi-Fi hotspots (Starbucks, Lotteria etc.) are available. While many tertiary institutions already provide Wi-Fi networking, the use of computers in class is constrained: firstly, institutional computer facilities are limited, and demand, notably at exam time, may exceed supply. Secondly, where computer labs are available, these tend to be desktop computers in fixed arrangements as in Figure 2 below, which likely springs from dated administrative perceptions that computing is a special kind of education, separable from general education, that takes place statically, and primarily in individual relationship to a teacher. But fixed computer labs do not enable the flexible groupings of students that typically occur in EFL classes, where students frequently alternate between whole-class activities and diverse individual, paired and group tasks as in Figure 3 below. Educational theorists promote the importance to pedagogy of connection, collaboration and networking, but the architecture of existing Korean educational computer facilities (and of regular language classrooms) discourages face-to-face collaboration and the integration of hybrid online computer-mediated learning with traditional learning that blended learning aims to achieve.

Thirdly, my surveys show that relatively few Korean students own laptop computers that they are willing to bring to class [18]. Fourthly, while smart phone usage in Korea is high, the small screen and keyboard size, limited applications and data

cost limit their intentional use in class. Fifthly, many native English-speaking teachers, faced with administrative wariness towards innovation, tend towards caution in
their approach to educational technology, and often lack the skill sets necessary to
successfully implement computer-mediated learning in their classrooms. This merely
reflects the somewhat commonplace unenlightened and reactionary administrative
policies of those who still ignore the realities of Web 2.0+ thinking, as described
in [5].

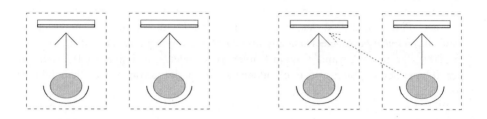

Fig. 2. Fixed desktop computer labs inhibit face-to-face pair and small group collaboration,
while encouraging cheating in online exams (as per dashed arrow on right)

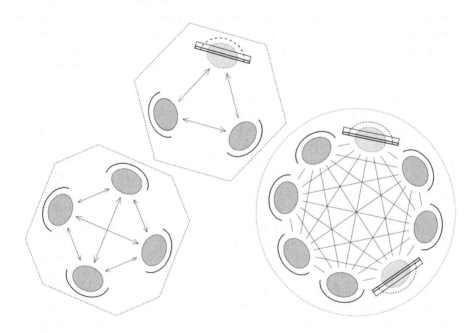

Fig. 3. Language students often alternate between whole-class activities and diverse individual,
paired and small group tasks, here shown formalized with and without computer use of online
resources and optional videoconferencing

But, as argued elsewhere [19], student adoption of new digital technologies is increasing exponentially and naturally is affecting expectations of how teaching and learning should occur. Simultaneously, EFL textbooks are merging with digital media [20]; with teachers starting to integrate online placement and progress tests, and web-hosted Learning Management Systems into their courses [21]. The relationship to digital media in the classroom is evolving from that of a precious Internet that can only be accessed as a specialized scarce resource, to that of the taken-for-granted Internet as constant companion. The iPad completely fits this new paradigm, as it popularizes it.

3 Implementing EFL Digital Literacy

3.1 Using Existing Facilities to Encourage EFL Digital Literacy

Elsewhere, working within a cognitive framework of using existing facilities more creatively, I recommend ways to encourage EFL student digital literacy in English [22]. In that paper, I encourage teachers to move from singular use of the traditional classroom to a more blended or hybrid form of education that combines traditional classroom instruction with computer-based language learning. Tasks can be computer-mediated, accomplished by students in their own time on computers in the university, at home, or in PC rooms, and submitted online. Classes can be held intermittently in existing computer labs. Quizzes and exams can then be set online, to be conducted in existing computer labs, using Internet-hosted exam writing and management services such as Cognero, an excellent full-featured online assessment system. About 1 in 4 computer classes is adequate for in-class tasks, quizzes and exams, care being taken to allow for potential server outages, loss of data online, and cheating through instant messaging, email, or cell phone SMS. Students are encouraged to make more intensive use of online resources in the target L2 language, here English (e.g. http://www.google.com and http://wikipedia.org in English). Teachers are advised to implement a computer-based Learning Management System, such as Moodle (http://moodle.org using hosting services such as http://ninehub.com) and to force that LMS to use English only [23]. (Moodle's strengths include that it is free and flexible; as open source software, users can freely adapt and modify it). These recommendations will all help develop desired L2 digital literacy skills in English.

3.2 The iPad as a Comprehensive Solution

But the new Apple iPad and iPhone OS 4.0 (renamed iOS 4) has provided a potential game-changer that will likely revolutionize education, and one particularly well suited to second language learning environments. The key advantage provided is that these developments put computing and the Internet firmly into the hands of their users [7], who in this context are EFL/ESL students and their teachers. The Internet has emerged as a fast-developing powerful educational tool; but it has been regarded as something special, that needs to be accessed indirectly: one locates a computer, makes sure that it is connected to the Internet and is multimedia-capable, and then rather self-consciously works on the computer on the Internet. However, the advent of mobile computing through smart phones like the iPhone, multitouch input, and increasingly portable laptops such as the MacBook Air, has signaled the transition to a new paradigm that the iPad fully recognizes and exploits: the Internet is already something that is no longer special, but something that is taken for granted, that conceptually is always available anywhere, anytime. The stored experience, knowledge, wisdom, and indeed folly of mankind is becoming immediately accessible, as through computer-mediated telecommunications, the distant is becoming proximal. Students should now be able to access learning content wherever they are, and whenever they want.

3.3   Advantages of a Comprehensive Solution

- Any classroom space can be used with the iPad; dedicated computer labs are not required. Students have no need for desktop computers, cabling or computer desks.
- As blended systems develop, textbooks are rapidly evolving from hard copy physical items that must be carried, and become obsolete every 3 to 5 years, to e-texts. This phenomenon needs to be taken into consideration when selecting textbook series [20]. E-texts are adopting the emerging e-Pub standard format, and can be downloaded onto the iPad as Apps and updated frequently. The student only needs to carry to class her iPad, on which has been installed all of her e-texts.
- Publishers can create hybridized content that draws from audio, video, interactive graphics in books, magazines and newspapers, whereas paper layouts are static. E-texts link to diverse multimedia digital resources and telecommunication services.
- Using the open e-Pub standard, institutions can therefore customize, and even create their own e-texts; these can be integrated into customized institutional, departmental and teacher-implemented learning management systems. Pages now allows export to ePub format; and the iPad is poised to make that format the lingua franca of electronic books, in the same way that the advent of portable digital music players, especially the iPod, made the MP3 format the de facto standard for audio.
- Student collaboration can be encouraged with tasks that include online components, to be undertaken both more formally in class and informally by loose groupings of students in libraries, unused classrooms, campus cafes, and off campus.
- Multimedia capability together with telecommunications means that students can engage within and outside class in L2 English videoconferencing locally and internationally (Skype video is free). EFL class students complain about the lack of other English speakers with whom to converse, so this may come to be regarded as a normal activity which substitutes for such demand through distance conversation.
- The iPad is simple to use. Once registered with the iPhones Store, software can be installed in class by running the App Store application, which does not bother the user with choices during the installation process; no DVDs, CDs or serial numbers are needed. User files such as documents are installed via synchronization to those stored on another computer, which for student files would be their teacher's class computer (to which would be connected a classroom printer for controlled student use). Once synchronized, documents simply appear for respective applications.
- Apps can be pre-installed by IT departments before giving the iPad to students. iOS 4 (which supports the iPad as well as the iPhone) provides institutions with the ability to manage iPads and distribute applications wirelessly without using iTunes.

The capacity to push applications from a central location via Wi-Fi or 3G means far less work, and less worry about iTunes being up to date (or even installed).

- Required iPad applications are cheap, so costs could in principle be absorbed by educational institutions. Apple's new productivity software suite iWork for iPad takes advantage of multitouch input: slides in Keynote, columns in Numbers, and text and graphics in Pages can be rearranged by tapping and dragging a finger. An on-screen keyboard appears when text needs to be typed. These apps import iWork '09 and Ms Office documents, and export iWork '09, PDF and ePub formats. iWork documents are synced between a Mac and the iPad using iTunes; the iPad works with a Dock Connector to VGA adapter so the iPad can connect to a projector to display Keynote slides on a screen. These apps cost just US$10 each.
- iPad/iPhone apps are sandboxed from one another. Data from an application isn't generally available to other applications, and storage can't be overwritten.
- The platform is simple, intuitive, and highly usable for tasks that include photos, music and movies, without needing to understand the underlying organization.
- Security problems common on PC platforms are unlikely to be encountered.

3.4 The Significance of iOS 4

iPhone OS 4 (now iOS 4) focused on seven "tentpole" features: multitasking, folders, Mail, iBooks, Enterprise features, Game Center, and iAd. The appearance of multitasking is achieved through a combination of app-switching features and background processes managed by the operating system itself. This meets student productivity needs to use several apps to perform a task, and to switch rapidly between them, e.g. using Safari to find and download images for a task in Pages. Apps can be frozen, and pick up right where they have been left. They are able to perform tasks in the background: e.g. the push-notification scheme; background audio; VOIP - so that Skype allows conversations to continue when switching to another app and to receive incoming calls; GPS tracking of location; Local Notifications; and task completion. These features satisfy issues of app switching, streaming audio, and location awareness.

3.5 Student Use of the iPhone with the iPad

The initial iPad lacks a camera, but Apple is expected to add a front-facing camera to the next generation iPad to offer FaceTime video conferencing across all their mobile devices. The portability of the iPhone and its popularity in Korea, and the compatibility of iPhone and iPad (which both run iOS 4), mean that the iPhone camera can readily be used to capture photos and video, and upload them to the iPad. A second front-facing camera in the newly released iPhone 4.0 also favors person-to-person videoconferencing, though small groups may prefer to view images together on the larger screen of the iPad, while using an iPhone for video capture.

4 Conclusion

The primary uses of English by non-native speakers will increasingly - and in my opinion, predominantly - be computer-mediated. These will be in the use of online resources, and in distance telecommunication with other mainly non-native speakers. Recognizing this, there is a critical need to strongly develop L2 Digital Literacy in English. Advantage should be taken of Korea's high level of broadband penetration by comprehensively upgrading Internet-connected computer facilities, to make them available to all students anywhere, anytime. This is effectively achieved by saturating campuses with Wi-Fi access; ensuring all classrooms have a teacher's computer for synching with high-speed Internet access, OHP and printer; and providing all students on enrollment with a Wi-Fi+3G iPad tablet with Apps and e-texts that are Wi-Fi managed by IT departments. Such a strategy, which would put the immense benefits of the Internet directly into the hands of both students and teachers, would greatly enhance Korea's competitiveness in the digital global community.

References

1. Meurant, R.C.: L2 digital literacy: Korean EFL students use their cell phone videocams to make an L2 English video guide to their college campus. In: Proceedings of IPC 2007, The 2007 International Conference on Intelligent Pervasive Computing, pp. 169–173. IEEE Computer Society, Los Alamitos (2007)
2. Meurant, R.C.: The key importance of L2 digital literacy to Korean EFL pedagogy: College students use L2 English to make campus video guides with their cell phone videocams, and to view and respond to their videos on an L2 English language social networking site. IJHIT: The International Journal of Hybrid Information Technology, SERSC 1(1), 65–72 (2008)
3. Cellan-Jones, R.: A multi-tasking moral panic. BBC News: dot.Rory (2010),
   http://www.bbc.co.uk/blogs/thereporters/rorycellanjones/
   2010/08/a_multi-tasking_moral_panic.html
4. Meurant, R.C.: L2 digital literacy in English in second language acquisition in Korea. In: Online Proceedings of the 2009 World Congress for Korean Politics and Society, Korean Political Science Association, Seoul (2009),
   http://www.kpsa.or.kr/congress2009/sub43.html
5. Meurant, R.C.: The Korean Need for L2 Digital Literacy in English. In: Proceedings of the KATE 2010 International Conference: Teaching and Learning English as a Global Language: Challenges and Opportunities, pp. 176–181. KATE, Seoul (2010)
6. Sutter, J.D.: Why Internet connections are fastest in South Korea (2010),
   http://edition.cnn.com/2010/TECH/03/31/
   broadband.south.korea/
7. Jobs, S.: Apple Special Event January 2010: Keynote Speech (2010),
   http://events.apple.com.edgesuite.net/1001q3f8hhr/
   event/index.html

8. Bradley, T.: iPad paves the way for coming tablet invasion. Macworld article, reprinred from PCWorld.com Net_Work blog (2010),
   http://www.macworld.com/article/153628/2010/08/
   ipat_tablets.html
9. Warschauer, M.: Technological change and the future of CALL. In: Fotos, S., Brown, C. (eds.) New Perspectives on CALL for Second and Foreign Language Classrooms, pp. 15–25. Lawrence Erlbaum Associates, Mahwah (2004)
10. Warschauer, M.: Laptops and Literacy. Paper presented at the Annual Meeting of the American Educational Research Association San Francisco, California (April 2006) (cited by permission of the author)
11. ACU: Mobile Learning: Our Progress (2010),
    http://www.acu.edu/technology/mobilelearning/
    progress/index.html
12. ACU: Solutions for the 21st Century University (2008),
    http://www.acu.edu/technology/mobilelearning/vision/
    solutions/index.html
13. Cohen, P.: Okla. university offers MacBooks to freshmen (2008),
    http://www.macworld.com/article/132346/2008/03/oc.html
14. Hamblin, M.: Oklahoma tech center eyes Kindles, iPhones for e-learning (2009),
    http://www.macworld.com/article/142990/2009/09/
    iphone_learning.html
15. Farivar, C.: Japanese university to track attendance with iPhone (2009),
    http://www.macworld.com/article/140854/2009/05/
    japanese_university_to_monitor_students_with_iphone.html
16. Mathis, J.: How schools are putting the iPad to work. Macworld.com article (2010),
    http://www.macworld.com/article/153672/ipaded.html?loomia_ow
    =t0:s0:a38:g26:r13:c0.002429:b36881800:z0
17. Foreign Policy magazine, quoted in English News at Chosubn Ilbo: Seoul named among World's most global cities (2010),
    http://english.chosun.com/site/data/html_dir/2010/08/17/
    2010081700406.html
18. Meurant, R.C.: Second survey of Korean college EFL student use of cell phones, electronic dictionaries, SMS, email, computers and the Internet to address L1:L2 language use patterns and the associated language learning strategies used in accessing online resources. In: Advances in Information Sciences and Services - Special Issue on ICCIT 2007, Advanced Institute of Convergence Information Technology (AICIT), Gyeongju, vol. 2, pp. 240–246 (2007)
19. Meurant, R.C.: Applied Linguistics and the Convergence of Information Communication Technologies: Collected refereed papers 2006–2009. The Opoutere Press, Auckland (2010)
20. Meurant, R.C.: EFL/ESL textbook selection in Korea and East Asia - Relevant issues and literature review. In: Tomar, G.S., Grosky, W.I., Kim, T.-h., Mohammed, S., Saha, S.K. (eds.) UCMA 2010. Communications in Computer and Information Science, vol. 75, pp. 89–102. Springer, Heidelberg (2010)
21. Meurant, R.C.: Computer-based Internet-hosted assessment of L2 literacy: Computerizing and administering of the Oxford Quick Placement Test in ExamView and Moodle. In: Communications in Computer and Information Science CCIS #60: Multimedia, Computer Graphics and Broadcasting, pp. 84–91. Springer, Heidelberg (2009)

22. Meurant, R.C.: The significance of second language digital literacy - why English-language digital literacy skills should be fostered in Korea. In: CPS Series Proceedings of ICCIT 2009: The Fourth International Conference on Computer Sciences and Convergence Information Technology, pp. 369–374. IEEE Computer Society, Los Alamitos (2009)
23. Meurant, R.C.: The use of computer-based Internet-hosted learning management systems, particularly Moodle, to develop critical L2 digital literacy. In: SICOLI 2009: The 2009 Seoul International Conference on Linguistic Interfaces, Seoul: Current Issues in Linguistic Interfaces, HankookMunhwasa, Seoul, vol. 2, pp. 213–225 (2009)

Dematerialization and Deformalization of the EFL/ESL Textbook - Literature Review and Relevant Issues

Robert C. Meurant

Director, Institute of Traditional Studies, Seojeong University College
Yongam-Ri 681-1, Eunhyun-Myeon, Yangju-Si,
Gyeonggi-Do, Seoul, Korea 482-863
Tel.: +82-10-7474-6226
rmeurant@me.com
http://web.me.com/rmeurant/INSTITUTE/HOME.html

Abstract. Rapid development and critical convergence of Information Communication Technologies is radically impacting education, particularly in second language acquisition, where the sudden availability of multimedia content and immediacy of distance communication offer specific advantage. The language classroom is evolving to integrate computer-mediated learning and communication with traditional schooling; digitization and the Internet mean the textbook is evolving from inert hard copy that is consumed, to dynamic e-texts that students participate in. The emergence of English as a Global Language, with the primary role of English on the Internet, means that the transition from fixed hard copy to fluid online digital environment is particularly evident in EFL/ESL. I review research, trace ways in which this transition occurs, and speculate on how, under the impact of ICTs and their convergence, the EFL/ESL textbook will reform, and may even disappear as a stand-alone entity.

Keywords: EFL/ESL, E-Text, E-book, L2 digital literacy, Second Language, SLA, Convergence, E-learning, blending learning, autonomous learning, Information Communication Technologies, Korea, East Asia.

1 Introduction

The early 21st century is characterized by radical transformations that are fueled by and supported by the rapid development of Information Communication Technologies and their critical convergence. Phenomena that include globalization, urbanization, mass transportation and communication, and the rapid evolution of the computer and the Internet are evidence of a profound revolution in human culture. But that developing global culture now faces the pressures of global warming, environmental pollution, economic crises and the mass extinction of species - to which all of these phenomena have in various ways contributed. Education is an integral part of this global transformation, as the computer and the Internet become integrated more fully into everyday life. Person-to-person interaction moves from a predominantly face-to-face mode to a predominantly computer-mediated mode. Global linguistic transformation is evident in the emergence of English as a Global Language, and its

T.-h. Kim et al. (Eds.): SIP/MulGraB 2010, CCIS 123, pp. 235–244, 2010.
© Springer-Verlag Berlin Heidelberg 2010

use, to a degree, as the *lingua franca* of computer-mediated communication. Inevitably, these various aspects and pressures of transformation are becoming evident in the EFL/ESL discipline and the classroom [1] - notwithstanding delayed actualization from reactionary resistance, lack of technological expertise, and financial and imaginative constraints.

Within the second language classroom and syllabus, the two major sources of exposure to the target language are the textbook and the native teacher. The first of these has important implications for the choice of a textbook series [2]. The popularity of the cell phone, smart phone, multitouch interface, and most recently the iPad tablet with its ecology of iOS 4, apps, iBookstore and e-texts is increasingly providing ubiquitous computing, putting the Internet firmly into the hands of its users, by making the Internet available to users anywhere, anytime [3]. The packaging of information is moving from relatively fixed inert bundles that are consumed, to dynamic interactive multimedia experiences that allow integration of individual, group, class, teacher, departmental and institutional user-created content and user-modifiable orientation and navigation as well as author updating, institutional wifi installation, maintenance and updating, and the integration of distance communication and online resource utilization with interpersonal communication and individual research, study and play.

The textbook evolves from a stand-alone object to an interactive node; from there it dissolves into non-distinguishable components of an online integrated learning management system that manages and delivers online learning content, tasks, applications, assessment, administration, interpersonal communication and social networking. Concomitant with this is a certain subjectivization, as language - and even reality itself - transforms from an essentially static resource to which one subscribes, to a more participatory immersion within which one navigates and of which one partakes.

2 A Survey of E-Textbook Literature Relevant to Korean and East Asian Native Teacher EFL/ESL Programs

In recent years, a significant literature has begun to emerge on the evolution of the E-book and more particularly the E-textbook. For pedagogical purposes that include helping integrate the e-book phenomenon into the EFL/ESL discipline, I review four papers on electronic books and texts, which can be accessed from the links provided.

2.1 Electronic Literacy in School and Home: A Look into the Future

Keith J. Topping, Director, Centre for Paired Learning,
Department of Psychology, University of Dundee.
Reading Online, May 1997.
http://www.readingonline.org/international/
inter_index.asp?HREF=/international/future/index.html

Topping [4] provides parents and teachers with a broad overview of electronic literacy in the school and at home, and includes practical links and references. Texts, hypertexts and hypermedia are defined, and changes in the definitions of reading and literacy are discussed. Current developments in electronic literacy are reviewed for various categories, and linking electronic activities between home and school considered, with access and equity issues and the practicalities of simpler alternative multimedia technologies in regard to international perspectives reviewed. The development of global electronic literacy from the home independent of the school is addressed, and future problems, opportunities and developments are reviewed, with implications for practitioners and researchers discussed, and issues of effectiveness emphasized.

Introduction: Texts, hypertexts, and hypermedia. While traditionally all writing systems have been linear, linear text is being replaced by hypertext in electronic literacy environments. Segments of linear text can be accessed by the user in any order via embedded structural electronic links, which can also link to other texts and materials. Thus hypertexts as large sets of parallel texts provide a network of possibilities, and exhibit ill-defined boundaries. Hypertexts thus provide a different way of organizing and linking symbolic information; and hypermedia go beyond textual symbols, typically offering much faster search and interactivity than regular books. Hypermedia may be characterized by organizational feature, variety, and immediacy of interactivity. These changes coupled with changes in the interaction between medium and user ensure that "reading" is no longer what it used to be.

What is reading, in the Electronic Age? Reading in a hypertext environment involves the reader in cognitive reconstruction to suit individual requirements and specifications. It is a search for personal relevance and salient information, involving more frequent and overt selectivity, and potentially more partial and deeper understanding. Hypertext reading is more like navigation than reading, requiring more active strategic management, with implications for metacognitive strategies and thinking skills.

Electronic literacy. As reading changes, so does literacy, as notions of literacy expand and develop as I elsewhere address in regard to the critical importance of enabling EFL digital literacy [9, 11, 12, 13]. Electronic literacy refers to literary activities that are delivered, supported and accessed through electronic means, rather than on paper.

Wangberg, cited by Topping, shows *Electronically supported reading* has been effective in raising literacy performance, as Reitsma shows adaptive computer programs scaffold and prompt successful reading. According to Salomon, Globerson and Guterman, inserting metacognitive prompts of questions, strategies and reminders results in superior reading comprehension and essay writing, while Reinking and Rickman suggest support for readers is more used and more effective when presented by computer. Davidson and Noyes, and Goodman, all cited by Topping, show that voice recognition, speech synthesis and digitized speech to give feedback promotes oral reading accuracy. Supported texts can extend the student's zone of proximal development. Citing Anderson-Inman and Horney [5], embedded resources can be *translational, illustrative, summarizing, instructional, enrichment, notational, collaborative,* and *general purpose* (see ¶2.2: *Types of Embedded Resources* below).

Electronically supported writing. Word processing, according to Bangert-Drowns, improves student writing; predictor programs scaffold and prompt the writing process, so student and machine develop an individualized, interactive and adaptive relationship combining human and artificial intelligence. Adaptive prompting spell-checkers engineer effective learning, while minimizing stress and sense of failure.

Electronic audiences. Electronic literacy can offer wide socialization and an expanded view of potential audiences, as students correspond and collaborate globally.

Electronic literacy assessment, feedback, and management. As programs for self-assessment of silent reading comprehension become more sophisticated, computers also assess reading; norm-referenced tests delivered, scored and interpreted by computer save teacher time, reduce student testing time, and generate prescriptive advice. *Electronic direct speech-text conversion* may even mean teaching writing will become of doubtful relevance, while teaching dictation skills becomes more important. *Miniaturisation and the virtual library.* Portable electronic text readers are supplanting books, as simultaneously public libraries become virtual libraries.

The electronic literacy home-school connection. For those with access, the potential for flow of worldwide information direct into the home is enormous; schools can help promote electronic literacy in the home, and electronic literacy activities can develop independently in the home irrespective of school involvement, given the international availability of information and communication.

Parental involvement in reading guided by the teacher accelerates reading achievement; while the *family literacy* movement embraces all aspects of literacy, and targets gains in literacy competence, motivation and self-image for all participants as family members help one another as collaborative partners and with the teacher.

Family electronic literacy. Topping, Shaw and Bircham argue for computer-based inter-generational literacy development leading to accelerated achievement gains.

Access and equity issues in electronic literacy. Critical practical access and compatibility issues redress the development of a digitally disposed information underclass.

Global electronic literacy from the home. While the hearth was the focal point of the family in the 19th century, and the television its focus in the 20th century, the computer may be focal in the 21st century. The stereotype of a school at the center of a village community is being replaced by the notion of a global electronic village; and parents and students are already interacting electronically independently of the school.

Electronic distance learning. Distance learning, with electronically scored and interpreted multiple-choice exams, is now delivered worldwide through the Internet [6].

Future opportunities. New methods should help students fulfill their potential by individualizing and enhancing learning, as the nature of learning changes in the face of the torrent of information. Transferable skills of selecting, processing, transforming, evaluating and adding to information are needed, rather than the traditional emphases on detailed knowledge and retention. Higher order thinking skills are increasingly needed, and information retained in the brain will be largely nodal such as signposts, frameworks, and strategies for navigating networks. Routine laborious tasks like writing will be taken over by machines, while electronic literacy offers the opportunity for expanding horizons while retaining privacy. Worldwide telecommunications allowing conversations with others in other parts of the world should generate a more global, less parochial view, and the seeing of foreigners as real people with similar concerns to oneself and one's peers. Print may not survive, and beyond hypermedia lies the world of virtual reality, as interactive virtual reality books are not "read" but are instead sensed and indeed lived. As communication reduces to digital electronic transfer, our crude symbols may no longer be needed: will any concept of literacy remain? *Effectiveness.* Issues of effectiveness are important in the research agenda for electronic literacy. If electronic literacy is to make a difference, inexpensive, simple, durable, and compatible hardware coupled with intelligent, interactive, and adaptive software and developmental menus of strategic practical options suitable for many different contexts and participants will be necessary.

2.2 Electronic Books: Reading and Studying with Supportive Resources

An adaptation of *"Electronic Books for Secondary Students"* Published in the Journal of Adolescent and Adult Literacy, 40(6), 486-491.

> Lynne Anderson-Inman, Director of the Center for Advanced Technology in Education and of the Center for Electronic Studying, University of Oregon; and Mark Horney, Research Associate, Center for Electronic Studying, University of Oregon.
> Reading Online, April 1999.
> http://www.readingonline.org/electronic/
> elec_index.asp?HREF=/electronic/ebook/index.html

Anderson-Inman and Horney [5] observe that the term *"electronic book"* means different things to different people; stringent criteria for deciding whether a piece of software is an electronic book are: 1. An electronic book must have electronic text and that text must be presented to the reader visually, though the delivery mechanism is unimportant. 2. The software needs to adopt the metaphor of a book in some significant way, so that the program is situated within a familiar genre of reading materials. 3. The software requires a focus or organizing theme, which rules out large networks of text-based information that are not thematically organized. 4. When media other than text are available, they are primarily used to support or enhance the text.

   Applying these four criteria in the selection of software results in a library of programs or websites that share important features. There is a strong textual thread and it is used to convey information or tell a story; the program has the "feel" of a book because features are labeled with terms traditionally applied to printed books. The focus of the book has some definable boundaries and a clear organizing theme; and other types of media, if available, are largely present to enhance and enrich the text.

Purposes. Despite considerable overlap, most electronic books can be matched with at least one of four general purposes: 1. Some are for reference, with small amounts of information on many topics. 2. Others have a strong instructional flavor, with information presented and supported to meet specific student needs. 3. Some are meant for in-depth studying, providing the reader with a large-scale network of interrelated documents, graphics and sounds, providing links across documents and text enhancements within documents. 4. Others are primarily for entertainment or pleasure reading. In evaluating an electronic book for classroom use, teachers need to understand what it was designed to do and whether this matches their instructional goals.

Advantages and Disadvantages. Advantages evolve from the specialized features inherent in most kinds of electronic documents; specifically, electronic documents are usually *searchable, modifiable,* and *enhanceable.*

Searchable means the reader can quickly locate key words or phrases simply by entering the desired word or words into the program's *find* or *search* function. *Modifiable* means that the electronic book can be changed during use to meet the needs of the reader. *Enhanceable* means that resources designed to enhance text use and comprehension can be embedded into an electronic book by the developers or by teachers. *Types of Embedded Resources.* The presence of embedded resources in electronic books has tremendous potential for improving student comprehension and promoting in-depth learning. Anderson-Inman and Horney categorize these types of resources by the function they perform in assisting readers to comprehend and learn from text:

Translational resources translate a word, phrase, or paragraph into something more comprehensible to the reader. *Illustrative resources* help readers understand the text by providing examples, comparisons, illustrations, and visuals. *Summarizing resources* provide an overview of the text without any encumbering or complicating detail. *Instructional resources* promote active processing of the text by suggesting activities for manipulating concepts and remembering information. *Enrichment resources* provide information directly related to, but not strictly necessary for, comprehension of the text. *Notational resources* are designed to enable interaction with an electronic book while supporting students' need to reinforce their learning by such actions as taking notes, outlining, diagramming, calculating, categorizing, summarizing, and collecting examples. *Collaborative resources* enable students to read and study collaboratively. *General purpose resources* are those that can be linked to an electronic book but were not developed to support the text in that specific document.

Evaluating and Selecting Electronic Books. Anderson-Inman and Horney advise the teacher and administrator to firstly, be very explicit about their purposes for selecting an electronic book or set of books; secondly, to look for features in the available electronic books that would support their purposes; and thirdly, to look for features that would enable a diverse body of learners to use the book effectively, attending specifically to translational, illustrative and instructional resources that support student comprehension of the text, particularly definitions for problematic words, explanations for difficult concepts, and graphics that clarify or supplement the text.

2.3   The Benefits and Advantages of Ebooks

Remez Sasson.
SuccessConsciousness.com.
http://www.successconsciousness.com/ebooks_benefits.htm

Sasson [7] provides a popularist, repetitive and dated list of the benefits of ebooks: they disseminate knowledge, so prices should be evaluated according to the usefulness, relevancy and uniqueness of the information contained, and the practical knowledge, inspiration, motivational tips and advice provided. E-books are purchased and delivered immediately, do not consume trees, do not require physical storage space, are highly portable, can be read anywhere, are more safely stored and carried, provide linkages, are searchable and interactive, are printable, and display resizable fonts.

2.4 "E-Book Flood" for Changing EFL Learners' Reading Attitudes

Chih-Cheng Lin and Irene Yi-Jung Lin.
Proceedings of the 17th International Conference on Computers in Education, 2009.
http://www.apsce.net/ICCE2009/pdf/C6/proceedings769-776.pdf

Lin and Lin [8] investigate the effects of using e-books in an extensive reading program (ERP) on EFL learners' attitudes toward reading in English. In Taiwan, 109 students from three junior high school classes were recruited in the ten-week ERP of e-books. Each class was introduced to a list of 140 selected e-books for the reading program; students were encouraged to read e-books after school, with weekly targets of reading four e-books each. The degree of changes in reading attitudes was assessed using Stokmans' reading attitudes scale, cited by Lin and Lin, before and after the e-book ERP; also, the teacher's class notes of the students' reading behaviors and reactions as well as their spontaneous oral or written feedback were analyzed for triangulation with quantitative data. The e-books had positive effects on the students' attitudinal changes in all dimensions of reading attitudes, namely utility, development, enjoyment and escape, as well as in all the cognitive, affective and cognative components. E-book features, especially oral reading, highlighting, animations and music/sound effects, were important in changing attitudes. The implementation of interaction and learner control in the e-books guaranteed positive attitudinal changes.

Introduction. High student computer literacy and keen interest in multimedia have inspired language teachers to convert the traditional teaching setting into an e-setting that students are constantly exposed to. Teachers are encouraged to use computer technology as an intervention strategy to overcome negative student attitudes to L2 reading. E-books change the nature of reading mainly through multimedia features of oral reading, highlighting, animations, music and sound effects. The motivation of the study is to find ways to utilize e-books effectively, reinforcing student reading attitudes by employing a literature-based reading program to encourage reading, spend more time reading, foster the love of reading, and develop long-term reading habits.

The E-book ERP. The positive effects of an Extensive Reading Program have been widely recognized, e.g. by Asraf and Ahmad, and by Davis, Elley and Mangubhai, all cited by Lin and Lin. ERP is one of the most effective ways to enhance reading comprehension and reading rate; but not all studies of ERPs or e-books have positive results, the key factors being the ways ERPs are implemented and e-books are utilized. E-books may be best implemented as a new medium in an ERP, according to Green, by introducing extensive reading within the purposeful interactive framework of the task-based language curriculum. *Characteristics of Successful ERPs* may be addressed through central philosophy, suggested material, and specific principles. ERPs that emphasize extrinsic motivation, excessive reading or comprehension tests result in student loss of reading pleasure and desire. ERPs need to be able to stimulate students to read, keep them reading, and induce them to read in large quantity, hence the importance of *reading enjoyment.* Students need an ample supply of short, easy and interesting books to choose from. ERP books must conform to the student reading comfort zone with not too many unknown words, and students must be provided with a wide range of books in terms of genres and contents. Specific tasks should be assigned to students so that they use rather than browse the Internet, as online browsing lacks direction and causes fatigue and boredom. With specific reading tasks in mind, students have clear goals to achieve, and are likely to really use Internet resources.

Conclusion: The study investigates the effects of e-books on EFL high school students' attitude toward reading in English and bears two central findings: E-books can effectively reinforce EFL learners' attitudes toward reading in English; and the features of e-books, such as oral reading, highlighting, animations, and music/sound effects, may reinforce reading attitudes. Participants had positive attitudinal changes because of the unique nature of e-books; and the process of reading has four dynamic stages: read increasingly, read easily, read happily and read regularly.

3   Discussion of E-Texts and EFL/ESL

3.1   Change in the Role of E-Texts: Does the EFL Textbook Have a Future?

With the rapid development of ICT, multimedia resources and e-texts in particular have undergone a change in role. Initially, they were regarded as adjunct resources that could in some sense merely provide peripheral assistance to traditional physical textbook instruction. Then, e-texts were considered as digital versions of traditional textbooks that essentially imitated physical texts, while embodying convenient substitute and supplementary tools. From there, the inherent potential of the new media became stronger, suggesting new avenues in which the media could function, as the e-text came to be realized as a powerful new educational tool in its own right that could complement traditional instruction. Finally, we are seeing the radical reformalization of the educational process, as the electronic realities of digital-based computer-mediated Internet-hosted EFL resources assume a primary role in EFL instruction. It is becoming normal for teachers and students to effortlessly shift between traditional face-to-face learning that uses physical texts, classroom objects and physically present teachers and students, and e-learning that exploits online resources and distance communication [9]. In language class, this shift has been presaged by students adopting bilingual e-dictionaries: initially stand-alone, and now as smartphone apps [10].

In this new hybrid reality, physical texts recede into helpful adjuncts to a digitally-delivered and participated-in education (in the transitional stage, those who still prefer printed texts could be offered them as alternatives; at present, access to e-texts that students might prefer is not an option). At a later stage, students who prefer physical texts could print their own, or arrange custom printing through a copy shop.

3.2   Dematerialization and Deformalization of the EFL/ESL Textbook

Texts as stand-alone entities may even eventually be completely discarded, as textbook boundaries dissolve into an integrated Learning Management System that delivers online learning content, tasks, assessment, grades, administrative content, interpersonal communication and social networking [11]. Moodle LMS allows integration of lessons that incorporate multimedia learning content into courses that include tasks, assessment quizzes, grades, administrative content, forums, blogs and so forth [12].

This new educational reality emerges as a subset of everyday reality, which moves effortlessly between on-line and off-line mode, while moving inexorably towards a predominantly online reality. In its ultimate realization, everyday face-to-face reality will for some, become computer-mediated by choice, as digital communication enables preferred enhancements such as augmented reality, that non-computer-mediated communication either does not provide or provides less effectively [13]. For others, traditional every-day face-to-face reality will remain the mode of choice.

4   Conclusion

The effects of the contemporary rapid development and convergence of Information Communication Technologies are radically impacting upon education and are - at least in potential - particularly evident in language education because of the sudden availability and tremendous potential of multimedia and of distance communication. These transformations will become pervasive (notwithstanding significant professional resistance in some quarters from those administrators and teachers who, apparently threatened by progress, prove incapable of accommodating to Web 2.0+ thinking [2]). In the author's opinion, language itself is being profoundly transformed, the depth and significance of this transformation extending well beyond those changes that linguists generally are willing to recognize. This transformation extends to reality itself, which is evolving from a consumerist model that is essentially received and to which one conforms, to a participatory model one confronts and with which one engages, as L2 speakers take control of their second language(s), abandoning native speakers to their restricted monolingual fate. Critical to these various interrelated transformations is the shift from traditional face-to-face reality to the adoption of computer-mediated reality and communication as being the taken-for-granted norm.

In accord with that shift, the role of the textbook in EFL/ESL is changing. The text is no longer to be regarded as a stand-alone physical object that contains received wisdom, but rather as a non-distinct hypermedia reality within which students engage and which they share and help shape. In language education, the EFL textbook is thus undergoing *dematerialization,* as it evolves through intermediary stages of digitization through e-book and beyond; it is simultaneously undergoing *deformalization,* as its traditional clear structure and shape becoming problematic as it merges with online Internet-hosted Learning Management Systems, to where it becomes merely a non-distinct aspect of that system. Ultimately, we may even see the complete abandonment of the stand-alone textbook, whether as physical object or as dedicated e-text.

That abandonment correlates with a profound shift in consciousness from viewing reality as an independent state to be conformed to, to an immediacy that might best be described, in poiesis, as the integration of the scholar with their field of study. Such a change in the nature of reality parallels the well-known aesthetic differentiation between the plasticity of object-centeredness, where separate identities stand in clear relationship (as in the Italian architecture of the Mediterranean), and the field perception of flow, in which phenomena merge into one another, and nothing stands still (as in Germanic forest landscapes or the Baroque). In the longer term, this immediacy itself may eventually exhibit a reformalization, as new modes of comprehension, being and communicating stabilize and become identifiable, understandable and describable.

References

1. Meurant, R.C.: Applied Linguistics and the Convergence of Information Communication Technologies. The Opoutere Press, Auckland (2010)
2. Meurant, R.C.: EFL/ESL textbook selection in Korea and East Asia - Relevant issues and literature review. In: Tomar, G.S., Grosky, W.I., Kim, T.-h., Mohammed, S., Saha, S.K. (eds.) UCMA 2010. Communications in Computer and Information Science, vol. 75, pp. 89–102. Springer, Heidelberg (2010)
3. Meurant, R.C.: Providing Every Student with an iPad as a Means of Helping Develop Korean EFL Digital Literacy. In: NCM 2010: 6th International Conference on Networked Computing and Advanced Information Management. IEEE/AICIT, Seoul (2010)
4. Topping, K.J.: Electronic Literacy in School and Home: A Look into the Future. Reading Online (1997),
   `http://www.readingonline.org/international/`
   `inter_index.asp?HREF=/international/future/index.html`
5. Anderson-Inman, L., Horney, M.: Electronic Books: Reading and Studying with Supportive Resources. Reading Online (1997),
   `http://www.readingonline.org/electronic/`
   `elec_index.asp?HREF=/electronic/ebook/index.html`
6. Meurant, R.C.: Computer-based Internet-hosted Assessment of L2 Literacy: Computerizing and Administering of the Oxford Quick Placement Test in ExamView and Moodle. In: CCIS 60: Multimedia, Computer Graphics and Broadcasting, pp. 84–91. Springer, Heidelberg (2009)
7. Sasson, R.: The Benefits and Advantages of Ebooks. SuccessConciousness.com (n.d.),
   `http://www.successconsciousness.com/ebooks_benefits.htm`
8. Lin, C.-C., Lin, I.Y.-J.: E-book Flood. for Changing EFL Learners' Reading Attitudes. In: Proceedings of the 17th International Conference on Computers in Education (2009),
   `http://www.apsce.net/ICCE2009/pdf/C6/proceedings769-776.pdf`
9. Meurant, R.C.: The Korean Need for L2 Digital Literacy in English. In: KATE 2010 International Conference: Teaching and Learning English as a Global Language: Challenges and Opportunities, pp. 176–181 (2010)
10. Meurant, R.C.: Using Cell Phones and SMS in Second Language Pedagogy. JCIT: Journal of Convergence Information Technology 2(1), 98–106 (2007)
11. Meurant, R.C.: How Computer-Based Internet-Hosted Learning Management Systems such as Moodle Can Help Develop L2 Digital Literacy. IJMUE: International Journal of Multimedia and Ubiquitous Engineering 5(2), 19–26 (2010)
12. Meurant, R.C.: Developing Critical L2 Digital Literacy through the Use of Computer-Based Internet-Hosted Learning Management Systems such as Moodle. In: CCIS 60: Multimedia, Computer Graphics and Broadcasting, pp. 76–83. Springer, Heidelberg (2009)
13. Meurant, R.C.: The Key Importance of L2 Digital Literacy to Korean EFL Pedagogy: College Students Use L2 English to Make Campus Video Guides with Their Cell Phone Videocams, and to View and Respond to Their Videos on an L2 English Language Social Networking Site. IJHIT: International Journal of Hybrid Information Technology 1(1), 65–72 (2008)

Real-Time Adaptive Shot Change Detection Model

Won-Hee Kim and Jong-Nam Kim

Div. of Electronic Computer and Telecommunication Engineering,
Pukyong University
{whkim07,jongnam}@pknu.ac.kr

Abstract. Shot change detection is the main technique in the video segmentation which is required real-time processing and automatical processing in hardware. In this paper, we propose the SCD model for real-time shot change detection in such as PMPs and cellular phones. The real-time SCD model determines shot change detection by comparing the feature value of current frame and a mean feature value on variable reference block. Proposed method can be used independently from the feature value of frame, can set automatic threshold using a mean feature value on variable reference block. We obtained better detection ratio than the conventional methods maximally by precision 0.146, recall 0.083, F1 0.089 in the experiment with the same test sequence. Therefore, our proposing algorithm will be helpful in searching video data on portable media player such as PMPs and cellular phones.

Keywords: Shot Change Detection, Real-time SCD Model, Adaptive Threshold, Variable Reference Block.

1 Introduction

Shot change detection is a technique to detect boundary between shots of the video data [3]. When camera is recording, shot consist of frames. They made of start and stop only one time. More comprehensive meaning of shot gives a name to scene. Scene is defined consist of shots and separated according semantics. Shot transition is separated abrupt transition and gradual transition. First thing is made up simple connection in two shots. Second thing is composed of slow change shots that is included image processing. Shot change detection is a basic technique and one of important things for management of efficient video data [1]. It is enable to reconstruction, and it has a function to divide and searching for video data [2].

Conventional related works are researched focus on accuracy of detection ratio. So it is not suitable to use in real application. Shot change detection technique should be operated by real-time detection to apply in real application. Real-time shot change detection should be satisfied no-delay, immediately response. In real application, automatic algorithm is required unnecessary user's command. It should be decided only using information of input video sequence.

In this paper, we propose shot change detection model which satisfy real-time and automatic processing. Proposed model determines shot change detection by comparing the feature value of current frame and a mean feature value on variable

T.-h. Kim et al. (Eds.): SIP/MulGraB 2010, CCIS 123, pp. 245–252, 2010.
© Springer-Verlag Berlin Heidelberg 2010

reference block. Proposed method can be used independently from the feature value of frame, can set automatic threshold using a mean feature value on variable reference block.

The remainder of this paper is organized as follow. In Section 2, we describe about conventional researches. The proposed method is presented in Section 3. In Section 4, we discuss our experimental results. Finally, we give our concluding remarks in Section 5.

2　Related Works

Shot change detection algorithms are approached by a various researched in spatial domain and frequency domain.

Zhang et al introduced pixel-wise comparison [5]. Pixel-wise comparison is a scheme to use difference intensity for each other pixel at continuous two frames. However, this method has the drawback of failing to distinguish between significant variations in a small section of video data or between slight variations in a large section of video data [6].

The block-based scheme focuses on local features that, otherwise, would be sensitive to camera and object activities [5]. Zhang et al introduced likelihood ratio comparison. This scheme decide for shot change's point through a frame by frame, each frame has pair of blocks. Shahraray proposed a scheme for mapping blocks with previous images after dividing a frame into 12 unoverlapping blocks and using brightness values [7]. Kasturi et al measured likelihood ratios through the use of means and variations for different blocks with the purpose of measuring variations in scene motion [8]. But these methods would be sensitive to camera and object activities [9].

The histogram based methods is widely used in detecting shot changes. Tonomura proposed the simplest scheme for detecting threshold-based scene changes through comparing among gray-level histograms [10]. Nagasaka et al proposed a x^2-test scheme for focusing on the difference value between two frames as well as on the motion of a camera or object [11]. However, the x^2-test scheme proved to be less powerful than the linear histogram comparison scheme proposed by Tonomura, and has the disadvantage of causing the amount of computations to increase. Ueda et al used the variation ratios of the color histogram to detect shot boundaries [12]. Gargi et al evaluated 3 histogram-based comparison methods through the use of 6 different color coordinate systems [13]. Shin et al proposed shot change detection method using local x^2-test on telematics [14].

A number of shot change detection methods have been proposed in frequency domain. In order to reduce the high complexity caused by IDCT, the approaches for shot change detection in MPEG compressed domain have been developed. Meng et al use the variance of DC coefficients and motion vectors to analyze shot changes [15]. Zhang et al count the number of valid motion vectors in predictive-coded pictures or P-pictures and bidirectionally predictive-coded pictures or B-picture [5]. Yeo et al use DC image sequences to analyze shot changes [16]. They experiment with pixel-based difference and histogram difference of DC images of successive frames. These approaches are faster than the frame-based approaches, but they are not still suitable for a low-powered PMP.

Fernando et al use the macro-block prediction statistics of B-pictures [17]. Seong et al simplify the Fernando's method to apply to a hard disk drive embedded digital satellite receiver [18]. However, they cannot be applied to the high quality video-stream which consists of only I-pictures and P-pictures without B-pictures. Sethi et al use the luminance histogram difference of the DC coefficients of successive I-pictures [19]. It is much faster than the above methods but less accurate. Kim et al implemented the shot change detection algorithm using reduced images of only intra pictures in MPEG-2 compressed domain and demonstrated the efficiency of algorithms in a commercial PVR [4].

Another most common method is commonly use edge comparison, normalized inner product based method, image segmentation based scheme, clustering based algorithms, Sendecor's F based method, T-Student based method [20][21][22]. Bescos et al introduced variety algorithms such as comparison based on pixel-wise, comparison based on statistics, comparison based on frequency domain [21].

3 The Proposed Method

For shot change detection, two factors is needed. First factor is defined FFV(frame feature value) to describe characteristic of each frame. Second factor is threshold. It is a standard to decide shot change or not change from FFV. In this paper, we define that FFV is F, threshold is TH. Therefore, to decide shot change detection become through Eq. (1).

$$F_i > TH_i \qquad (1)$$

F_i is FFV in ith frame, TH_i is a threshold in ith frame. Namely, if FFV of random frame is higher than threshold, the random frame is shot change frame.

The FFV can become every factor which represent characteristic of the frame. For example, mean of intensity of frame or histogram are used commonly in other researches. Conventional shot change detection algorithms are generated dependently from the FFV. Namely, these algorithms are proposed for shot change detection according FFV. In this case, one algorithm is operated only about one FFV. If we want to use other FFV, we should be adjusted the algorithm. It is a disadvantage point on conventional methods.

Proposed real-time SCD model is designed independently from FFV. If frame characteristic is induced scalar value, we can apply the FFV in proposed SCD model. The reason of independent of proposed method from FFV is as follows. In SCD model, FFV is only representation values which represent one frame. So our model does not use meaning of FFV to decide shot change. Just it uses value of FFV to judge shot change. For example, we suppose that we use two FFV mean of frame and mean of histogram. They have different meaning and frame characteristic each other, but they take equal meaning which is only representation value of frame in proposed method. Therefore, proposed real-time SCD model can do independent from FFV. So our method can be applied various application, because proposed model can be used with FFV all scalar value in spatial domain and frequency domain. Moreover, our method is easy to implement in hardware platform which is required small computation amount. Because proposed model can decide shot change detection through simple computed FFV.

Next, considering factor is a threshold. To decide shot change detection is only a role of the threshold from FFV. Therefore, optimal threshold make correct shot change detection algorithm.

Fixed threshold based methods is used commonly. It detects shot change using some value, iterates adjustment of threshold until best results. This method does not have logical basis about threshold setting. One threshold can use only certain video sequence. Another method is threshold setting by whole video sequence analysis. This method is not used in real application, because it demands iteration of whole video sequence analysis more than minimum 2. It require much computation time, so it is not suitable in real application.

Therefore, threshold setting scheme demand 3 conditions in real-time shot change detection. First, it should be possible adaptively about every video sequence. Second, it must have the sequence like as replay of frame. Third, it should not be taken a delay time during processing.

We propose the solutions about above three conditions. First, our method use input information of video sequence itself to set adaptive threshold for every video. Second, we use only information of previous frames from current frame to take the sequence. Third, proposed method use already computed value to reduce computation amount.

We propose adaptive threshold setting scheme satisfied three solutions. It is presented Eq. (2).

$$TH_i = \frac{\sum_{j=s_p+1}^{s_i-1} F_j}{s_i - s_p + 1} \times w_i \tag{2}$$

S_i is ith frame, S_p is previous shot change frame, W_i is threshold weighting in ith frame. That is to say, threshold TH_i in ith frame is defined W_i times of mean of F_j from $p+1$th frame to $i-1$the frame. Here, we can know importance of W_i. If we do not use weighting, namely, case of $W_i=1$, the result is $TH_i \approx F_i$. The reason is the phenomenon because of TH_i consist of mean of F values. It makes similarly threshold and FFV, so many false alarm occurred. It decreases reliability of induced threshold TH_i. Therefore, we should be applied weighting to increase reliability of threshold.

In this paper, we propose to decide weight W_i using variation between successive frames. It use ratio between variation and non-variation. To apply the variation is various according kind of FFV. Eq. (3). and (4). present weight determine equation, when we use histogram.

$$NZM_i = \frac{\sum_{j=S_p+1}^{S_i-1} NZ_j}{S_i - S_p + 1} \tag{3}$$

$$W_i = \frac{512}{NZM_i} \tag{4}$$

NZM_i is mean of NZ on variance reference block in ith frame, NZ_j is a number of non-zero bins in different histogram of jth frame. This method can be possible complete automatic threshold setting without user's command. Also, we obtain improved detection ratio better than fixed weighting.

4 Experimental Results and Analysis

The proposed SCD model was tested following conditions. Environment of PC is CPU Core Duo 2.4GHz and RAM 2G. A detection program is implemented using VC++ .net 7.1. We used MATLAB 7.3 for comparison detection results and real shot change frames. In the experiment, we use ten video data in 4 genres by random choice. We generated 9000 frame sequence, 176×144 size, YUV(4:2:0) format. Table 1 is presented title of video at experiment and a number of shot change frame each videos. Table 2 is shown distribution and kind of shot change frame. In table 2, we can know about more than 98% in abrupt transition. The frequency of gradual transition is little in most of real video sequence. Therefore, in this paper, we do not use different scheme for gradual shot change detection.

Table 1. Video title and a number of shot change frame

Genre	Title	The number of Shot change frame
TV-Drama	War of money	52
	The coffee-prince	93
	Prison break	88
Movie	Freedom writers	111
	The highway star	32
	The old miss diary	52
News	KBS news	69
	SBS news-tracking	63
Entertainment	Infinite Chanllenge	98
	Gag-Concert	78

Table 2. A number of shot change frame

	Whole SCD frame	Abrupt transition	Gradual transition
A number of frame	736	723	13
Ratio(%)	100	98.2	1.8

Proposed method is evaluated by precision and recall, F1 [25]. They are represented eq. (8)~(10). N_c is a number of correctly detected shot change, N_f is a number of false detected shot change, N_m is a number of missed detected shot change. Precision means the ratio in the correct detection in the shot change frame by algorism. It is measure for the standard in false alarm. Recall means the ratio in the correct detection in the real shot change frame. It is measure for the standard in missed detection. *F1* means total detection ratio and is defined harmonic mean for precision and recall. *F1* is presented higher value when the both measures have higher value.

$$precision = \frac{N_C}{N_C + N_f} \tag{8}$$

$$recall = \frac{N_C}{N_C + N_m} \tag{9}$$

$$F1 = \frac{2 \times precision \times recall}{precision + recall} \tag{10}$$

The proposed method was tested two experiments. Firstly, we compared proposed method with conventional methods through same feature value. We used 4 methods features which are pixel-wise comparison, histogram comparison, chi-square test, variance of frame comparison. In the experiment, conventional 4 methods are applied fixed threshold. They have results through correction of threshold of 30 times or more per method to search optimal result. We experiment and obtain result until we search the best result about ten video sequences. Table 3 shows results of first experiment. The proposed SCD model is improved with 0.013~0.071 in precision, 0.023~0.052 in recall, 0.018~0.062% in F1, which may vary according to comparison methods. If same feature value is used, proposed SCD model have improved detection ratio than conventional methods.

Table 3. Comparison results of precision, recall, and F1(conventional algorithms)

Methods		precision	recall	F1
Conventional Method	Pixel-wise[5]	0.845	0.845	0.845
	Histogram[10]	0.935	0.932	0.933
	Chi-square[11]	0.955	0.957	0.956
	Variance[1]	0.884	0.880	0.882
Proposed Method	Pixel-wise	0.891	0.895	0.893
	Histogram	0.990	0.963	0.977
	Chi-square	0.968	0.980	0.974
	Variance	0.955	0.932	0.944

Next, we compared proposed method with adaptive threshold based methods. We obtained better detection ratio than three other conventional methods maximally by precision 0.146, recall 0.083, F1 0.089 in the experiment. Therefore, we known that proposed real-time SCD model is more suitable than conventional adaptive threshold based methods for correct shot change detection.

On the basis of two experimental results, we verified that proposed real-time SCD model can correct shot change detection than conventional methods. Proposed method can describe certainly difference between successive frames on shot changed. Because it compute only FFV of frames which have same characteristic with current frame using variable reference block. Also, adaptive threshold setting scheme and application of weighting can separate effective shot change or not shot change.

Table 4. Comparison results of precision, recall, and F1(adaptive algorithms)

Methods	precision	recall	F1
Kim's method[4]	0.897	0.880	0.888
Cheng's method[23]	0.861	0.943	0.900
Ko's methods[24]	0.844	0.990	0.911
Proposed method(histogram)	0.990	0.963	0.977

5 Conclusions

In this paper, we propose the SCD model for real-time shot change detection. The real-time SCD model determines shot change detection by comparing the feature value of current frame and a mean feature value on variable reference block. Proposed method can be used independently from the feature value of frame, can set automatic threshold using a mean feature value on variable reference block. We obtained better detection ratio than the conventional methods maximally by precision 0.146, recall 0.083, F1 0.089 in the experiment with the same test sequence. We verified real-time operation of shot change detection by implementing our algorithm on the PMP from some company of H. Therefore, our proposing algorithm will be helpful in searching video data on portable media player such as PMPs and cellular phones.

Acknowledgments. This research was financially supported by MEST and KOTEF through the Human Resource Training Project for Regional Innovation, and supported by LEADER.

References

1. Cotsaces, C., Nikolaidis, N., Pitas, I.: Video Shot Detection and Condensed Representation. IEEE Signal Processing Magazine 23, 28–37 (2006)
2. Smoliar, S.W., Zhang, H.J.: Content-Based Video Indexing and Retrieval. IEEE Multimedia 1(2), 62–72 (2006)
3. Yu, J., Srinath, M.D.: An dfficient method for scene cut detection. Pattern Recognition Letters 22, 1379–1391 (2001)
4. Kim, J.R., Suh, S.J., Sull, S.H.: Fast scene change detection for personal video recorder. IEEE Transaction on Consumer Electronics 49, 683–688 (2003)
5. Zhang, H.J., Kankamhalli, A., Smoliar, S.W.: Automatic partitioning of full-motion video. In: ACM Multimedia Systems, New York (1993)
6. Hampapur, A., Jain, R., Weymouth, T.: Digital Video Segmentation. In: Proc. ACM Multimedia 1994, pp. 357–364 (1994)
7. Shahraray, B.: Scene Change Detection and Content-Based Sampling of Video Sequences. In: Proc. in Digital Video Compression: Algorithms and Technologies, vol. SPIE-2419, pp. 2–13 (1995)
8. Xiong, W., Lee, J.C.M., Shen, D.G.: Net Comparison: An Adaptive and Effective Method for Scene Change Detection. In: SPIE (1995)
9. Lee, J.C.M., Li, Q., Xiong, W.: Automaeiv and Dynamic Video Manipulation. Research and Development in Information Retrieval (1998)

10. Tonomura, Y.: Video handing based on structured information for hypermedia system. In: Proc. ACM International Conference Multimedia Information Systems, pp. 333–344 (1991)
11. Nagasaka, A., Tanaka, Y.: Automatic Video Indexing and Full Video Search For Object Appearances. In: Proceedings of the IFIP TC2/WG 2.6 Second Working Conference on Visual Database Systems II, pp. 113–127 (1991)
12. Ueda, H., Miyatake, T., Yoshizawa, S.: ImPACT: An Interactive Natural-motion-picture Dedicated Multimedia Authoring System. In: Proceedings of CHI, pp. 343–350 (1991)
13. Gargi, U., Kasturi, R., Antani, S.: Evaluation of video sequence indexing and hierarchical video indexing. In: Proc. SPIE Conf. Storage and Retrieval in Image and Video Databases, pp. 1522–1530 (1995)
14. Shin, S.Y., Sheng, G.R., Park, K.H.: A Scene Change Detection Scheme Using Local x^2-Test on Telematics. In: International Conference on Hybrid Information Technology, vol. 1, pp. 588–592 (2006)
15. Meng, J., Juan, Y., Chang, S.F.: Scene change detection in a MPEG compressed video sequence. In: Digital Video Compression: Algorithms and Technologies, vol. SPIE-2419, pp. 14–25 (1995)
16. Yeo, B., Liu, B.: Rapid scene analysis on compressed video. IEEE Transactions on Circuits and Systems for Video Technology 5(6), 533–540 (1995)
17. Fernando, W.A.C., Canagarajah, C.N., Bull, D.R.: scene change detection algorithms for content-based video indexing and retrieval. Electronics and Communication Journal 13(3), 117–126 (2001)
18. Seong, Y.K., Choi, Y., Park, J., Choi, T.: A hard disk drive embedded digital satellite receiver with scene change detector for video indexing. IEEE Transactions on Consumer Electronics 48(3), 776–782 (2002)
19. Sethi, I.K., Patal, N.: A statistical approach to scene change detection. In: Storage and Retrieval for Image and Video Databases III, vol. SPIE-2420, pp. 329–338 (1995)
20. Gargi, U., Kasturi, R., Strayer, S.H.: Performance Characterization of Video-Shot -Change Detection Methods. IEEE Transactions on Circuits and Systems for Video Technology 10(1), 1–13 (2000)
21. Bescos, J., Cisneros, G., Menendez, J.M.: Multidemensional Comparison of Shot Detection Algorithms. In: Proceedings of International Conference on Image Processing, vol. 2, pp. II-404–II-404 (2002)
22. Bescos, J., Cisneros, G., Menendez, J.M., Cabrera, J.: A unified model for techniques on video-shot transition detection. IEEE transaction on Multimedia 7, 293–307 (2005)
23. Cheng, Y., Yang, X., Xu, D.: A Method for Shot Boundary Detection With Automatic Threshold. Proceedings of IEEE TENCON 1, 582–585 (2002)
24. Ko, K.C., Rhee, Y.W.: Video Segmentation using the Automated Threshold Decision Algorithm. KSCI Journal 10(6), 65–73 (2005)
25. Boccignone, G., Chinaese, A., Moscato, V., Picariello, A.: Foveated Shot Detection for Video Segmentation. IEEE Transactions on Circuits and Systems for Video Technology 15, 365–377 (2005)

Search for Effective Distance Learning in Developing Countries Using Multimedia Technology

Sagarmay Deb

Central Queensland University, 400 Kent Street, Sydney 2000
NSW Australia
debsc@idx.com.au

Abstract. Although the developments of multimedia technology and internet networks have contributed to immense improvements in the standard of learning as well as distance learning in developed world, the developing world is still not in position to take advantage of these improvements because of limited spread of these technologies, lack of proper management and infrastructure problems. Unless we succeed in solving these problems to enable people of developing countries to take advantages of these technologies for distance learning the vast majority of the world population will be lagging behind. In this paper we take stock of the current situation and suggest some future directions in the resolution of these problems.

Keywords: Distance learning, multimedia technology, developing countries.

1 Introduction

The concepts of distance learning are prevalent in developing countries for last few decades and it is very much in vogue in developed countries [1],[7]. In developing countries it started like many other countries did with correspondence courses where printed learning materials used to be despatched to the students at regular intervals and students were expected to read the materials and answer questions. The basic philosophy was teachers would be physically away from the students and have to conduct the teaching process from distance [2].

With the development of computer industry and internet networks during the last three decades things have changed and global communication has reached an unprecedented height [1]. With these developments immense scopes have come to the surface to impart learning in a much more efficient and interactive way. Multimedia technology and internet networks have changed the whole philosophy of learning and distance learning and provided us with the opportunity for close interaction between teachers and learners with improved standard of learning materials compared to what was existing only with the printed media. It has gone to such an extent to create a virtual class room where teachers and students are scattered all over the world. Although some of these facilities are expensive still the developed world is in a position to take advantage of these facilities to impart much better distance-learning to students residing in the developed countries. But for developing countries the story is different as computerization and network connections are still very limited compared

T.-h. Kim et al. (Eds.): SIP/MulGraB 2010, CCIS 123, pp. 253–259, 2010.

to the developed world. In this paper we focus our attention on defining the problems of using these technologies for much more improved and extensive distance-learning and suggest how we could possibly reach these vast majority of people from the developing countries with the improved quality of distance-learning provided by multimedia and internet networks.

Section one gives an introduction of the area. Section two presents the problems in developing countries to make use of these technologies. Section three presents statistical data to show the status of internet usage in developing and developed countries. Section four suggests future directions of research to solve these problems. We put our concluding remarks in section five.

2 Analyses of Works Done

In a distance learning, institutional implementation, administrative and organizational resources will have a heavier role than other elements such as individual course design techniques. In an individual delivery unit, instead, course design and management techniques will have a key role [1].

The open-universities which started functioning by late sixties and early seventies of last century, reaching off-campus students delivering instruction through radio, television, recorded audio-tapes and correspondence tutoring. Several universities particularly in developing countries still use educational radio as the main instructional delivery tool [1].

With the extended application of information technologies (IT), the conventional education system has crossed physical boundaries to reach the un-reached through a virtual education system. In the distant mode of education, students get the opportunity for education through self-learning methods with the use of technology-mediated techniques. Efforts are being made to promote distance education in the remotest regions of developing countries through institutional collaborations and adaptive use of collaborative learning systems [2].

Initially, computers with multimedia facilities can be delivered to regional resource centers and media rooms can be established in those centers to be used as multimedia labs. Running those labs would necessitate involvement of two or three IT personnel in each centre. To implement and ascertain the necessity, importance, effectiveness, demand and efficiency, an initial questionnaire can be developed. Distributing periodical surveys among the learners would reflect the effectiveness of the project for necessary fine-tuning. After complete installation and operation of a few pilot tests in specific regions, the whole country can be brought under a common network through these regional centers [2].

The sad reality is that the participation of many developing countries in the global information society remains insignificant. This is attributable to many causes, including perceived incompatibilities between cultures and technologies, an idealistic preference for self-reliance, and simple lack of economic or human resources to acquire and utilize the technology [3].

In developed economies, newer versions of technology are often used to upgrade older versions, but in developing economies where still older versions of technology are often prevalent (if they exist at all), the opportunities for leapfrogging over the successive generations of technology to the most recent version are that much greater [3].

The invisibility of the developing countries owes much to the lack of telecommunications infrastructure, though satellite services have the potential to change that. In 1995, just four (of 55) African countries had an Internet presence. By October 1998, all but four countries had Internet connections of 64kbps or faster, Eritrea, the Congo, Libya and Somalia relying on dial-up connections. Many of these links were via INTELSAT, which owns and operates a global communications satellite system providing voice/data, Internet and video services to over 200 countries and territories [9].

In the conventional view, (i.e. as seen by technology developers and donors), developing countries passively adopt technology as standard products which have been developed in industrialized countries and which can be usefully employed immediately. However, successful use of IT requires much more than mere installation and application of systematized knowledge. It also requires the application of implied knowledge regarding the organization and management of the technology and its application to the contextual environment in which it is to be used. This implied IT knowledge often represents experience with the deployment of previous technology accumulated over time, such experiences contributing towards the shaping of new technology [3].

In addition to purely technological issues, the development of appropriate human resources skills are required, i.e. extensive training of the people who are going to use (and train others how to use) the resources. Training is seen as particularly important as this is not technology just a few people to benefit from, but for many. As Pekka Tarjanne, Secretary General of the ITU, made clear at Africa Telecom '98, "communication *is* a basic human right" (original emphasis). Nelson Mandela, at Telecom 95 in Geneva, urged regional co-operation in Africa, emphasizing the importance of a massive investment in education and skills transfer,thereby ensuring that developing countries also have the opportunity to participate in the information revolution and the "global communications marketplace"[3].

In the 1990s, Egypt's Information and Decision Support Center (IDSC), started a series of projects to support the establishment of Egypt's Information Highway. The project involved diffusion of decision making based on databases developed, maintained and directed by the Governorates [10]. The PC-based systems staff and developers were predominantly women from the local areas. It has been generally observed, however, that such developments may cause friction in the local communities, where the role of women may be downplayed by more fundamentalist groups.

The stated intention of Malaysia, as articulated by the former Prime Minister, Dr Mahathir Mohamad, is to become a "fully-developed, matured and knowledge-rich society by 2020".To this end, it has embarked on the Multimedia Super Corridor (MSC) project. Malaysia plans to leapfrog into the 21st century and the Information Age with an intellectual and strategic leadership, smart cities and smart lifestyles. To achieve this vision, however, growth at 7% per annum is required. In the present economic situation, this growth rate is unrealistic, though the government sees the MSC as a key stimulator of growth for the future.

Canada's International Development Research Centre (IDRC) runs a number of developing country projects that involve technology leapfrogging. The Pan Asian Network (PAN) was set up to fund ICT infrastructure and research projects in developing countries across Asia. Individuals, development institutions, and other organizations should all be able to use the infrastructure so as to share information [3].

PAN works with Bangladesh's world famous grassroots Grameen Bank. One service here is a "telecottage", where network services can be obtained. The technology and the material will be tailored to meet the needs of Grameen's typically poorly educated clients. One of PAN's objectives is gender equity. Women, who constitute some 95% of Grameen's borrowers, will be prominent among PAN users in Bangladesh [3].

PAN is also responsible for linking Laos to the Internet. The Science, Technology and Environment Organization (STENO) of the Lao Government invited some Laotian IT professionals living and working overseas to return home and share their experiences with their colleagues in the country. STENO collaborated with PAN in designing an 18-month long project to build the necessary infrastructure for a dial-up e-mail service. Among the pioneer users were "researchers working on agriculture and aquaculture projects; journalists managing national news agencies and newspapers; lawyers consulting on international legal issues; travel agents planning business trips; computer resellers tracking down suppliers and obtaining pricing information; and about 20 others in both the public and private sectors" [11].

Enns and Huff (1999) describe the development of Mongolian telecommunications networks from a historical perspective, focusing on the country's first ISP (Datacom) sponsored by IDRC's PAN. Mongolia's infrastructure was deteriorating when the project was initiated, with a tightly controlled media and a one-party political system that effectively isolated the country. However, Mongolia's complex transition to a market economy and its requirement for information resources provided project planners with the opportunity to study technical challenges that would be common in other countries with similar backgrounds. Sierranet is a telephone-based computer network, funded by IDRC and created by university students in Sierra Leone to support research. Two functional e-mail networks – for the university community and for health professionals - were established. Although most users have fled the country in the wake of current military activity, the network environment is reported to be still intact. Once the carnage in Sierra Leone has passed, it is hoped that Sierranet will play a role in the reconstruction of the country and the restoration of Democracy [12].

The Navrongo Health Research Centre (NHRC) in Ghana is part of an international initiative funded by IDRC and other aid agencies called HealthNet. The system uses lowearth-orbit satellites and telephone-based computer networks for the exchange and transfer of health-related information in the developing world. NHRC started in 1988 as a field site for Vitamin A studies. Within ten years, it had become a world-class health research facility. It leads the way in applying epidemiological methodologies and is a forerunner in the adoption of new ICTs. The NHRC exerts an influence on Ghanaian and WHO intervention programs and policies, training researchers from Asian and African countries, attracting expatriate Ghanaian scientists to work in Ghana.   Most significantly, the research undertaken at NHRC has resulted in dramatic reductions in child morbidity and mortality, as well as significant national improvements in standards of health care [3].

Over about the last ten years 'conventional' e-learning has been exemplified technologically by the rise of virtual learning environments (VLEs), such as WebCT and Blackboard, and the demise of computer-assisted learning 'packages', by expectations of ever increasing multi-media interactivity, power, speed, capacity, functionality and bandwidth in networked PC platforms. Pedagogically, we have seen the rise of social constructivist models of learning over previous behaviorist ones. All this is however only really true for Europe, North America and East Asia. In sub-Saharan Africa the term 'mobile learning' is recognized but as something grafted onto a tradition of open and distance learning and onto different pedagogic traditions, ones that have concentrated on didactic approaches rather than discursive ones. Mobile learning in these parts of the world is a reaction to different challenges and different limitations – usually those of infrastructure, poverty, distance or sparsity [4].

3 Experimental Evidences

In the developed world, Cunningham et al. (2000) referred in their report that "notwithstanding the rapid growth of online delivery among the traditional and new provisions of higher education, there is as yet little evidence of successful, established virtual institutions." However, in a 2002 survey of 75 randomly chosen colleges providing distance learning programs, results revealed an astounding growth rate of 41% per program in the higher education distance learning (Primary Research Group, 2002). Gunawardena and McIsaac (2003), in their *Handbook of Distance Education* , has inferred from the same research case that, "In this time of shrinking budgets, distance learning programs are reporting 41% average annual enrollment growth. Thirty percent of the programs are being developed to meet the needs of professional continuing education for adults. Twenty-four percent of distance students have high-speed bandwidth at home. These developments signal a drastic redirection of traditional distance education." According to an estimate, IT-based education and the e-learning market across the globe is projected at $11.4 billion (United States dollars) in 2003) [2].

In the developing world, studies show only a tiny percentage of Africans enjoy Internet connectivity (Amoako, 1998), perhaps one in ten thousand outside South Africa. These people are effectively invisible in an electronic world. As Roche and Blaine (1997) have observed, if one measures the IT capacity of countries in terms of millions of instructions per second (MIPS), then it has been estimated that most of the developing world suffers from a "MIPS gap ratio" in the order of something like 1:26 with the developed world. Another estimate of the disparity suggests that developing countries, whilst representing around 80 percent of the world population, account for only 2 percent of the total global expenditure on informatics (Hanna, 1991) [3].

The major challenges are related to lack of time and to the unstable infrastructure, causing some of the registered students to drop out without completing the course. The lack of infrastructure and access to modern technology is often argued against the strategy for offering higher education to target groups in developing countries through internet. Statistics now show, however, that the situation is changing drastically. From data collected in around 2008 show in Africa internet usage is 4.7% of the population, in Asia it is 12.4%, in Middle East 17.4% and in Latin Am/Carib it is 20.5% of the population [14].

4   Suggested Future Directions of Research

In Section 2, we presented various efforts made to make distance learning effective in developing countries. Presentation of course materials through multimedia in remote locations where in villages there could be school structures where those presentations could be made is feasible. Of course learning materials must be self-explanatory and not boring. Using multimedia facilities like videos, audios, graphics and interesting textual descriptions, it is possible to reach the remote locations of the world where computer technology has not reached yet. As the areas not covered by computer and internet technology is still profoundly vast in the world this approach seems to be very constructive and should be pursued.

Wherever possible distance learning through multimedia should be imparted through internet as internet and networks are the vehicles of multimedia. But since bandwidth connection is still very limited in vast areas of Asia, Africa and Latin America it would still take long time to reach major part of the population of the above-mentioned regions with multimedia and web.

Mobile technology offers a very hopeful way to reach the vast population of the developing countries as it does not require bandwidth connections. We have to develop distance learning using multimedia through mobile technology. This seems to be the most viable way to reach billions living in the rural areas of the developing countries. Hence considerable research efforts must be dedicated to this line. The author plans to work on how to use mobile technology to provide distance learning in both developed and developing countries in an efficient way using advanced multimedia tools.

5   Conclusion

In this paper we studied the problems of imparting distance learning through multimedia in developing countries. More research needs to be carried out to tap the vast opportunity of reaching to billions in developing countries through mobile technology and gearing up multimedia technology to be easily transported to those locations.

References

1. Passerint, K., Granger, M.J.: A Developmental Model for Distance Learning Using the Internet, Computer & Education, vol. 34(1) (2000)
2. Rahman, H.: Interactive Multimedia Technologies for Distance Education in Developing Countries - Introduction, Background, Main focus, Future trends, Conclusion (2000), http://encyclopedia.jrank.org/articles/pages/6637/Interactiv e-Multimedia-Technologies-for-Distance-Education-in-Developing-Countries.html
3. Davison, R., Vogel, D., Harris, R., Jones, N.: Technology Leapfrogging in Developing Countries – An Inevitable Luxury? Journal of Information Systems in Developing Countries (2000)

4. Traxler, J.: Mobile learning in 'developing' countries – not so different (2008), http://www.receiver.vodafone.com/mobile-learning-in-developing-countries
5. Vuksanovic, I., Zovko-Cihlar, B., Boras, D.: M-Learning in Croatia: Mobile multimedia systems for distance learning (2008), http://ieeexplore.ieee.org/xpl/freeabs_all.jsp?tp=&arnumber=4418832&isnumber=4418780
6. World Resources Institute Digital Dividend – Lesson, from the field: Distance education, http://www.digitaldividend.org/pubs/pubs_03_overview_distance_education.htm
7. Ruth, S., Giri, J.: The Distance Learning Playing Field:Do We Need Different Hash Marks? (2001), http://technologysource.org/article/distance_learning_playing_field/
8. Verma, R.: Distance Education in Technological Age, p. 419. Anmol Publications, New Delhi (2005), http://tojde.anadolu.edu.tr/tojde18/reviews/review2.htm
9. Semret, N.: Intelsat Brings Internet and New Technologies Summit to Africa, Email Message sent to the AFRIK-IT Discussion List March 24 (1999)
10. Nidumolu, R., Goodman, S., Vogel, D., Danowitz, A.: Information Technology for Local administration Support. The Governorates Project in Egypt, MIS Quarterly 20(2), 197–224 (1996)
11. Nhoybouakong, S., Ng, M.L.H., Lafond, R.: (1999), http://www.panasia.org.sg/hnews/la/la01i001.htm
12. Enns, H.G., Huff, S.L.: Information Technology Implementation in Developing Countries: Advent of the Internet in Mongolia. Journal of Global IT Management 2(3), 5–24 (1999)
13. Jurich, S.: Before the e-mail there was the p-mail: Distance learning by postal correspondence. TechKnowLogia I(1) (1999), http://www.techknowlogia.org/TKL_active_pages2/TableOfContents/main.asp?IssueNumber=1
14. Hauge, H., Ask, B.: Qualifying University Staff in developing countries for e-learning (2008), http://www.eife-l.org/publications/proceedings/ilf08/contributions/Supporting%20Lifelong%20Learning%20and%20Employability/Hauge_Ask.pdf

Scheduling and Planning Software for Corrugation Process

C. Chantrapornchai and T. Sathapanawat

Department of Computing, Faculty of Science, Silpakorn University
Thailand, 73000
ctana@su.ac.th

Abstract. In this paper, we are interested in developing a software for production planning for the corrugation process. We particularly study the scheduling problem of paper roll feeding in the mills. In the real industry, many constraints are post in order to assign paper rolls to the mill roll stand while the factory has several goals to meet such as saving paper cost and balancing worker loads. Such problems are known to be a difficult problem. We develop a heuristic to find the schedule under the goals to minimize waste, to balance the roll switching, and minimize butt rolls. Then, the web-based application is developed to create such a schedule and simulate the schedule results to verify the solutions.

Keywords: Corrugation Process, Scheduling, Planning.

1 Introduction

It is known that planning and scheduling in a factory is a difficult problem. This problem is usually modeled as an optimization problem. Mathematic models are needed to find an optimal solution. To find the best solution, it always takes a long time. Heuristic is often needed to find an approximate solution.

In this paper, we study a production planning problem for corrugation process. In the paper factory, the mill stand is fed with the sequence of the paper rolls for corrugation. For a large factory, there are several mill stands and several workers. The worker needs to feed the roll to the mill stands. The sequence of rolls to be fed needs to be determined in advance. We provide a scheduling algorithm which yields the sequence of roll feeding. Our schedule provides a sequence of feeding rolls to each stand under the constraints and maximizing the production goals. The algorithm is integrated in the corrugation software and the simulation to verify the schedule.

Several works have been done in cutting stock, scheduling and production planning. Most of them are solved using heuristic or linear programming approach. Nevertheless, very few of these has focused on the corrugation process. Also, existing software for corrugation process or production planning is commercial and very expensive.

For example, the worked by Dyckhoff (1981) presented the use of linear programming to solve 1-D cutting stock which is NP-complete problem. Afshar (2008) also used linear programming model to solve 1-D cutting stock with trim loss. Renio et.al. (2006) has given a good literature review on 1-D cutting stock problem.

T.-h. Kim et al. (Eds.): SIP/MulGraB 2010, CCIS 123, pp. 260–266, 2010.

For the scheduling problem, Pandelis G. Ipsilandis (2007) uses MOP in scheduling in project management. His model attempts to minimize the overall time, idle time as well as cost. Sang M. Lee et. al. (2003) used linear programming to solve job-scheduling and compared with the genetic algorithm. Chan et. al. (1998) proposed the machine scheduling using linear programming and compared the approach with list scheduling.

The closest work in the paper mill is the worked by Robert Hasseler (1980,1981,1990) which attempted to solve the trim loss for the corrugation process. He modeled using 0-1 linear programming. The trim loss problem is not considered in the original scheduling model. However, he did not consider the amount of butt rolls in the inventory and the selection of the paper roll to balance the mill feeding.

Several heuristics have been studied for scheduling problem. Kops et.al. (1994) presented a scheduling algorithm for single operation job flow. The approach used both linear programming and dynamic programming and considered the due date and minimizing the work in process. Haessler (1971) presented a heuristic to roll scheduling for cutting to minimize the setup cost, trim loss and total time based on the random solutions. Cruz et.al. (2002) proposed the technique for corrugating for corrugate sheet. Darley et.al.(2008) used agents for corrugate boxes in the factory. The agent analyzed the proper process to reduce inventory stock and considered the due date.

Many heuristics on the similar problem such as cutting stock problem are such as Shen et.al. (2007) which used particle swarm. Chiong and Beng (2007) compared genetic algorithm and evolutionary algorithm. Many commercial software exists for cutting stock problem. The closet ones are from SolarSoft (2008) which is for corrugate packing and from Mccullough (2007) from CTI company which presented a DSS system for a corrugating box for the whole production cycle.

In our work, we consider a scheduling problem in the paper factory. Particularly, for a corrugation process, we consider the paper roll feeding problem. In our problem, several goals are concerned such as the minimize butt rolls, waste, paper roll switching. The scheduling heuristic is developed. The software to perform the scheduling and simulate the results is presented. In the future, the linear programming model will be presented and compared to the heuristic solutions. Also, the comparison to existing scheduling approach will be done.

2   Problem Backgrounds

In a corrugation box factory, corrugating is the most important process to save the cost. Normally, there are two types of corrugating boxes: Single wall and double wall corrugated boards as shown in Figure 1. In order to create the single wall board, 3 layers are needed while for the double wall board, 5 layers are needed.

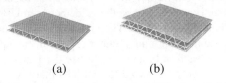

(a)   (b)

Fig. 1. Types of corrugating boards (a) Single wall (b) Double wall

In details the process will take the paper liner together with the paper medium and put together with the starch to become a piece of boards. Each paper is from the inventory which is in the roll as shown in Figure 2 (a). Each roll has a length of 4,000-6,000 meters. The unused roll is called full rolls while the already used one is called, butt rolls.

(a)   (b)

Fig. 2. (a) Paper rolls (b) Mill stands

Each paper roll is put on the mill roll stand in Figure 2(b). For each machine, there is two paper roll heads. The other one is the spare roll using when the first roll is run out. Thus, while the first roll is feeding, the worker needs to prepare the paper in the other roll and preparing to lengthen the paper.

Each paper has attributes such as the paper grade which implies the paper weight in grams and the width the paper which implies by the arrow in Figure 2(a). To calculate the paper size, the dimension of the paper box is used together with the total boxed needed plus the size to spare.

Thus, in our work, to create the schedule, we consider the following.

1.   Total waste should be minimized. In the factory, the roll with less than 300 meters is a waste since it cannot be used on the mill stand anymore.
2.   The butt rolls should be minimized. We will attempt to minimize the use of the new rolls.
3.   To feed the rolls, we consider the small rolls (the rolls with the less paper, more than 300 meters but not more than 1,000 meters) in the first priority.
4.   For the worker on each mill stand, the work on roll changing should be balanced. If the workers are assigned with the small rolls, he will need to changes the rolls very often.
5.   For the rolls with the attributes, we will select the older rolls. We will attempt to use the old one in the stock first to prevent the inventory aging.

3 Algorithms

Assume that LB is the lower bound length in meters and UB is the upper bound lengths in meters. The waste is assumed to be the value between UB and LB. The size of small rolls or large rolls is given by ML. The major goal is to minimize the waste. The minor goals are 1) to use the minimum number of rolls in each plan. 2) to balance the number of rolls used for each mill stand. 3) to use the paper in the order of the aging. The constraints are 1) between two different jobs that needs the same roll but on the different mill stand, the two job must be five jobs apart so that the paper roll can be moved on time. 2) For each mill stand, it is to avoid two small paper rolls in sequence, since the worker cannot setup the paper roll on time. 3) If there is no required paper in stock, the mill must stop, log report is generated.

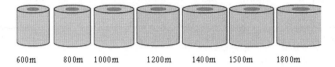

600m   800m   1000m   1200m   1400m   1500m   1800m

Fig. 3. Rolls examples

For example, UB = 300, LB = 10, ML = 1200 and the total order size is 3,000. Consider the example rolls in Figure 3. If we feed the rolls in sequence, 1200 -> 600 -> 1400, then we get the total length 3,200 and the waste is 200 for the roll 1,400. But if we feed the rolls, 1,200-> 1,800 , the total length is 3,000 with no waste. It is also seen that the problem is similar to the subset sum problem which is NP-hard.

Thus, our algorithm is based on the greedy approach as following.

1. read inputs : stock balance, inventory.
2. loop for each job, each mill stand:
   a. Select the paper rolls that meets the constraints: quality, size, the aging priority.
   b. If there are more than one eligible roll, consider the same roll as the previous job. If the same one is used, calculate the total switching for each mill. Also, consider the small rolls first. Calculate the total switching for each mill for each case. Then, select the one that most balance the roll switching.
   c. Also, for the selected roll, check if the current job use the same paper roll as any previous job, check the minimum distance between the two jobs. If the distance is not possible, go back to (a)-(b) to try on another eligible roll. If still not possible, then insert the blank job (idle).
   d. Accumulate the number of switching for each mill.
   e. Update used length on each roll in the inventory.

Note that for the roll selection, first we need to consider the quality of the paper that meets the criteria first. If none is available, the reorder needs to be processed. Otherwise, we will select the rolls for that quality according to the required size. The size may be considered as first fit, best fit etc. Next, the aging of the paper is considered. Both size and aging may be considered as one selection function. Then, we consider to balance the number of roll switching and minimize the butt rolls. Next the distance on the same roll usage is checked.

4 Software Interface

Figure 4 shows the inventory of the paper rolls. Column "SKU: Grade" shows the type of the paper, Column "width" is the width of the roll, Column "Roll ID" is the paper roll reference, Column "Length (m) before" is the original length of the roll.

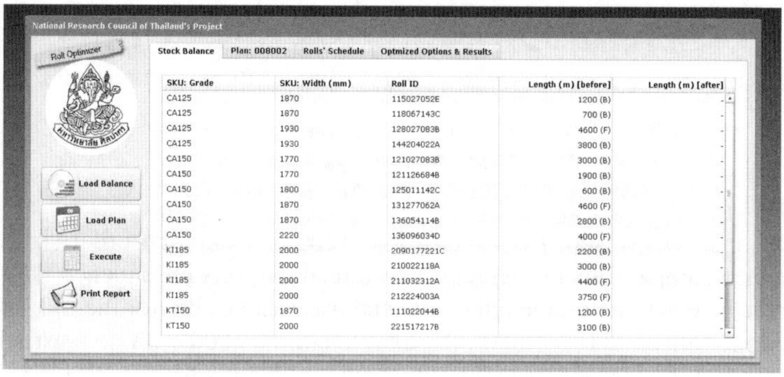

Fig. 4. Inventory of the paper rolls

Figure 5 shows the schedule results. The results are read as follows: Column "No" shows the job number. Thus, the same row implies the same job. Column "Width" is the width of the roll. 5 mill stands will require the same width since they process the same job. Column "Grade#1" shows the paper grades needed for mill stand #1. Column "len#1" is the length required on mill stand #2 and so son. The empty entry in column "Grade" means there is no feeding at that sequence and the job uses the same paper grade as in the above row. For example, job no.3 at grade#2 "CA125" uses the same paper until job no.5. The whole length is 5,310 meters.

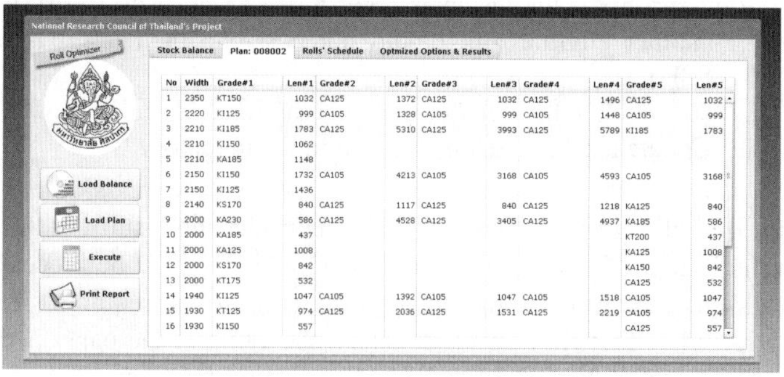

Fig. 5. Roll feeding plan

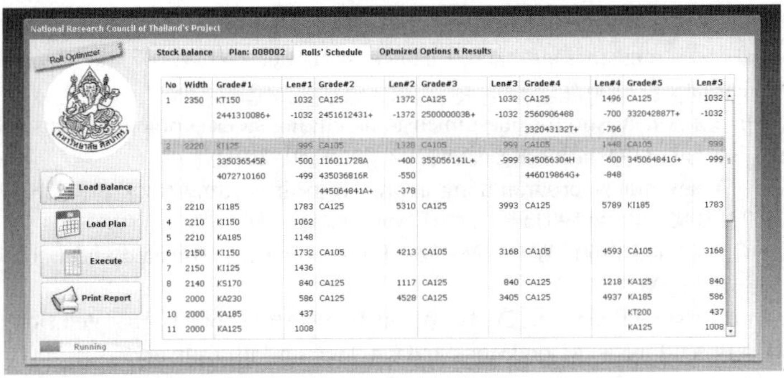

Fig. 6. Plan simulation

Figure 6 shows the plan execution (simulation) to test the possibility of the plan. The current highlight shows the current job executing. While executing, the length of the roll is decreased. In Column "Grade#1" under KI125, shows the two rolls that are used for this job. " 335036545R" is used 500 meters and "4072710160" is used 499 meters for the total of 999 meters.

5 Conclusion

In this paper, we present the software for the corrugation process. Corrugation process is important in a corrugating factory. The waste of paper should be minimized to save the cost. The heuristic for creating a schedule plan for roll feeding is developed. The algorithm considers to minimize the waste, the butt rolls, and to balance the roll switching. It also considers the aging of the paper and the distance for the same roll usage. The software includes the inventory management and the simulation of the plan to verify the schedule.

In the future, the mathematical models for the problem will be completed. The linear programming solution will be compared with the solution from the heuristic. The usability of the software will be tested. The performance of the schedule will be tested against the real data and satisfactory reports will be included.

Acknowledgement

We would like to thank National Research Council of Thailand for the funding support for this work.

References

Affshar, A., Amiri, H., Eshtelardian, E.: An improved linear programming model for one dimensional cutting stock problem. In: 1st Intl. Conf. on Construction in Developing Countries (ICCIDE-1), Pakistan (August 2008)

Chan, L.M.A., Muriel, A., Simchi-Levi, D.: Parallel Machine Scheduling, Linear Programming, and Parameter List Scheduling Heuristics. Operations Research 46(5), 729–741 (1998)

Chiong, R., Beng, O.: A comparison between genetic algorithms and evolutionary programming based on cutting stock problem. Engineering Letter 14(1) (2007) (advance online publication: February 12, 2007)

Cui, Y., Yang, Y.: A heuristic for one-dimensional cutting stock problem with usable leftover. European Journal of Operation Research 204, 245–250 (2010)

Dyckhoff, H.: A new linear programming approach to the cutting stock problem. Operations Research 29, 1092–1104 (1981)

Santa Cruz, C.D., Sandstrom, W.R.: Method for bending a corrugated sheet, United States Patent 6354130 (2002)

Darley, V., von Tessin, P., Sandler, D.: An Agent-Based Model of a Corrugated Box Factory: Tradeoff between Finished-GoodStock and On-Time-in-Full Delivery, http://www.santafe.edu/~vince/pub/CorrugatedBoxFactory.pdf (accessed February 20, 2008)

Goldratt, E.M., Cox, J.: The Goal. North River Press, 3rd Revised edn. (2004)

Haessler, R.W.: A heuristic programming solution to a nonlinear cutting stock problem. Management Science 17, 793–802 (1971)

Haessler, R.W.: Production Planning and Scheduling for an Integrated Container Company. Automatica 21(4), 445–452 (1985)

Haessler, R.W., Brian Talbot, F.: A 0-1 Model For Solving The Corrugator Trim Problem, Working Paper No. 205, Division of Research, Graduate School of Business Administration, The University of Michigan (February 1980)

Haessler, R.W.: Short Term Planning and Scheduling in a Paper Mill, Working Paper No. 268, Division of Research, Graduate School of Business Administration, The University of Michigan (May 1981)

Haessler, R.W., Sweeny, P.E.: Evaluating the Impact of Small Changes in Ordered Widths on Trim Yields, Working Paper No. 640, Division of Research, Graduate School of Business Administration, The University of Michigan (July 1990)

Ipsilandis, P.G.: Multiobjective Linear Programming Model for Scheduling Linear Repetitive Projects. Journal of Construction Engineering and Management 133(6), 417–424 (2007), doi:10.1061/(ASCE)0733-9364(2007)133:6(417)

Kops, L., Natarajan, S.: Time partitioning based on the job-flow schedule — A new approach to optimum allocation of jobs on machine tools. International Journal of Advanced Manufacturing Technology 9(3), 204–210 (1994)

Lee, S.M., Asllani, A.A.: Job scheduling with dual criteria and sequence-dependent setups: mathematical versus genetic programming. Omega 32(2), 145–153 (2004)

Mccullough, J.: Improved Planning Results in Success, Softwar Technology Case Studies. Corrugated Technologies Inc., (January-February 2007)

Pinedo, M.L.: Scheduling: Theory, Algorithms, and Systems, 3rd edn. Springer, Heidelberg (2008)

Renildo, G., et al.: Linear Programming Models for the One-dimensional Cutting Stock Problem, INPE ePRint:sid.inpe.br/ePrint@80/2006/12.18.16.52 v1 2006-12-19 (2006)

Shen, X., et al.: A Heuristic particle swarm optimization for cutting stock problem based on cutting pattern. In: Shi, Y., van Albada, G.D., Dongarra, J., Sloot, P.M.A. (eds.) ICCS 2007. LNCS, vol. 4490, pp. 1175–1178. Springer, Heidelberg (2007)

SolarSoft Co. Ltd., http://www.solarsoft.com/industries/packaging/corrugated_packaging.htm

Hierarchical Data Structures for Accessing Spatial Data

Rituparna Samaddar[1], Sandip Samaddar[1],
Tai-hoon Kim[2,*], and Debnath Bhattacharyya[2]

[1] Heritage Institute of Technology
Kolkata-700107, India
{rituparna.snh,Samaddar.sandeep}@gmail.com
[2] Department of Multimedia
Hannam University
Daejeon, Republic of Korea
taihoonn@empal.com, debnathb@gmail.com

Abstract. Spatial data is data related to space .In various application fields like GIS, multimedia information systems, etc., there is a need to store and manage these data. Some data structures used for the spatial access methods are R tree and its extensions where objects could be approximated by their minimum bounding rectangles and Quad tree based structures where space is subdivided according to certain rules. Also another structure KD Tree is used for organizing points in a k dimensional space. In this paper we have described an algorithm for insertion of points in a Quad tree based structure, known as PR Quad Tree. Another data structure called K-D Tree is also described and the insertion procedure is defined. Then the comparison between the two structures is drawn.

Keywords: Spatial Data, Quad Tree, PR Quad Tree, KD Tree.

1 Introduction

Conventionally, database systems were designed to manage and process alphanumeric data represented by character strings, numerical values, date and Boolean or logical expressions There was no provision in conventional relational database systems to support the storage and processing of spatial data represented by points, lines, polygons and surfaces. Spatial Databases were used which supports spatial Data Types.

Spatial information system requires an efficient spatial data handling [9, 10, 14]. The large amount of information and complexity of characteristics of data have given rise to new problems of storage and manipulation.

Modern applications are both data and computationally intensive and require [11] the storage and manipulation of voluminous traditional (alphanumeric) and nontraditional data sets, such as images, text, geometric objects, time series, audio, video. Examples of such emerging application domains are: geographical information systems (GIS), multimedia information systems, time-series analysis, medical

---

* Corresponding author.

T.-h. Kim et al. (Eds.): SIP/MulGraB 2010, CCIS 123, pp. 267–274, 2010.
© Springer-Verlag Berlin Heidelberg 2010

information systems, on-line analytical processing (OLAP) and data mining. These applications impose diverse requirements with respect to the information and the operations that need to be supported. Therefore from the database perspective, new techniques and tools need to be developed towards increased processing efficiency.

Some data structures adopted in most research on spatial access methods, are R trees [6,11] and their extensions R+ trees and R* trees [7], where objects could be approximated by their minimum bounding rectangles [8]. The other approach where space is subdivided according to certain rules is the quad tree data structures.

Another data structure for storing a finite set of points from a k-dimensional space is KD -Tree. It was examined in detail by J.Bentley [1].

2   Introduction to Quad Tree

Quad tree is a 2-D spatial data structure that successively partitions a region of space into 2^2 quadrants or cells. Cells are successively subdivided into smaller cells. The main idea of quad tree structure is repeatedly divide a geometrical space into quadrants and the strategy it follows is similar as divide and conquer strategy. A quad tree has four children; each corresponds to each direction, namely NE, NW, SW and SE. The most studied quad tree approach to region representation is termed as region quad tree, which is proposed by Klinger. The region quad tree is based on a regular decomposition of the space of equal size. An object is put into a quad tree node corresponding to a quadrant if and only if it is inside that quadrant but is not inside any of its children. It is a variation of Maximum block representation where the blocks must be disjoint and must have a standard size, which is in power of 2. It is developed with the intension of representing homogeneous parts of an image in a systematic way. As one property of decomposition is to partition the space infinitely, so any desired degree of resolution is achieved. It is helpful in the field of image processing and database management. If points represent the objects, i.e. zero sized elements e.g. the locations of all the cities in a particular state, and then either point quad tree or PR quad tree is used. In the case of PR quad tree the total space is decomposed into quadrants, sub quadrants, until the number of points in each quadrant is within a certain limit. This limit is known as node capacity of the PR quad tree. But, in case of Point quad tree, the space is decomposed into quadrants, sub quadrants, till a point exists in a quadrant.

3   PR Quad Tree

3.1   Introduction to PR Quad Tree

PR quad tree (P for point and R for region): It was proposed by Hanen Samet [11,12,5,13]. The PR quad tree is organized in the same way as the region quad tree. The difference is that leaf nodes are either empty or contain a data point and its coordinates. A quadrant contains at most one data point. Actually, the search is for the quadrant in which the data point says A, belongs (i.e., a leaf node). If the quadrant is already occupied by another data point with different x and y coordinates, say B, then the quadrant must repeatedly be subdivided (termed splitting) until nodes A and B no longer occupy the same quadrant.

In effect, each node of a PR quad tree represents a particular region in a 2D coordinate space. Internal nodes have exactly 4 children (some may be empty), each representing a different, congruent quadrant of the region represented by their parent node.

Internal nodes do not store data. Leaf nodes hold a single data value. Therefore, the coordinate space is partitioned as insertions are performed so that no region contains more than a single point. PR quad trees represent points in a finitely-bounded coordinate space.

3.2 Insertion in PR Quad Tree

Data points are inserted in a PR Quad Tree in a manner analogous to Point Quad Tree. The algorithm for insertion is as follows. Data points are inserted in a PR Quad Tree in a manner analogous to Point Quad Tree.

The Algorithm for insertion is as follow:
A is the point to be inserted.
Root refers to the root node of the tree.
Current refers to the node most recently traversed in the tree
sqdt is the sub-quadrant in which the point to be inserted lies

Suppose a point A is to be inserted.
Step1: (Check for null tree)
        If root =null i.e tree is empty
           then set Root <–A
           go to Step 9
      End if
Step2: current = root
Step3: if current is not a leaf node
          sqdt = sub-quadrant in which A lies w.r.t the quadrant
             represented by the current node.
      Step 3a: if child node of current corresponding to the sqdt sub-quadrant is
            empty then go to step8.
         End if
      Step 3b: set parent=current
           set current=child node of current corresponding to sqdt sub-
             quadrant. Go to step 3
      End if
Step 4: Temp=current
Step 5: sqdt=subquadrant of parent in which current lies
Step 6: Create a node (n1) referring the sqdt subquadrant.
      Make n1 the child node of parent corresponding to sqdt subquadrant
           Set current=n1
           Partition the quadrant corresponding to current node equally
           into 4 subquadrants
           set sqdt = subquadrant in which A lies after partition

Step 6a: if sqdt is same as subqquadrant corresponding to temp node's
data
then parent=current
go to step6.
End if
Step7: Insert temp as the NE or SE or SW or SE child of current node based on the
sub-quadrant current node in which temp's data lies.
Step8: Insert A as the NE or SE or SW or SE (based on sqdt) child of current node.
Step9: End

From the above procedure we can conclude that:
a.   The shape of the tree is entirely independent of the order in which the data
elements are added to it.
b.   In the worst case it may happen that the points are so closely spaced that the
number of partitions increases and the tree level goes on increasing and results in a imbalanced skewed tree.

4   The KD-Tree (Organization of Point Data)

4.1   Introduction to KD Tree

A KD-tree is a data structure for storing a finite set of points from a k-dimensional
space. It was examined in detail by J.Bentley [1]. In [1] Bentley described the KD-
tree as a k-dimensional binary search tree. A node in the tree (Figure below) serves
two purposes: representation of an actual data point and direction of a search. A discriminator D (P), whose value is between 0 and k-1 inclusive, is used to indicate the
key on which the branching decision depends. A node P has two children, a left son L
(P) and a right son R (P). If the discriminator value of node P is the jth attribute (key),
then the jth attribute of any node in the L(P) is less than jth attribute of node P, in Fig.
1 and the jth attribute of any node in the H(P) is greater than or equal to that of
node P.
The basic form of the KD-tree stores K-dimensional points. This section concen-
trates on the two-dimensional (2D) case shown 2. Each internal node of the KD-tree
contains one point and also corresponds to a rectangular region.

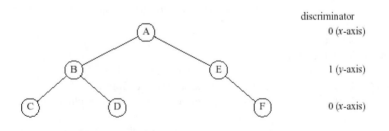

Fig. 1. The structure of a KD-tree

The root of the tree corresponds to the whole region of interest. The rectangular region is divided into two parts by the x-coordinate of the stored point and by the y-coordinate alternately. A new point is inserted by descending the tree until a leaf node is reached. At each internal node the value of the proper coordinate of the stored point is compared with the corresponding coordinate of the new point and the proper path is chosen. This continues until a leaf node is reached. This leaf also represents a rectangular region, which in turn will be divided into two parts by the new point.

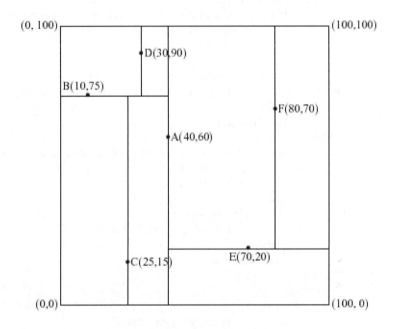

Fig. 2. The planar representation of KD Tree

In [2, 3] also it is described how the nearest neighbor algorithm is applied to KD trees. KD trees are most often utilized as a primary storage structure, but the extended K-d tree [18, 15] has been modified for secondary storage.

A second K-d tree hybrid, called the K-D-B tree [16, 17, 4] has evolved for very large indexes KD- B trees combine features of B-trees and K-d trees.

A third K-d tree hybrid 1s defined as an optimized K-d tree for finding best matches [19].

4.2 The Algorithm for Insertion in a KD Tree

In 1975 Bentley described the algorithm for insertion as follows. A node P is passed which is not in the tree.

Step1: (Check for null tree)

    If root =null i.e tree is empty

        Then set Root <-A

        Go to Step 7

    End if

Step2: Set current = root

Step3: Set Discriminator=x

Step4: if current is not null

    If discriminator=x then

        If x coordinate of P less than x coordinate of current node

            Parent=current

                Current=left child (current)

        Else

            Parent=current

            Current=right_child (current)

        Endif

        Discriminator=y

    Else

        If y coordinate of P less than y coordinate of current node

            Parent=current

                Current=left_child (current)

        Else

            Parent=current

            Current=right_child (current)

        Endif

        Discriminator=x

    End if

    Go to step 4

  End If

Step 5: Create a new node ndnew representing the point P

Step 6: if discriminator =x then

    If y coordinate of P less than y coordinate of parent node

        Set left child (parent)= ndnew

    Else   Set right_child (parent)= ndnew

    End if

  Else

    If x coordinate of P less than x coordinate of current node

        Set left child (parent)= ndnew

    Else   Set right_child (parent)= ndnew

    End if

  End If

Step 7: End

From this algorithm we can conclude that:

  a. the shape of the tree depends on the order in which the insertion is taking place.

  b. In the worst case the number of levels in the tree becomes equal to the number of points that is inserted.

5 Comparative Study

The following points can be highlighted by comparing PR Quad Tree and KD Tree:

- PR Quad Tree is a tree data structure used to partition a 2 dimensional space by recursively subdividing it into 4 quadrants or regions.
  Whereas a KD Tree is a k dimensional tree data structure used for organizing points in a k dimensional space.

- In a PR Quad Tree each node of a PR quad tree represents a particular region in a 2D coordinate space.
  Whereas every node in a KD Tree is a k dimensional point.

- In a PR Quad Tree non-leaf nodes do not store data. Internal nodes have exactly 4 children (some may be empty), each representing a different, congruent quadrant of the region represented by their parent node.
  Whereas in a KD Tree even non-leaf nodes is a k dimensional point. Every non leaf node generates a splitting hyper plane that divides the space into 2 subspaces.

- In PR Quad tree only leaf nodes contain data points.
  Whereas in a KD Tree every node is a k dimensional point.

- In a PR Quad Tree the structure of the tree is determined entirely by the data values it contains, and is independent of the order of their insertion.
  Whereas in a KD Tree the shape of the tree depends on the order in which the points are inserted.

- In a PR Quad Tree dense data regions require more partitions and therefore the quad tree will not be balanced in this situation.
  Whereas In a KD Tree in the worst case the number of levels in the tree becomes equal to the number of points that is to be inserted.

6 Conclusion

We have shown in this paper the methods of the storage and manipulation of voluminous spatial data sets, which were really a problem in conventional Database. Among spatial data set we have considered only point data and its storage and access methods. We have focused mainly on one Quad Tree based structure, PR Quad Tree and another Hierarchical Data structure KD Tree used for organizing point data. There are also other structures like point Quad Tree and some extended version of KD Tree, R Tree which are not discussed in detail.

We have mainly stressed on the insertion procedure of these structures. The deletion procedure is also equally important which is not covered in this paper.

Acknowledgement

This work was supported by the Security Engineering Research Center, granted by the Korea Ministry of Knowledge Economy.

References

1. Bentley, J.L.: Multidimensional binary search trees used for associated searching. Comm. Of the ACM 18, 509–517 (1975)
2. Moore, A.W.: An introductory tutorial on KD trees. Technical Report No. 209, Computer Laboratory, University of Cambridge (1991)
3. Overmars, M.H., van Leeuwen, J.: Dynamic Multi-Dimensional data structures based on Quad and KD tress. Acta Informatica 17, 267–285 (1982)
4. Güting, R.H.: An Introduction to Spatial Database Systems. The VLDB Journal — The International Journal on Very Large Data Bases 3(4), 357–399 (1994)
5. Samet, H.: The Quadtree and Related Hierarchical Data Structures. ACM Computing Surveys (CSUR) 16(2), 187–260 (1984)
6. Guttman: R-Trees: A Dynamic index structure for spatial searching. In: Proceedings ACM SIGMOD, pp. 47–57 (1984)
7. Beckmann, N., Kriegel, H.P., Schneider, R., Seeger, B.: The R* Tree-An efficient and Robust Access Method for points & Rectangles. In: Proceedings ACM SIGMOD 1990 International Conference on management of data (1990)
8. Brinkhoff, T., Kriegel, H.P., Seeges, B.: Efficient Processing of spatial join using R-Trees. In: Proceedings ACM SIGMOD 1993 International Conference on Management (1993)
9. Chang, S.K., Jungert, E., Li, Y.: The Design of Pictorial Database based upon the theory of symbolic projections. In: Proceedings 1st International Symposium on large Spatial databases, Santa Barbara, USA, pp. 303–323 (1989)
10. Nardelli, E., Projetti, G.: Time & space efficient secondary memory representation of Quadtrees. Information System 22(1), 25–37 (1997)
11. Papadopoulos, A.N., Manolopoulos, Y.: Nearest Neighbor search, A Database Perspective. Springer, Heidelberg (January 2005)
12. Samet, H.: The Design and Analysis of Spatial Data Structures. Addison-Wesley, Reading (1990)
13. Samet, H.: Applications of Spatial Data Structures. Addison-Wesley, Reading (1990)
14. Finkel, R.A., Bentley, J.L.: Quad Trees: A Datastructure for retrieval of composite keys (April 8, 1974)
15. Overmars, M.H., Van Leeuwen, J.: Multikey retrieval from K-d trees and QUAD-trees. In: International Conference on Management of Data, pp. 291–301 (1985)
16. RobInson, J.T.: The K-D-B-Tree A Search Structure for Large, Multidimensional Dynamic Indexes. In: 4CM-SIGMOD 1981 International, Conference 0" Management -o-f Data Association (1981)
17. Van Leeuwen, J., Overmars, M.H.: Stratified Balanced Search Trees. Acta, Informatica 18, 345–359 (1983)
18. Chang, J.M., Fu, K.S.: Dynamic Clustering Techniques for Physical Database, Design. In: International Conference on Management of Data, pp. 188–199 (1980)
19. Friedman, J.H., Bentley, J.L., Finkel, R.A.: An Algorithm for Finding Best Matches in Logarithmic Expected Time. ACM Transactions on Mathematical Software 3(3), 209–226 (1977)

OSI Layer Wise Security Analysis of Wireless Sensor Network

Shilpi Saha[1], Debnath Bhattacharyya[2], and Tai-hoon Kim[2,*]

[1] Computer Science and Engineering Department
Heritage Institute of Technology
Kolkata, India
shilpisaha7@gmail.com
[2] Department of Multimedia
Hannam University
Daejeon, Korea
debnathb@gmail.com, taihoonn@empal.com

Abstract. The security in wireless sensor networks (WSNs) is a critical issue due to the inherent limitations of computational capacity and power usage. While a variety of security techniques are being developed and a lot of research is going on in security field at a brisk pace but the fields lacks a common integrated platform which provides a comprehensive comparison of the seemingly unconnected but linked issues. In this paper, we have tried to analyze some attacks and their possible countermeasures in OSI layered manner.

Keywords: WSN, security, Sybil, wormhole, spoofing, eavesdropping, selective forwarding.

1 Introduction

Wireless sensor networks (WSN) are quite useful in many applications since they provide a cost effective solution to many real life problems. But it appears that they are more prone to attacks than wired networks. They are susceptible to a variety of attacks, including node capture, physical tampering and denial of service [1]. An attacker can easily eavesdrop on, inject or alter the data transmitted between sensor nodes. Security allows WSNs to be used with confidence and maintains integrity of data. Without security, use of WSN in any application domain would result in undesirable consequences. Particularly in military based projects where a compromise in security can lead to disastrous consequences. Thus security must be addressed in such critical sensor applications. It turns out that providing security in wireless sensor networks is pivotal due to the fact that sensor nodes are inherently limited by resources such as power, bandwidth, computation and storage. Efficiency is thus a crucial issue as sensors are usually deployed in remote area for a long time. Although a lot of progress has been made for the past few years, the field remains fragmented, with contributions scattered over seemingly disjoint yet actually connected areas. As for

---

* Corresponding author.

T.-h. Kim et al. (Eds.): SIP/MulGraB 2010, CCIS 123, pp. 275–282, 2010.
© Springer-Verlag Berlin Heidelberg 2010

example, key management only makes sure the communicating nodes possess the necessary keys, at the same time protecting the confidentiality, integrity and authenticity of the communicated data. However, it only assures a sense of security in one layer whereas the security of the network can be ruptured in other layers as well like network layer, physical layer, etc.

In this paper, we have explored various security issues of wireless sensor network. Our contribution is therefore to identify and describe the threats and the preferable modes of countermeasures according to the OSI layer.

2   WSN Architecture

In a typical WSN (shown in Fig. 1), we can see the following components:

- Sensor motes (Field devices): Sensor motes are mounted in the process and must be capable of routing packets. In most cases they characterize or control the process or process equipments. A router is a special type of field device that does not have process sensor or control equipment and as such does not interface with the process itself.
- Gateway or Access points: A gateway enables communication between host application and field devices.
- Network manager: A network manager is responsible for configuration of the network, scheduling communication between devices, management of the routing tables and monitoring and reporting the condition of the network.
- Security manager: The security manager is responsible for the generation, storage and management of keys.

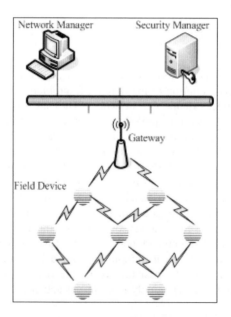

Fig. 1. WSN Architecture

3 Issues in WSN Security

Security mechanisms in WSN are developed in view of certain constraints. Among these some are predefined security strategies; whereas some are direct consequences of the hardware limitations of sensor nodes. Some of the issues described here pave way for the guidelines in the next section:

a. Energy efficiency: The requirement for energy efficiency suggests that in most cases computation is favored over communication, as communication is three orders of magnitude, more expensive than computation [2]. The requirement also suggests that security should never be overdone – on the contrary, tolerance is generally preferred to overaggressive prevention [3]. More computationally intensive algorithms cannot be used to incorporate security due to energy considerations.

b. No public key cryptography: Public key algorithms remain prohibitively expensive on sensor nodes both in terms of storage and energy [4]. No security schemes should rely on public key cryptography. However, it has been shown that authentication and key exchange protocols using optimized software implementations of public key cryptography is very much viable for smaller networks [2].

c. Physically tamper able: Since sensor nodes are low-cost hardware that is not built with tamper-resistance in mind, their strength has to lie in their number. Even if a few nodes go down, the network survives. The network should be resilience of attacks [3, 1].

d. Multiple layers of defense: Security becomes an important concern because attacks can occur on different layers of a networking stack (as defined in the OSI model). Naturally it is evident that a multiple layer of defense is required, i.e. a separate defense for each layer [3]. The issues mentioned here are in general applicable to almost all sorts of domain irrespective of their traits.

4 Security Requirements

WSN Security Requirements can be as follows:

Availability: Sensors are strongly constrained by many factors, e.g. limited computation and communication capabilities. Additional computations or communications consume additional energy and if there is no energy, data will not be available. Energy is extremely limited resource in large scale wireless sensor network. A single point failure will be introduced while using the central point scheme. This greatly threatens the availability of the network. The requirement of security not only affects the operation of the network, but also is highly important in maintaining the availability of the whole network [5]. Moreover, wireless sensor networks are vulnerable to various attacks. The adversary is assumed to possess more resources such as powerful processors and expensive radio bandwidth than sensors. Equipped with more rich resources, the adversary can launch even more serious attacks such as DoS attack, resource consumption attack and node compromise attack.

Confidentiality: Data confidentiality is the most important issue in network security. Confidentiality, integrity and authentication security services are required to thwart the attacks from adversaries. These security services are achieved by cryptographic primitives as the building blocks. Confidentiality means that unauthorized third parties cannot read information between two communicating parties. A sensor network should not leak sensor readings to its neighbors. Especially in military application, the data stored in the sensor node may be highly sensitive [5]. In many applications, nodes communicate highly sensitive data, e.g. key distribution; therefore it is extremely important to build a secure channel in a wireless sensor network. Public sector information, such as sensor identities and public keys, should also be encrypted to some extent to protect against traffic analysis attacks. Generally, encryption is the most widely used mechanism to provide confidentiality.

Integrity and Authenticity: Confidentiality only ensures that data cannot be read by the third party, but it does not guarantee that data is unaltered or unchanged. Integrity means the message one receives is exactly what was sent and it was unaltered by unauthorized third parties or damaged during transmission. Wireless sensor networks use wireless broadcasting as communication method. Thus it is more vulnerable to eavesdropping and message alteration [1]. Measures for protecting integrity are needed to detect message alteration and to reject injected message. Authentication ensures that the sender was entitled to create the message and the contents of the message have not been altered. In the public key cryptography, digital signatures are used to seal a message as a means of authentication. In the symmetric key cryptography, MACs are used to provide authentication. When the receiver gets a message with a verified MAC, it is ensured that the message is from an original sender. Digital signature is based on asymmetric key cryptography, which involves much more computation overhead in signing/decrypting and verifying/encrypting operations. It is less resilient against DoS attacks since an attacker may feed a victim node with a large number of bogus signatures to exhaust the victim's computation resource for verifying them [3].

Data Freshness: Data freshness means that the data is recent and any old data has not been replayed. Data freshness criteria are a must in case of shared key cryptography where the key needs to be refreshed over a period of time. An attacker may replay an old message to compromise the key.

Self Organization: Due to the ad hoc nature of WSNs it should be flexible, resilient, adaptive and corrective in regards to security measures.

5   Various Types of Attacks in Different Layers and Preferable Modes of Countermeasures

Security attacks in sensor networks can be broadly classified into Passive attacks and Active attacks. Passive attacks are in the nature of eavesdropping on or monitoring of transmissions. The motive of the attacker is to obtain information that is being transmitted. Two types of passive attacks are release of message contents and traffic analysis. Active attacks involve some modification of the data stream or the creation of a false stream and can be subdivided into four sub categories: masquerade, replay, modification of services and denial of service. We are mainly looking at two types of protection: protection from denial-of-service (DoS) attacks and protection of the secrecy of information. Multiple defenses, each for one layer of the networking stack should be implemented. One layer is discussed at a time.

5.1 Physical Layer

The physical layer refers to mechanical, electrical, functional and procedural characteristics to establish, maintain and release physical connections between data link entities. This layer defines certain physical characteristics of the network, e.g. frequency, data rate, signal modulation and the spread spectrum scheme to use. DoS attacks on the physical layer are radio jamming. Well known countermeasures to radio jamming include adaptive antenna systems, spread spectrum modulations, error correcting codes and cryptography.

In Table 1, we describe physical layer attacks and possible countermeasures in case of wireless sensor network.

Interference: interfering into communication between to sensor nodes;

Jamming: the transmission of a radio signal that interferes with the radio frequencies being used by the sensor network;

Sybil: a single node pretends to be present in different parts of the network;

Tampering: nodes are vulnerable to physical harm or tampering.

Table 1. Physical Layer Attacks and Countermeasures

Attack	Countermeasure
Interference	Channel hopping and Blacklisting
Jamming	Channel hopping and Blacklisting
Sybil	Physical protection of devices
Tampering	Protection and changing of key

5.2 Data Link Layer

Data Link layer: The data link layer defines how data are encoded and decoded, how errors are detected and corrected, the addressing scheme as well as the medium access scheme.

In Table 2, we describe data link layer attacks and possible countermeasures in case of wireless sensor network.

Collision: with fake routing information more than one packet destined for different locations can collide

Exhaustion: injecting fake identifying information in the network

Spoofing: injecting fake routing control packets in the network

Sybil: a single node pretends to be present in different parts of the network

De-synchronization: injecting fake neighbor information

Traffic analysis: traffic analysis attacks are forged where the base station is determinable by observation that the majority of packets are being routed to one particular node. If an adversary can compromise the base station then it can render the network useless

Eavesdropping: by listening to the data the adversary can easily discover the communication content

Table 2. Data Link Layer Attacks and Countermeasures

Attacks	**Countermeasures**
Collision	CRC and Time diversity
Exhaustion	Protection of Network ID and other information that is required to joining device
Spoofing	Use different path for re-sending the message
Sybil	Regularly changing of key
De-synchronization	Using different neighbors for time synchronization
Traffic analysis	Sending of dummy packet in quite hours and regular monitoring the WSN network
Eavesdropping	Key protects DLPDU from Eavesdropper

5.3 Network Layer

The International Standards Organization (ISO) model for Open Systems Interconnection (OSI) states that network layer provides functional and procedural means to exchange network service data units between two transport entities over a network connection depending upon parameters such as latency or energy. It provides transport entities with independence from routing and switching considerations. There are two types of routing protocols for WSNs: ID-based protocols, in which packets are routed to the destination designated by the ID specified in the packets themselves and data-centric protocols [6], in which packets contain attributes that specify what kinds of data are being requested or provided.

In Table 3, we describe network layer attacks and possible countermeasures in case of wireless sensor network.

Wormhole: tunnels packets from one part of the network to another

Selective forwarding: compromised node might refuse to forward packets. It is more dangerous when compromised node forward selected packets

DoS: It is any event that diminishes or eliminates a network's capacity to perform its expected functions

Sybil: a single node pretends to be present in different parts of the network

Traffic analysis: traffic analysis attacks are forged where the base station is determinable by observation that the majority of packets are being routed to one particular node. If an adversary can compromise the base station then it can render the network useless

Eavesdropping: by listening to the data the adversary can easily discover the communication content

Table 3. Network Layer Attacks and Countermeasures

Attacks	Countermeasures
Wormhole	Physical monitoring of Field devices and regular monitoring of network using Source Routing. Monitoring system may use Packet Leach techniques
Selective forwarding	Regular network monitoring using Source Routing
DoS	Protection of network specific data like Network ID etc. Physical protection and inspection of network
Sybil	Resetting of devices and changing of session keys.
Traffic analysis	Sending of dummy packet in quite hours and regular monitoring the WSN network
Eavesdropping	Session keys protect NPDU from Eavesdropper

6 Conclusion

Security in wireless sensor network is vital to the acceptance and use of sensor networks. In particular, wireless sensor network product in industry will not get acceptance unless there is a full proof security to the network. In this paper, we have made an attack analysis to the wireless sensor network and suggested some countermeasures. Link layer encryption and authentication mechanisms may be a reasonable first approximation for defense against mote class attackers, but cryptography is not enough to defend against laptop class adversaries and insiders: careful protocol design is needed as well.

Acknowledgement

This work was supported by the Security Engineering Research Center, granted by the Korea Ministry of Knowledge Economy.

References

1. Perrig, A., Stankovic, J., Wagner, D.: Security in wireless sensor networks. Commun. ACM 47(6), 53–57 (2004)
2. Wander, A.S., Gura, N., Eberle, H., Gupta, V., Shantz, S.C.: Energy analysis of public key cryptography for wireless sensor networks. In: Third IEEE Conference on Pervasive Computing and Communications (PERCOM 2005), pp. 324–328. IEEE Computer Society Press, Los Alamitos (2005)
3. Yang, H., Luo, H., Ye, F., Lu, S., Zhang, L.: Security in Mobile ad hoc Networks: Challenges and Solutions. IEEE Wireless Communications 11(1), 38–47 (2004)
4. Carman, D., Matt, B., Balenson, D., Kruus, P.: A communications security architecture and cryptographic mechanisms for distributed sensor networks. In: DARPA SensIT Workshop. NAI Labs, The Security Research Division Network Associates, Inc. (1999)
5. Xiao, Y. (ed.): Security in Distributed, Grid and Pervasive Computing. Auerbach Publications, CRC Press (2006)
6. Ganesan, D., Cerpa, A., Yu, Y., Estrin, D.: Networking issues in wireless sensor networks. Journal of Parallel and Distributed Computing (JPDC), Special issue on Frontiers in Distributed Sensor Networks
7. Karlof, C., Wagner, D.: Secure routing in wireless sensor networks: Attacks and Countermeasures. In: First IEEE International Workshop on Sensor Network Protocols and Applications, pp. 293–315 (May 2003)
8. Sharma, K., Ghose, M.K., Kuldeep: Complete Security Framework for Wireless Networks. International Journal of Computer Science and Information Security 3(1) (2009)
9. Ganeriwal, S., Srivastava, M.B.: Reputation Based Framework for?High Integrity Sensor Networks. In: Proceedings of the 2nd ACM Workshop on Security of ad hoc and sensor networks, pp. 66–77. ACM Press, New York (2004)

Business Planning in the Light of Neuro-fuzzy and Predictive Forecasting

Prasun Chakrabarti[1], Jayanta Kumar Basu[2], and Tai-hoon Kim[3,*]

[1] Sir Padampat Singhania University,
Rajasthan, Pin-313601, India
prasun9999@rediffmail.com
[2] Bengal Institute of Technology and Management
West Bengal, Pin-731236, India
basu.jayanta@yahoo.co.in
[3]Hannam University,
Daejeon 306-791, Republic of Korea
taihoonn@empal.com

Abstract. In this paper we have pointed out gain sensing on forecast based techniques.We have cited an idea of neural based gain forecasting. Testing of sequence of gain pattern is also verifies using statsistical analysis of fuzzy value assignment. The paper also suggests realization of stable gain condition using K-Means clustering of data mining. A new concept of 3D based gain sensing has been pointed out. The paper also reveals what type of trend analysis can be observed for probabilistic gain prediction.

Keywords: gain sensing, neural, fuzzy value, K-Means clustering, trend analysis.

1   Introduction

Business gain depends significantly on the proper realization of factors associated. Neural computing environment facilitates gain estimation. The weights should be properly adjusted and based on accumulated output, gain class is defined. Statistical and fuzzy means can also be applied in this perspective. Information retrieval by artificial intelligence and data mining based techniques also plays a pivotal role in this context.

2   Neural Approach of Forecasting Gain

2.1   Fuzzy Based Analysis

The fuzzy values assigned to the parameters on which gain of a business organization depends fall in the range of rare fuzzy or frequent fuzzy ($0.5 < \text{ff} < 1$) as per pre-computation. ($0 < \text{fr} < 0.5$)

---

* Corresponding author.

T.-h. Kim et al. (Eds.): SIP/MulGraB 2010, CCIS 123, pp. 283–290, 2010.
© Springer-Verlag Berlin Heidelberg 2010

The weights are identical and desired gain will match with optimum condition if the value assigned is upper value of crisp (1).

Let production (P), cost (C), market-competition (M), quality (Q) and risk level (R) bear weights of 0.2

Aopt = 1 which reveals that WP=WC=WM=WQ=WR=0.2 and

PV=CV=MV=QV=RV=1

f(Aopt)= PV * WP + CV * WC + MV * WM + QV * WQ + RV * WR =1

2.2 Learning Strategy Analysis

If the actual gain bears a significant relation with the input value of parameter, in that case by adjusting weight of that particular parameter (maintaining $\sum_{i=1}^{n} wi = 1$, where n being total number of factors or parameters) the deviation in gain comparison can be minimized. Using this supervised learning pattern, the probability of achieving optimum gain is enhanced. In case of stochastic deviation of parameter value, the same can be compensated by weights and values of remaining parameters.

2.3 Perceptron Based Analysis

After sensing the parameters and assigning their values, the associator unit extracts the features and finally the class of gain is analyzed [1].

$$gti = w/n \sum_{i=1}^{j} Pv,j$$

where gti = gain at ith instant of time

n = number of trials

w = identical weight = 0.2

j = total number of parameters

Pv,j = fuzzy mapping of production value at timing instant i

The gain production can also be done using auto-associator

Step a: Weights are initialized to store gain pattern.

Step b: For (testing of input i.e) each parameter of business planning, prefer step c to e.

Step c: set bias weight w as per gain class.

Step d: adjust other weights such that $\sum_{i=1}^{n} wi = 1$, n being number of parameters

Step e: Analyze activation value (Av) and realize gain_class (Gc)

If(0.5<Av<0.6), Gc= marginal; If(0.6<Av<0.7), Gc= satisfactory;

If(0.7<Av<0.8), Gc= good;

else If(0.8<Av<1), Gc= excellent.

Step f: Terminate testing condition.

3 Statistical Realization and Fuzzy Approach of Gain

3.1 Range Estimation Using Statistical Means

Assume that the gain of a business organization is linearly related [2][3] to its respective year in the form of $g = a + by$,
where g and y are variables while a, b are constants.

Gain realization is curved out in the period (y_i to y_j); j>i. The minimum gain observed is gmin in ymin while gmax denotes maximum gain in year ymax.

For b>0 , gmax = a + b ymax
$\qquad\qquad$ gmin = a + b ymin
Therefore , gmax - gmin = b(ymax - ymin)
Hence , Range(gain) = b Range(year)   (1)

For b<0 , gmax = a + b ymin
$\qquad\qquad$ gmin = a + b ymax
Therefore , gmax - gmin = -b(ymax - ymin)
Hence , Range(gain) = -b Range(year)   (2)

3.2 Gain Analysis Using Fuzzy Membership Value Assignment

As cited in previous section , the activation value of gain of a business organization can be realized by a fuzzy rule. As per human intuition[4],which involves contextual and semantic knowledge, gain prediction can be done. Linguistic truth values in terms of gain analysis of a business is involved in this context. For achievement of desired gain , parameter status on the basis of quantitative nature is to be sensed . Neuro-fuzzy inference systems are systematic and facilitate by enhancing prediction accuracy level.

The steps in gain prediction for a particular business organization in the light of fuzzy logic control are as follows-

Identify the factors on which gain depends viz problem, quality, cost, market-competition, risk-level.
Partition universe of discourse.
Assign fuzzy values as per scaling in case of factors.
Input the respective weights of each factor such that summation of the weights is 1
Aggregate fuzzy outputs.
Apply defuzzication to form a crisp gain of a business organization.

In case of optimum parameter selection , trials are carried out and the average of the best possible ones are taken into consideration. The features of these types of non-adaptive fuzzy based gain estimation of a business organization are as follows:

Uniform weight

Av is excellent ($0.8 < Av < 1$)

Fixed membership function

Low accuracy level

Hierarchical rule structure

3.3   Business Status in the Light of Knowledge Management

The information (I) resident in any organization is proportional to the order (O) of the organization [5] i.e.

$I = CO$

where C = constant of proportionallity

Information on Manager's Desk X Manager's Mobility = Constant

Again, O * Manager's Mobility = Constant

This means that the order of organization is high, the manager's mobility is low. Thus Torn Stonier's Theory is acceptable to this extent .

Efficiency estimation is represented by $W = P - (P * OPE)$

where W = loss of business,

       P = process cost

   OPE = Overall process efficiency of any business with applicable of KM, OPE will inarcase and that will cause the loss W to decrease.

Volume of information or knowledge available on manager's desks * volume of mobility of manager's = constant.This means more the knowledge sharing or information available on manager's dasks, the less is the movement required by managers. So, Km = Invention + Innovation.

RKI (Relative knowledge index) = (Outflow in bits – Inflow in bits) / Total flow in bits.

Over a period of assessment, if a knowledge organization (P) inputs information from other organization (Q) over a time of t seconds using a link of c bps (bits per second) and outputs information for that other organization over a time of T seconds using a link of C bps, then the said organization's (p's) knowledge index with respect to Q will be

$PKIPQ = (CT - ct) / (CT + ct)$

PKI may range from -1 to +1

4   Feature Extraction of Data Mining Based Gain Analysis

On the basis of feature extraction of data mining , gain prediction towards optimum condition can be analyzed.

Table 1. Association of key factors against each gain

GAIN	Key factors associated i.e. optimum parameter selection
G1	SK1 = (K1,K3,K4,K6)
G2	SK2 = (K3,K5)
G3	SK3 = (K4,K5,K6)
G4	SK4 = (K2,K3,K5)
G5	SK5 = (K1,K2)
G6	SK6 = (K1,K2,K3,K6)

Table 2. Determination of count and value

Key factors	Initial value	Count	Value	(Value)2
K1	0.1	3	0.3	0.09
K2	0.2	3	0.6	0.36
K3	0.3	4	1.2	1.44
K4	0.4	2	0.8	0.64
K5	0.5	3	1.5	2.25
K6	0.6	3	1.8	3.24

Now CF = (x , y , z)
where x = number of elements , y = linear sum of the elements and z = sum of the square of the elements
CF1 = (4 , 4.1 , 5.41)
CF2 = (2 , 2.7 , 3.69)
CF3 = (3 , 4.1 , 6.13)
CF4 = (3 , 3.3 , 4.05)
CF5 = (2 , 0.9 , 0.45)
CF6 = (4 , 3.9 , 5.13)

So CFnet = accumulation of maximum of each tuple = (4 , 4.1 , 6.13)
Based on the above tuple , fuzzy value assignment of parameters towards optimum condition is to be performed .

5 K-MEANS Clustering Based Gain Analysis

In this section we have cited a new approach of gain realization on the basis of K-Means clustering[6]. It is to be noted that our proposed approach facilitate in the process of classification of gain for a particular business after proper sensing of its gain after nth timing instant. Based on the gain estimates , we can analyze the gain ranges .

Let G = { G1,G2,.….Gn} be the set of gain estimates in years Y = {Y1,Y2,.…..Yn}.

We assign G1 and G2 as the two initial means and accordingly form clusters in initial timing instant such that K1={G1,...} and K2={G2,....}.It is to be noted that elements of G1 will be nearest to G1 rather than G2. Similar is the case for elements of K2.

This process will be helpful for business gain analysis as after certain iterations the K clusters for subsequent stages will be identical which reveal saturation gain point. Hence that will be considered as optimum gain range for that particular business.

6 Gain Analysis Using Graphical Notation

Gain can also be sensed on the basis of graphical study. We propose a new approach of realizing a gain as a co-ordinate point(x,y,z) in 3-dimensional space where x is the real gain estimate , y is the deviation from base value and z is the deviation from previous gain. The concept is as follows-

Step1- Analyze the estimate of return after investment of a particular business as Rb and denote that as no gain no loss condition
Step2 - Sense the first gain .
Step 3- Sense the second gain. Plot in 3D.
Step3- Repeat this for (n-2) times, if we assume that n number of gain observations is to be taken.

The plotting signifies the following-

If $y > 0$, it is gain
If $y < 0$, it is loss
If $y = 0$,then it is base case.
If $z = 0$,then it signifies stable condition.
If $y > 0$ and $z < 0$,then it is relative loss but actual gain
If $y > 0$ and $z > 0$, then it is relative gain as well as actual gain.

The limitation of the approach is that we cannot plot the first sensed gain. The reason is that the z value is calculated based on the previous gain. If by default we denote as 0,then it will also seem as if it is stable condition.

7 Gain Prediction Based on Trend Analysis

In this section we have shown gain prediction on the basis of statistical theory of trend analysis. We consider gain sensing upto k timing instants (k>1). Hence using this theory there can be some prediction from (k+1)th timing instant.

Let G = {G1,G2,G3,......., Gk} be the set of gain sensed.
The following conditions are to be studied:
If G1>G2>G3>...........>Gk, then it is secular gain with upslope.
If G1<G2<G3<............<Gk, then it is downslope.

If $(Gi - G i+1)$ is not equal to 0 and $(G i+1 - G i) = (G i+2 - G i+1)$ such that $Gi \in G$, $1 <= i <= k-2$, in that case it will signify secular trend in arithmetic progression form.
If gain redundancy occurs after definite period of time, then it is seasonal trend.
If a series of n number adjacent gains (n<k) occurs m times (m>=1),then it is m-cycle trend.

We now represent the trend analysis on the basis of a numerical example.

The following table shows corresponding percentages gain of various businesses observed over 10 adjacent years –

Table 3. Percentage gain of business over 10 years

	y1	y2	y3	y4	y5	y6	y7	y8	y9	y10
G1	2	4	6	8	10	12	10	8	6	4
G2	8	6	9	4	11	17	8	6	9	10
G3	3	5	4	6	5	7	4	6	22	5
G4	6	7	8	6	11	9	6	5	3	6

It is seen that the trends are as follows-

i) In G1, we can reveal that from y1 to y6 it is A.P. series in upscale form, y6 onwards it is A.P. downscale and peak gain is in y6. This is normal secular trend.
ii) In G2, the gains in y1, y2 y3 again repeat in succession from y7 onwards. This reflects cyclical variation.
iii) In G3, it is observed that huge fluctuation occurs at y9 and y10. So it is irregular trend
iv) In G4, it is seen that redundancies of percentage gain 6 occurs at y1,y4,y7,y10 which means that after regular span of time. It signifies seasonal gain.

8   Conclusion

In this paper the basic concept of neural computing has been applied for gain estimation. Fuzzy, learning strategy and perceptron based analysis is depicted. Statistical based relation investigation and knowledge management based business status have been pointed out with quantitative measures. Feature extraction based data mining in gain forecasting for a business organization has also been shown in this paper. The paper also reveals stable gain condition and actual gain realization using 3D graph plotting. The paper also suggests gain inspection on the basis of trend analysis

References

1. Chakrabarti, P., De, S.K., Sikdar, S.C.: Statistical Quantification of Gain Analysis in Strategic Management. IJCSNS, Korea 9(11), 315–318 (2009)
2. Giri, P.K., Banerjee, J.: Introduction to Statistics. Academic Publishers, New York (2001)

3. Spiegel, M., et al.: Probability and Statistics. Tata McGraw Hill (2004)
4. Rajasekaran, S., Pai, G.A.-V.: Neural Networks, Fuzzy Logic and Geretic Algorithm. PHI Pvt. Ltd. (2003)
5. Bhunia, C.T.: Introduction to Knowledge Management. Everest Publishing House (2003)
6. Pujari, A.K.: Data Mining Techniques. University Press, New Haven (2001)

Identification of Plant Using Leaf Image Analysis

Subhra Pramanik[1], Samir Kumar Bandyopadhyay[3],
Debnath Bhattacharyya[2], and Tai-hoon Kim[2,*]

[1] Heritage Institute of Technology
Kolkata, India
subhra.dumdum@gmail.com
[2] Department of Multimedia
Hannam University
Daejeon, Korea
debnathb@gmail.com, taihoonn@empal.com
[3] Department of Compuer Science and Engineering
University of Calcutta
Kolkata, India
skb1@vsnl.com

Abstract. The trees are basically identified by their leaves. There are different varieties of trees grown throughout the world. Some are important cash crop. Some are used in medicine. The tree identification is very important in day to day life. Their identifications had been studied using various laboratory methods. The morphological and genetically characteristics were employed to classify different leafs. However, the presence of wide morphological varieties through evolution among the various leaf cultivars made it more complex and difficult to classify them. Therefore manual identification as well as classification of these leaves is a tedious task. During the last few decades computational biologists have studied various diversities among leaf due to huge number of evolutionary changes. Leaf structures play a very crucial role in determining the characteristics of a plant. The broad and narrow shaped leaves, leaf arrangement, leaf margin characteristics features which differentiate various leaf of a tree. This project proposed the methods to identify the leaf using an image analysis based approach.

Keywords: edge detection, image processing, recognition, segmentation.

1 Introduction

The Plants are basically identified by their leaves. There are different varieties of trees grown throughout the world. Some are important cash crop. Some are used in medicine. The tree identification is very important in day to day life. Their identifications had been studied using various laboratory methods [2, 3, 4]. The morphological and genetically characteristics were employed to classify different leafs. However, the presence of wide morphological varieties through evolution among the various leaf

---

* Corresponding author.

T.-h. Kim et al. (Eds.): SIP/MulGraB 2010, CCIS 123, pp. 291–303, 2010.

cultivars made it more complex and difficult to classify them. Therefore manual identification as well as classification of these leaves is a tedious task. During the last few decades computational biologists have studied various diversities among leaf due to huge number of evolutionary changes [1, 5, 6]. Leaf structures play a very crucial role in determining the characteristics of a plant. The broad and narrow shaped leaves, leaf arrangement, leaf margin etc. characteristics features which differentiate various leaf of a tree. This project proposed the methods to identify the leaf using an image analysis based approach.

The Leaf structure: To identify a leaf, first we have to know the structure of the leaf. The following is the description of the structure of the leaf. An example of a sample jpeg image of a leaf structure is shown in Fig. 1.

An entire leaf has the following parts:-

a. The Leaf Apex: The Apex is the tip of the leaf.
b. The Leaf Blade: The blade is the entire leaf unit. Sometimes this is made up of several smaller leaflets.
c. The Leaf Margin: The margin is the term used to describe the edge of the leaf.
d. The Leaf Base: Base is the name given to the part of the blade that is closest to the stem.
e. The Petiole: The leaf parts that is closest to the stem. The Petiole is the thin stalk that connects the leaf blade to the stem.
f. The Stipule: The leaf parts that is closest to the stem. Stipules are tiny leaf-like structures that may or may not be present at the base of the petiole.
g. Petiolate and Sessile: Leaves which have petioles are called Petiolate. If a leaf has no petiole it is called Sessile.

Importance of Identifying Leaf:

Identify Plants: If we can able to identify leaf, we can easily able to identify plants.

Disease identification: The structure of the leaf will be changed, if the leaf is affected by some disease. So, we can quickly identify the disease of the plant [7, 8, 9]. If the disease is early detected then we can take the step to prevent the diseases that occur in the plant. In this way we prevent the production fall in agriculture system. In this way we can achieve green revolution which helps our country to grow rapidly.

Plant leaves classification: If we can able to identify leaf, we can easily able to classify the plants.

Pollen leaves classification: If we can able to identify leaf, we can easily able to classify the pollen [10]. It is found that pollen grain of some plant families is responsible for producing allergy or asthma of human being. So this type of pollen should be identified.

Seed classification: If we can able to identify leaf, we can easily able to classify the seed [11]. This classification helps the farmer very much.

Crop Productivity: If we can able to identify the leaves, it helps to identify the different varieties of seeds, diseases, changing the color of leaves in heat, etc. This helps in crop productivity.

Identification of individual crop plants in the field: If we can able to identify the leaves, it helps to identify of individual crop plants in the field and locating their exact position of the crop [12]. Thus it helps in modern automated farming.

Recreation: Many people enjoy gardening or home landscape work. They need to identify which plants are useful and which ones are weeds. Other people simply like being able to name the plants they see when they walk through a city park or down a country road.

Food and Fiber: It is useful to be able to recognize plants which are sources of food (fruits, vegetables, spices, etc.) and fiber (lumber, firewood, paper pulp, etc.). This ability can also be profitable.

Career Skill: The ability to identify plant materials is an important skill for people who are interested in horticulture, forestry, landscape architecture, parks and recreation, botany, and a number of other career areas.

2 Previous Works

In this paper several methods [9] for recognizing of leaf are proposed. They are described below:-

- Leaf Blade Shape
- Simple and Compound Leaf Grouping
- Leaf Arrangement
- Leaf Venation
- Leaf Margin
- In this paper we will discuss all the above points.

2.1 Leaf Blade Shape

The first step in identifying a plant is to look at its leaves and determine their shape, and grouping.
One way to identify a plant is by the shape of its leaf blade (Fig. 2). The main part of a leaf is the blade. There are nine major shape categories:

- Linear: Narrow, with parallel or nearly parallel sides, (Fig. 3).
- Lanceolate: Lance shaped, longer than broad and tapering to a point at the tip (Fig. 4).
- Oblong: Much longer than wide, with nearly parallel sides (Fig. 5).
- Elliptic: Shaped like an ellipse, wider at the center and tapering to equal tips (Fig. 6).

- Ovate: Egg-shaped, broader at the base than the tip (Fig. 7).
- Cordate: Heart-shaped, either the leaf shape or base (Fig. 8).
- Reniform: Kidney-shaped (Fig. 9).
- Spatulate: In the shape of a spatula or spoon (Fig. 10).
- Orbicular: Round (Fig. 11).

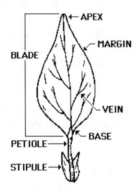

Fig. 1. Entire leaf structure

Fig. 2. Leaf blade shape

Fig. 3. Leaf with linear blade shape

Fig. 4. Leaf lanceolate blade shape

Fig. 5. Leaf with oblong blade shape

Fig. 6. Leaf elliptic blade shape

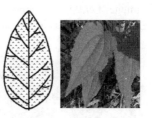

Fig. 7. Leaf with ovate blade shape

Fig. 8. Leaf cordate blade shape

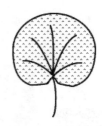

Fig. 9. Leaf with reniform blade shape

Fig. 10. Leaf spatulate blade shape

Fig. 11. Leaf with orbicular blade shape

2.2 Leaf Grouping

After determination of the leaf shape, we need to examine how the leaves are grouped on the stem. This is called Leaf Grouping. There are two types of leaf grouping.

Simple Grouping: Some leaves are simple, meaning they appear alone (Fig. 12). There are three main types of Simple leaf groupings as follows:

- Entire
- Palmately lobed
- Pinnately lobed

Compound Grouping: Other leaves are compound, meaning that they appear in groups and are made up of leaflets. Three kinds of compound leaves are as follows:

- Palmate: Here leaflets form a fan shape (Fig. 13).
- Pinnate: Leaflets are opposite each other on the stem (Fig. 14).
- Bipinnate: Leaflets are in pairs of pinnate groups (Fig. 15).

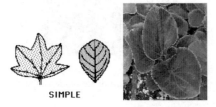

Fig. 12. Leaf with simple grouping

Fig. 13. Leaf with palmate compound grouping

Fig. 14. Leaf with pinnate compound grouping

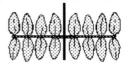

Fig. 15. Leaf with orbicular blade shape

2.3 Leaf Arrangement

After we have determined the shape and grouping of the plant's leaves, we need to examine the arrangement of the leaves on the stem. Here are the three principle ways leaves are arranged on the stem.

- Opposite: Leaves are directly opposite each other on the stem (Fig. 16).
- Alternate: Leaves are on both sides, but not directly across from one another (Fig. 17).
- Whorled: Leaves are arranged alternately around the stem.

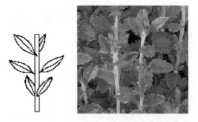

Fig. 16. Opposite leaves arrangement

Fig. 17. Alternate Leaves arrangement

2.4 Leaf Venation

The lines that appear on the surface of a leaf and look like blood vessels are called Veins. The two main types are Palmate and Pinnate (Fig. 18 and Fig. 19, respectively).

2.5 Leaf Margin

After we have determined the shape and grouping of the plant's leaves, the arrangement of the leaves on the stem, and the structure of the leaf veins, we need to examine the leaf margin characteristics.

The shape of the edge or margin of a leaf is another way it can be identified. Three kinds of margins are Entire, Toothed, and Lobed.

VEINS

Fig. 18. Leaf with palmate leaf venation

Fig. 19. Leaf with pinnate leaf Venation

Entire Margins: Margin smooth no bumps (Fig. 20).

Toothed Margins: A number of margins are named by the kinds of "teeth" they have (Fig. 21 and Fig. 22). Leaf Margins are as follows:

- Toothed: shallow bumps;
- Dentate: pointing outward;
- Serrate: pointing to the leaf's tip;
- Crenate: broad and round; and
- Incised: deeply cut with sharp, irregular teeth.

Fig. 20. Leaf with smooth margin

Fig. 21. Toothed leaf margin

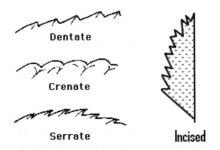

Fig. 22. Leaves with various types of toothed margins

Lobed Margins: Leaves with lobed margins are as follows (Fig. 23):

- Lobed: deep indents or sinuses;
- Sinuate: very wavy margin; and
- Undulate: wavy margin, but not as wavy as a sinuate margin.

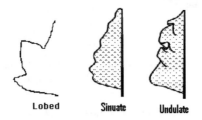

Fig. 23. Leaves with lobed margins

3 Our Work

In this paper we applied some methods and algorithms to identify the leaf. These are explained as follows:

3.1 Algorithms for Database Creation

The leaf samples are stored in the database. The images are in .bmp format. Here we consider only the leaf margins. There are nine types of leaf margins. These nine types of leaves are stored in the database. Algorithm is stated hereunder:

Step 1: Start
Step 2: List all .bmp files from the folder
Step 3: Initialize the variable Numfids with total no. of listed .bmp files
Step 4: Declare an array to store the loaded .bmp files
Step 5: Read the .bmp files from the folder

Step 6: Store the loaded .bmp file in an array for future use
Step 7: Stop

3.2 Algorithms for Preprocessing of Images in the Database

The leaf samples which are stored in the database are scaled in standard size and then converted into gray scale image. From these gray scale images edge are detected using 'sobel' technique. Algorithm is given hereunder:

Step 1: Start
Step 2: Initialize the variable   Numfids with total no. of listed .bmp files
Step 3: Read the images from the stored array
Step 4: Scale the images with size 186,200
Step 5: Convert the color images to gray scale images
Step 6: Identify the edge of the gray images using sobel technique
Step 7: Store the properties of the detected edge in an array
Step 8: Stop

3.3 Finding Out the Mathematical Characteristics of the Stored Image in the Database

The leaf samples which are stored in the database edges are identified. These edges are logical in nature. The edges area will store '1' in the array. From that array we calculate the correlation coefficient for each image and store in another array. The correlation coefficient is calculated with the following formula mentioned below.
The correlation coefficient of x and y is:

Correlation coefficient$=cov(x,y)/\sigma_x * \sigma_y$
Where $cov(x,y)$ is covariance of x and y,
σ_x and σ_y are standard deviation of the values of x and y respectively.

$$\sigma^2_x = 1/n \sum x^2_i - (1/n \sum x_i)^2 \text{ and } \sigma^2_y = 1/n \sum y^2_i - (1/n \sum y_i)^2$$

The covariance of x and y is

$$Cov(x,y) = 1/n \sum_{i=1}^{n} (x_i - \overline{X}) (y_i - \overline{Y})$$

where $\overline{X}$ and $\overline{Y}$ are the arithmetic means of the value assumed by x and y respectively,
$\overline{X} = 1/n(x_1 + x_2 + \ldots\ldots\ldots + x_n)$

and $\overline{Y} = 1/n(x_1 + x_2 + \ldots\ldots\ldots + x_n)$

The algorithm is given hereunder:

Step 1: Start
Step 2: Initialize the variable Numfids with total no. of listed .bmp files
Step 3: Read the array which store the detected edge properties of the stored
        images

Step 4: Find out the correlation coefficient from the array
Step 5: Store the correlation coefficient value in an array
Step 6: Stop

3.4   Algorithms for Preprocessing of Input Image and Mathematical Characteristic

Same algorithms are used to process the input image for identification, from the algorithm stated in 3.2 to 3.3.

3.5   Algorithms for Comparison of Characteristics of the Input Image with the Stored Images

After the mathematical calculation we get the value of correlation coefficient of the stored images and the input images. Then we calculate the matching percentage with the following formula:

Percentage = (Correlation coefficient of each stored image/ Correlation coefficient of the input image)*100;

The matching percentage if greater than 80% then we say 'image is fully matched'. If it is greater than 60 but less than 80 then we say 'image is partially matched' else we say that 'image is not matched'.

Step 1: Start
Step 2: Initialize the variable   Numfids with total no. of listed .bmp files
Step 3: Read the array which store the correlation coefficient value of the stored images
Step 4: If Correlation coefficient of the stored image=correlation coefficient of the input image?
Then Print "Image is found" Else Print "Image is not found"
Step 5: Stop

3.6   Another Algorithms for Comparison of Characteristics of the Input Image with the Stored Images

Step 1: Start
Step 2: Initialize the variable Numfids with total no. of listed .bmp files
Step 3: Read the array which store the correlation coefficient value of the stored images
Step 4: Calculate Matching percentage= (Correlation coefficient of the individual stored image/correlation coefficient of the input image)*100
Step 5: Store the matching percentage in an array
Step 6: Initialize the variable Per with the maximum matching value from the array which stores the matching percentage value
Step 7: if per>=80 then Print "image is fully matched"
Step 8: if (per>=60) && (per<80) Print "image is partially matched, Else Print "image is not matched"
Step 9: Stop

4 Result and Analysis

Table 1, shows the data of input file and Table 2 shows the detail experimental result. Fig. 24 to Fig. 27 shows the Algorithm output. Fig. 28 shows the Matlab implementation shot.

Table 1. Detail information of input leaf

Serial no	Name of input file	Value of the correlation coefficient of the input file
1.	toothed_large.bmp	-0.0911

Table 2. Detail of output information after matching

Serial no	Name of stored files	Value of the correlation coefficient of the stored file	Matching percentage with the input image file	Output
1.	Crenate.bmp	0.0891	-97.7624	image is not found
2.	Dentate.bmp	0.0619	-67.9869	image is not found
3.	Entire.bmp	0.0344	-37.7459	image is not found
4.	Irregular.bmp	0.1592	-174.8101	image is not found
5.	Lobed.bmp	0.0159	-17.4670	image is not found
6.	Serrate.bmp	0.1006	-110.4112	image is not found
7.	Sinuate.bmp	-0.1114	81.8079	image is not found
8.	Undulate.bmp	0.2544	-279.2548	image is not found
9.	Toothed.bmp	-0.0911	100	Image is matches up to 100 % image is fully matched

Fig. 24. Original Toothed Leaf   **Fig. 25.** Toothed Leaf after scaling

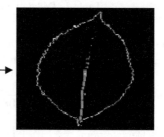

Fig. 26. Leaf after gray conversion   **Fig. 27.** After Edge Detection

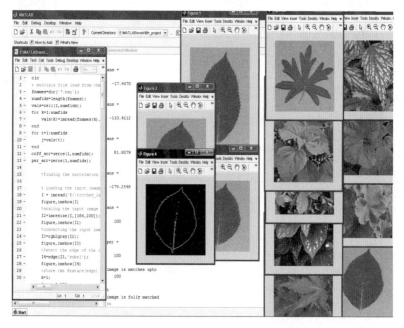

Fig. 28. Matlab Implementation Screen shot

5   Conclusion

This approach is useful in MLP classifier that helps in identification of various plant based on its morphological or genetically characteristics. Since the leaf structure varies in different cultivars, the outcome of the present work clearly shows the exact characteristics present in each variety of leaves. The figure of percentage required for the presence of similar characteristics of different leaves. For a more detailed and accurate classification other morphological features like fruits, stem, flower, and root can be considered for further processing.

Acknowledgement

This work was supported by the Security Engineering Research Center, granted by the Korea Ministry of Knowledge Economy.

References

1. Sanyal, P., Bhattacharya, U., Bandyopadhyay, S.: Analysis of SEM Images of Stomata of Different Tomato Cultivars Based on Morphological Features. In: IEEE International Conference, AMS 2008 (2008)
2. Sabino, D.M.U., da Costa, L.F., Rizzatti, E.G., Zago, M.A.: A texture approach to leukocyte recognition. Real-Time Imaging 10, 205–216 (2004)
3. Soille, P.: Morphological image analysis applied to crop field mapping. Image and Vision Computing 18(13), 1025–1032 (2000)
4. Stojanovic, R., Mitropoulos, P., Koulamas, C., Karayiannis, Y., Koubias, S., Papadopoulos, G.: Real-time vision-based system for textile fabric inspection. Real-Time Imaging 7(6), 507–518 (2001)
5. Polder, G., van der Heijden, G.W.A.M., Young, I.T.: Hyperspectral Image Analysis For Measuring Ripeness Of Tomatoes. ASAE International Meeting (2000)
6. Tzionas, P., Papadakis, E.S., Manolakis, D.: Plant leaves classification based on morphological features and a fuzzy surface selection technique. In: Fifth International Conference on Technology and Automation, Thessaloniki, Greece, pp. 365–370 (2005)
7. Zhao-yan, L., Fang, C., Yi-bin, Y., Xiu-qin, R.: Identification of rice seed varieties using neural network. Journal of Zhejiang University SCIENCE, 1095–1100 (2005)
8. Neuman, M., Sapirstein, H.D., Shwedyk, E., Bushuk, W.: Wheat grain color analysis by digital image processing: I. Methodology. J. Cereal Sci. 10, 175–182 (1989)
9. Damian, M., Cernadas, E., Formella, A., Sa-Otero, P.M.: Pollen classification of three types of plants of the family Urticaceae,
http://trevinca.ei.uvigo.es/~formella/inv/aaa/
formella-2002-pollen.pdf,
http://cas.psu.edu/docs/CASDEPT/Hort/LeafID/Arrangement.html
10. Slaughter, D.C., Harrell, R.C.: Discriminating fruit in a natural outdoor scene for robotic harvest. Trans. Of the ASAE 32(2), 757–763 (1989)
11. Tian, L., Slaughter, D.C., Norris, R.F.: Machine Vision Identification Of Tomato Seedlings For Automated Weed Control,
http://www.age.uiuc.edu/faculty/lft/papers/tomato.pdf
12. Sanyal, P., Bhattacharya, U., Parui, S.K., Bandyopadhyay, S.K.: Color Texture Analysis of Rice Leaves for Detection of Blast Disease. In: Proceedings of the 20th CSI Conference, pp. 45–48 (2006)

Using Incentives to Reduce Active Network Privacy Threats and Vulnerabilities

Maricel Balitanas and Tai-hoon Kim[*]

Multimedia Engineering Department, Hannam University,
133 Ojeong-dong, Daeduk-gu, Daejeon, Korea
jhe-c1756@yahoo.com, taihoonn@hnu.kr

Abstract. This paper scrutinize an active network scenario and proposes an efficient dual authentication key exchanged. The scheme protects personal privacy of identity information. It also provides an effective method to protect against DOS attacks with the scope information of initiator's random number table sent by the responder.

Keywords: Active Networks, Security, Identity Management System.

1 Introduction

In current generation information society has been governed by a collection of huge amounts of information and services that provides convenience to people. However, IMS (Identity Management Systems) have been crucial as the information society is getting bigger. IMS provides a description of the infrastructure within one or between several organizations that have agreed upon a mutual model of trust in managing and using identities. Identity management or ID management is a broad administrative area that deals with identifying individuals in a system (such as a country, a network or an organization) and controlling the access to the resources in that system by placing restrictions on the established identities.

In the real-world context of engineering online systems, identity management can involve three perspectives: A general model of identity can be constructed from a small set of axiomatic principles, for example that all identities in a given abstract namespace are unique and distinctive, or that such identities bear a specific relationship to corresponding entities in the real world. An axiomatic model of this kind can be considered to express "pure identity" in the sense that the model is not constrained by the context in which it is applied.

In general, an entity can have multiple identities, and each identity can consist of multiple attributes or identifiers, some of which are shared and some of which are unique within a given name space. The Figure1 below illustrates the conceptual relationship between identities and the entities they represent, as well as between identities and the attributes they consist of.

---

[*] Corresponding author.

T.-h. Kim et al. (Eds.): SIP/MulGraB 2010, CCIS 123, pp. 304–316, 2010.
© Springer-Verlag Berlin Heidelberg 2010

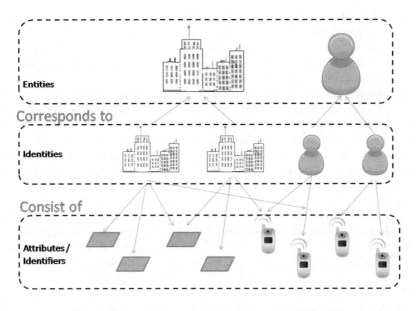

Fig. 1. Conceptualize relationship in pure identity [5]

In most theoretical and all practical models of digital identity, a given identity object consists of a finite set of properties. These properties may be used to record information about the object, either for purposes external to the model itself or so as to assist the model operationally, for example in classification and retrieval. A "pure identity" model is strictly not concerned with the external semantics of these properties.

The most common departure from "pure identity" in practice occurs with properties intended to assure some aspect of identity, for example a digital signature or software token which the model may use internally to verify some aspect of the identity in satisfaction of an external purpose. [19] To the extent that the model attempts to express these semantics internally, it is not a pure model.

Identity management, then, can be defined as a set of operations on a given identity model, or as a set of capabilities with reference to it. In practice, identity management is often used to express how identity information is to be provisioned and reconciled between multiple identity models.

Today [update], many organizations face a major clean-up in their systems if they are to bring identity coherence into their influence. Such coherence has become a prerequisite for delivering unified services to very large numbers of users on demand — cheaply, with security and single-customer viewing facilities.

The Diffie-Hellman key exchange scheme makes use of difficulty in computing discrete logarithms over a finite field. Since this scheme does not authenticate the participants while exchanging messages, it is vulnerable to man-in-the-middle attacks. For this reason, various authenticated key exchange schemes based on the Diffie- Hellman have been studied by many researchers [10, 11, 12]. These schemes can be categorized into two kinds of classes. The first class employs 'certificates' in its key exchange protocol, which foil man-in-the-middle attacks. Certificate-based schemes require additional cost and complexity in key exchange that they are not widely accepted in the market.

The other class proposes its authenticated key exchange protocol with an assumption that a pre-shared secret password or a secret key exists between two communication parties. Most of these authenticated key exchange schemes are not efficient because they use a public key cryptography mechanism which requires high computing power. Recently proposed ones like the IKE [2, 8, 20] consider privacy of personal identity and DOS attacks, which require much more computing power. Recently, mobile computing environment requires low computing power and small memory space even for security service. That is, authenticated key exchange schemes that do not use certificates and public key cryptography are preferable to the mobile environment. This paper proposes an efficient authentication and key exchange scheme that does not use certificates and public key cryptography, while protecting against man-in-the-middle attacks, replay attacks, DOS attacks and privacy intrusion. Characteristics of our scheme are as follows. First, it uses a symmetric block cipher with using a one-way hash function, but without using certificates for dual authentication and key exchange. Since symmetric block cipher requires smaller computing amount and memory space, our scheme is more adaptable to modern distribution environment, such as in ubiquitous and mobile computing. Next, due to the authentication key's one-time property used at each session, our scheme can detect various attacks, such as DOS attacks and man-in-the-middle attacks, without severe computing and memory overhead which overcomes the weakness of Diffie-Hellman. In addition, it solves the problem of identity privacy as well as perfect forward secrecy for future data confidentiality.

Wireless Internet networks security has become a primary concern in order to provide protected communication between mobile nodes in a hostile environment. Unlike the wire-line networks, the unique characteristics of wireless Internet networks pose a number of nontrivial challenges to security design, such as open peer-to-peer network architecture, shared wireless medium, stringent resource constraints, and highly dynamic network topology. These challenges clearly make a case for building multifence security solutions that achieve both broad protection and desirable network performance. The unreliability of wireless links between nodes, constantly changing topology due to the movement of nodes in and out of the networks, and lack of incorporation of security features in statically configured wireless routing protocols not meant for wireless Internet environments all lead to increased vulnerability and exposure to attacks. Security in wireless Internet networks is particularly difficult to achieve, notably because of the limited physical protection of each node, the sporadic nature of connectivity, the absence of a certification authority, and the lack of a centralized monitoring or management unit [2][3].

2   Wireless Communication

2.1 Scenario

2.1.1   The Security of Active Network: The active network should provide the solution for authentication, authorization and integrity to support the basic security service. In the Discrete Approach, the authentication of program code sender and the secret and integrity of the program code itself are the essential security points. If the program code is modified on bad purpose or it has the potential problem, it will become the unexpected error, so not only low performance of the entire active nodes but also a big security problem will be raised. In addition, if the authentication of program code is not performed, the hacker will modify the program code, and it will be a serious security problem. Now many projects of active network security, such as SANE, Seraphim, PLAN and Safety-Net are ongoing, but they cannot assure the basis of safety in the active network. Therefore, new security system that removes weak points is strongly necessary.

A review was done in an existing study, the security model that provides the basic security solutions such as authentication, authority, integrity is necessary. If the basic security problem is ignored, the performance of the entire active network node will be lowered, and the privacy violation and network congestion will be caused.

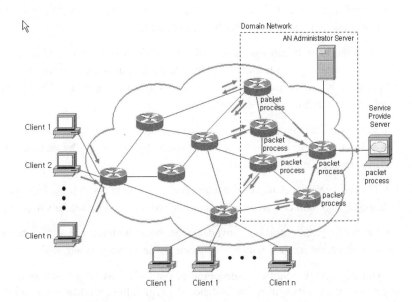

Fig. 2. Active network topology [16]

To resolve these security threats, we should authenticate active node users on the screte Approach. To authenticate active node users, we can restrict the access of hacker who tends to transmit the offensive program code, and block the forgery of program code. Also, we can reduce the deterioration of performance that the program reinstallation in active node causes through the management of frequently using program code. The active network structure that we propose is shown at Figure 2.

This study utilizes this scenario and use the discrete approach for the authentication proposed below. The active node management server authenticates and manages program codes, too. The active node server authenticates clients and the clients register the program code at the active node management server. The proposed system focuses on the authentication of middle node and the safe transmission of program code in active node.

2.2 Attacks in Wireless Communication

Many reasons are presented why wireless internet network are at risk, from a security point of view. Wireless Internet networks, do not have centralized machinery such as a name server, which if present as a single node can be a single point of failure. Wireless links between nodes are highly susceptible to link attacks, which includes the following listed below [4]

- Physical Attack: It gets rid of temper-defense-package in chip and then explains main information to put on prove on IC (Integrated Circuit) chip. We analyze electron-wave which emits from attacking prove, communication devices and computer.

- Denial of Service: It is a mean of attack which emits obstructive wave having special frequency for normally not to operate.

- Message loss: It can lose a part of communication method which reciprocates between tag and leader, cause by intention of attacker or error of system. Spoofing: It is a method which passes the authentication-process that individual which is unfair, deceives like to fair.

- Location Tracking: Attacker (invader) or leader, who is sinister, perceives position of tag. So it is a type which disturbs user's privacy by method which grips moving path of tag- owner.

- Traffic Analysis: Attacker (invader) analyse contents which get from eavesdropping and then can predicts tag's answer which is about leader's inquiry.

- Eavesdropping: Attacker (invader) can hear without big effort because communication method which is between tag and leader, is wireless.

Attacks typically involve only eavesdropping of data whereas active attacks involve actions performed by adversaries, for instance the replication, modification and deletion of exchanged data. External attacks are typically active attacks that are targeted to prevent services from working properly or shut them down completely. Intrusion prevention measures like encryption and authentication can only prevent external nodes from disrupting traffic, but can do little when compromised nodes internal to the network begin to disrupt traffic. Internal attacks are typically more severe attacks, since malicious insider nodes already belong to the network as an authorized party and are thus protected with the security mechanisms the network and its services offer. Thus, such compromised nodes, which may even operate in a group, may use the standard security means to actually protect their attacks [6, 7, 8].

As a summary a malicious node can disrupt the routing mechanism employed by several routing protocol in the following ways: [9]

Attack the Route Discovery Process by:

- Changing the contents of a discovered route.
- Modifying a route reply message, causing the packet to be dropped as an invalid packet.

- Invalidating the route cache in other nodes by advertising incorrect paths.
- Refusing to participate in the route discovery process.

Attack the Routing Mechanism By:
- Modifying the contents of a data packet or the route via which that data packet is supposed to travel.
- Behaving normally during the route discovery process but drop data packets causing a loss in throughput.

Launch DoS Attacks By:
- Sending a large number of route requests. Due to the mobility aspect of MANET's, other nodes cannot make out whether the large numbers of route requests are a consequence of a DoS attack or due to a large number of broken links because of high mobility.
- Spoofing its IP and sending route requests with a fake ID to the same destination, causing a DoS at that destination.

· Active Scanning / Probing Threat:
The most common threat of wireless networks is doing attack by Active software like Net Stumber (for Windows) and Dstumber (for Unix/ Linux). These software works on the method of active scanning. Attacks transmit the probe request to find any access point. If any access point is available, it will transmit probe response for that request. This response frame consists of SSID, Source/Destination MAC Address. Once attack captures this response frame, he/she has all the necessary information to enter in the network. Hence, if there is not any strong authentication mechanism, attackers may easily enter in the network.

In another scenario, if access point is using open system authentication, then also the attacker has no problem to join the network. On the third scenario, if access point is using 'Shared Key Authentication', which is based on encrypted challenge-response mechanism, the job of attacker becomes a bit tough, but not impossible.

· Spoofing Threat:
Another major threat in wireless networks is 'MAC Address Spoofing' which alters the manufacture assigned MAC address to any other value. This is conceptually different than traditional IP address spoofing where an attacker sends data from any arbitrary source address and does not expect to see a response to their actual source IP address. An attacker may choose 'MAC Address Altering' for several reasons, e.g. to bypass access control list, to impersonate an already authenticated user or disguising his/her presence on the network.

· 802.11 Beacon Flood Threat:
This technique requires generating thousands of counterfeit/fake 802.11 beacon frames and then transmits them on the network. Beacon frame contains the information about SSID of the network. Hence, it becomes difficult for the client to choose correct SSID to find a legitimate AP. There are several tools available to generate and

transmit the fake beacon frames. The famous tool for such a activity is Fake-AP (for both Windows and Linux).

• Authentication/ De Authentication Flood Threat:

In this, the attacker broadcasts the association or authentication request frames from the fake addresses to either access point (infrastructure mode) or to clients (ad-hoc mode). So, access point or client sends reply and keep the information about that request for some time in memory and wait for response, which is never going to come. Thus they are loaded with false authentication/ de-authentication requests and legitimate entities are put on hold for sometime, hence denying services to them. Air Jack and Void 11 are the tools which are used to achieve this effect.

Threat from Unauthorized Devices: In case of wireless networks, unauthorized access are not only limited to clients, but it is also applicable to access points. Sometimes, an authorized person, due to intruder/malicious users does not plant these access points. Once planted, this rouge access point is configured to operate on higher broadcasting power and poses itself as a valid access point. Sometimes, the legitimate users plant access point to improve their coverage. Attackers use wireless networks analyzing tools for this purpose. If the access point is established within firewalled network, it creates a backdoor within that network. Jamming Threat:

2.3 Malignant Code and Worm

Among the various types of system threatening codes such as virus, Worm and Trojan virus, internet worm is the most dominating system damaging factor. In DOS age, Worm was treated as non harmful code even though it copy and reproduced by itself continuously, it never contaminate the existing other files and system. As the development of network and internet, Worm also evolved to produce damages to system but its original l characteristics are never changed at all. The original worm of DOS ages is call as just Worm but current worm is called as I-Worm (Internet Worm).

Worm of prototype just create so many useless trash files by copy itself continuously and it is not so harmful to system but I-Worm decrease the system speed seriously by attempting copy through the network. During past few years, many different types of I-Worm were created. As a result of it, I-Worms are classified by two different types such as Network Worm and Internet Worm according to its propagation ways. If it is propagated through local network, it is called as Network Worm and if it is propagated through global network like internet, it is called as Internet Worm. Internet Worm is classified into three categories according to PC infection method. First group of Internet Worm is activated by just reading e-mails. Second group is activated by opening attached files of e-mail. Third group is activated by itself without any PC user's action. Also E-mail Worm is classified as Slow mass-mailers and Fast mass-mailers depending on its dissemination speed. Slow mass-mailers Worm is transferred at the same time when the infected PC users send an e-mail and Fast massmailers Worm is disseminated to many e-mail users at once. E-mail Worm use the email client such as Microsoft Outlook and Outlook Express to disseminate the worm to other PC users and it is transferred at the same time to all the users whose e-mail addresses are listed in specific mail client. On that way, if one is infected by e-mail worm then so many other PC users whose e-mail addresses are stored in an infected PC have possibilities of infection. This chain reaction can cause great amount of PC infections and damages in very short time. Current trends of e-mail worm such as Loveletter and Navidad, use very sensitive words which stimulate the PC user or use the title that lewd photos or video files are attached. Moreover, recent e-mail worm disguise that updated virus vaccine files are attached. These methods evoke the PC user's curiosities to open attached files or e-mail without any doubt. Network Worm is disseminated by local network system and is consist of next three steps.

- Find a Shared Drive
- Mapping Drive
- Copying Worm and Execute

3 Preceding Authentication Method

When looking at privacy concerns in addition to the integrity service, we may consider that each entity authorized to access to an information and possibly to change it must be able to sign the new version of this information in such a way that its identity has to be indistinguishable from the identities of all the user entitled to access the database in writing.[18]

There is an existing hash based authentication method Figure 3 which is proposed by Henrico and Muller[10]. This method is a protocol which prevents location tracking by updating ID based on hash. Manufacturer constructs database which can save h(ID), ID, TID, LST, AE and save ID, TID and LST in TAG. The TAG which received query increases 1 of TID and calculates h(ID), T=h(TID xor ID), TID and transmits to READER. The database searches ID with h(ID) and calculates T' which is added pertinent TID to ΛTID. In [Fig 1], If T and T' are same in Behaving of (3), Database calculates and transmits Q and xor calculates randomly generated R for updating ID. The tag which received (5) also calculates Q' and compares with Q. If both Q' and Q are same, the tag updates its ID. AE is designed safe from errors in system or losing messages by attacker. Because AE has previous ID information.

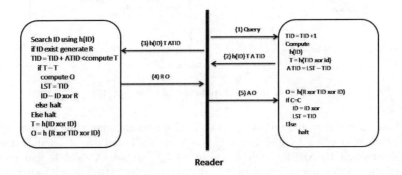

Fig. 3. Authentication Protocol based on hash [11]

In the perspective of the location tracking this method is safe because the ID updates when authentication process is over. In case of abnormal authentication processing between TAG and database that is attacker send query to TAG for attack, The attacker can do location tracking attack to TAG because The TAG always replies corresponding h(ID). If the TAG transmit (2) in session with database before opening normal authentication session, database is authenticated to attacker as normal TAG. Attacker give R that continuous character string which consists of 0 in the value of (5) that READER transmits to TAG in the middle of session. And then, attacker transmits T instead of Q. Therefore, the TAG can't notice error. When next authentication processing, server can find existed ID with h(ID). However, there is an disadvantage that TAG can't receive authentication, existing ID about LST is not corresponded with saved TAG and database. There is another method that READER generates random value S with Pseudo random number generator and query to TAG previously [5]. However, these methods have an disadvantage. If the 3rd person send spoofing query to TAG as READER, the TAG can't notice normal user or not. Of course, several advanced methods are proposed to solve the disadvantage. But they can't solve original problem.

4 Suggested Authentication Method

As suggested in [11] a new authentication method which is safe against spoofing attack and reducing hash time that 2 of hash function time reduce one in tag calculation time. This is shown in Figure 4. This method is similar to ID transformation protocol based on advance hash [12].

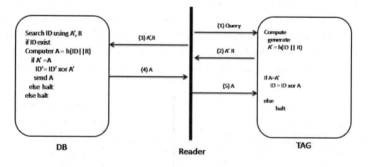

Fig. 4. Suggested Authentication Method [12]

This is done as of the following, first TAG makes random value R from pseudo random number generator and then, creation A'=h(ID||R) and then, A' and R transmit to DB through READER. DB is searching ID through A' and R and creating A=h(ID||R). DB compares A with A'. If both are same, DB authenticates the right TAG and updating XOR calculating ID to A' then transmit it to TAG. TAG compares A with A' and if both are the same, TAG calculates ID=ID xor A.

The authentication process is summarize in the following:

```
query to TAG                          /*Reader*/
generate R;                    /*Tag*/
compute A' = h(ID‖R);
send (A',R) to READER
bypass to DB;                         /*Reader*/
search ID using (A',R)                /*DB*/
if ID exist
    computer A=h(ID‖R);
    if A'=A
            ID'=ID' XOR A';
            Send A to READER
bypass to TAG;                 /*Reader*/
if A=A' then ID=ID=XOR A;      /*Tag*/
```

5 Proposed Key Exchange in Authentication Mechanism

The key exchange mechanism proposal in [8][15] has three phases. The first is a preliminary setup phase. In this phase, public information and random number matrix of an initiator are delivered to its responder. In the second phase, the initial seed is generated which is used to create a shared secret key for the data communication session. The third phase performs dual authentication between communication parties and creates a data encryption key that are shared.

5.1 Model and Notation

An entity which initiates a key exchange mechanism is called an initiator, and an entity which responds to the initiator's request is called a responder. Both are kinds of user. Another type of entity which is not a user but an attacker is called an adversary. An adversary is a Polynomial-Time Machine that attacks the secrecy of key exchange mechanism.

5.2 One-Time-ID(OID)

The OID is an identity that can be used only once for identifying a user. One-time-ID can be used to protect DOS attacks and man-in-the-middle attacks. To prevent DOS and man-in-the-middle attacks, OIDI(i) is attached at every i-th message transmission

$$OIDI(i) = h([m,n], MX[m,n], j)$$

OIDI(i) is a hash function of [m,n], MX[m,n], and j, where [m,n] is a random position among the random numbers assigned to the initiator within the random number matrix, MX[M,N], and j is just a random number in i-th message transaction. By using [m,n], we can detect DOS attacks and also decide who the initiator is. Man-in-themiddle attacks are detected by checking whether the transferred hash value is correct or not. For message integrity, we attach the hash value of all of the parameters transferred together at each message transmission.

5.3 Phases

In the preliminary set up phase, preliminary number are generated and delivered to the opposite side for the next phase. For the Diffie-Hellman key exchange, the initiator side generates $X_1(1)$, $g_1(1)$, $p_1(1)$ and deliverits public values to the responder. Fro the purpose of protecting it from DOS attack and Man-in-the-middle attacks using One-time-ID, the responder generates a MxN random number matrix and assigns them to the initiator. Be careful, however, that the same cell should not be assigned to different initiator. The responder should save the cell assignment information. This information is used for protecting against DOS attacks.

The authentication and key exchange phase is in the sth session. Each party has the same seed that can be used to create the shared secret key. At the first step the initiator sends $OID_1(1)$ that includes $AK_1(S)$ as a member of hash input. Using this value $OID_1(1)$, the responder can know whether the sender has the correct shared secret key or not, authenticating the initiator. The working key WK_S is used as the shared data encryption key during the sth session.

6 Analysis

A safe authentication method presented in Section 5 and particularly by using the proposed key-exchange mechanism generates a scheme that provides exceeding security to an unsecure communication. The key exchange method [15] was proven to provide dual-authentication key exchange mechanism as well as data integrity and data confidentiality. The existing method IKE and P-SIGMA are based on the fixed seed of shared key like One-Time –ID and an authentication key. Accordingly, if the culprit knows the fixed secret information, one can impersonate the initiator in the future session.. in contrast, the proposed scheme can regenerate the initial seed dynamically depending on the current client environment or adapting to the change in the security level. Because the proposed system does not use public key cryptography like Deffie-Hellman and RSA, which requires much computing power, it can be used for thin clients like mobile or ubiquitous computing devices. Moreover, the proposed system is so efficient as to finish within two messages round for authenticated key exchange. Aside from that, the scheme provides more concrete protection against DOS attack and Man-in-the middle attacks.

7 Conclusion

Active Networks architecture is composed of execution environment. This differ from the traditional network architecture which seeks robustness and stability by attempting to remove complexity and the ability to change its fundamental operation from underlying network component. Active networking allows the possibility of highly tailored and rapid "real-time" changes to the underlying network operation. This paper proposed an efficient dual authentication key exchanged in an active network scenario. The proposed system does not require the public key cryptography like Diffie-Hellman and RSA and certificates. The scheme protects personal privacy of identity information. It also provides an effective method to protect against DOS attacks with the scope information of initiator's random number table sent by the responder.

Acknowledgment. This work was supported by the Security Engineering Research Center, granted by the Korean Ministry of Knowledge Economy.

References

[1] Gross, R., Acquisti, A., Heinz, J.H.: Information revelation and privacy in online social networks. In: Proceedings of the 2005 ACM Workshop on Privacy in the Electronic Society, pp. 71–80 (2005), doi:10.1145/1102199.1102214

[2] Koodli, R., Perkins, C.: Fast Handover and context Relocation in Mobile Networks. ACM SIGCOMM Comp. Commun. Rev. 31 (October 2001)

[3] Balazinska, M., Castro, P.: Characterizing Mobility and network usage in a Corporate Wireless Local Area Network. In: Int'l. Conf. Mobile Systems, Apps, and Services (May 2003)

[4] Yoo, S., Kim, K., Hwang, Y., Piljoong Lee, H.: Satus-Based RFID Authentication Protocol. Journal of The Korean Institute of Information Security and Cryptology 14(6), 57–67 (2004)

[5] http://Wikipedia.org

[6] Pack, S., Choi, Y.: Pre-Authenticated Fast Handoff in a public Wireless LAN based on IEEE 802. 1x Model. IFIP TC6 Pers. Wireless Commun. (October 2002)

[7] Nakhjiri, M., Perkins, C., Koodli, R.: Context Transfer Protocol, Internet Draft: drafti-etfseamoby-ctp01.txt (March 2003)

[8] Perlman, R.: An Algorithm for Distributed Computation of a Spanning Tree in an Extended LAN, pp. 44–53 (1985)

[9] Min, B.-M., Cho, S.-P., Kim, H.-j., Lee, D.C.: System Development of Security Vulnerability Diagnosis in Wireless Internet Networks. In: Gervasi, O., Gavrilova, M.L., Kumar, V., Laganá, A., Lee, H.P., Mun, Y., Taniar, D., Tan, C.J.K. (eds.) ICCSA 2005. LNCS, vol. 3481, pp. 896–903. Springer, Heidelberg (2005)

[10] Henrici, D., Muller, P.: Hash based enhancement of location privacy for radio frequency identification devices using varying identifiers. In: PerSec 2004, pp. 149–153 (March 2004)

[11] Ko, H., Sohn, B., Park, H., Shin, Y.: Safe Authentication Method for Security Communication in Ubiquitous. In: Gervasi, O., Gavrilova, M.L., Kumar, V., Laganá, A., Lee, H.P., Mun, Y., Taniar, D., Tan, C.J.K. (eds.) ICCSA 2005. LNCS, vol. 3481, pp. 442–448. Springer, Heidelberg (2005)

[12] Hwang, Y., Lee, M., Lee, D., Lim, J.: Low-Cost RFID Authentication Protocol on Ubiquatous. In: CISCS 2004, pp. 120–122 (June 2004)

[13] Krawczyk, H.: The IKE-SIGMA Protocol, Internet Draft (2001)

[14] Imamoto, K., Sakurai, K.: A Design of Diffie-Hellman Based Key Exchange Using One-Time ID in Pre-shared Key Model. In: AINA 2004. IEEE, Los Alamitos (2004)

[15] Lee, Y., Choi, E., Min, D.: An Authenticated Key Exchange Mechanism Using One-Time Shared Key. In: Gervasi, O., Gavrilova, M.L., Kumar, V., Laganá, A., Lee, H.P., Mun, Y., Taniar, D., Tan, C.J.K. (eds.) ICCSA 2005. LNCS, vol. 3481, pp. 187–194. Springer, Heidelberg (2005)

[16] Kim, J.-M., Han, I.-s., Ryou, H.-b.: An Active Node Management System for Secure Active Networks. In: Gervasi, O., Gavrilova, M.L., Kumar, V., Laganá, A., Lee, H.P., Mun, Y., Taniar, D., Tan, C.J.K. (eds.) ICCSA 2005. LNCS, vol. 3481, pp. 904–913. Springer, Heidelberg (2005)

[17] Oatley, G., Mcgarry, K., Ewart, B.: Prioritizing of Offenders in Networks. In: 6th WSEAS International Conference on Simulation, Modelling and Optimization 2006, September 22-24, pp. 144–146 (2006)

[18] Akimana, R., Markowitch, O.: Data and Code Integrity in Grid Environments. In: WSEAS International Conference on Simulation, Modelling and Optimization 2006, September 22-24, pp. 677–682 (2006)

[19] Pervez, S., Ahmad, I., Akram, A., Swati, S.U.: A Comparative Analysis of Artificial Neural Network Technologies in Intrusion Detection Systems. In: WSEAS 2006, September 22-24, pp. 84–89 (2006)

[20] Mahfoudhi, A., Bouchelligua, W., Abed, M., Abid, M.: Towards a new approach of model-based HCI Conception. In: 6th WSEAS International Conference on Multimedia, Internet & Video Technologies, September 22-24, pp. 117–125 (2006)

Security Encryption Scheme for Communication of Web Based Control Systems

Rosslin John Robles and Tai-hoon Kim[*]

Multimedia Engineering Department, Hannam University,
133 Ojeong-dong, Daeduk-gu, Daejeon, Korea
rosslin_john@yahoo.com, taihoonn@hnu.kr

Abstract. A control system is a device or set of devices to manage, command, direct or regulate the behavior of other devices or systems. The trend in most systems is that they are connected through the Internet. Traditional Supervisory Control and Data Acquisition Systems (SCADA) is connected only in a limited private network Since the internet Supervisory Control and Data Acquisition Systems (SCADA) facility has brought a lot of advantages in terms of control, data viewing and generation. Along with these advantages, are security issues regarding web SCADA, operators are pushed to connect Control Systems through the internet. Because of this, many issues regarding security surfaced. In this paper, we discuss web SCADA and the issues regarding security. As a countermeasure, a web SCADA security solution using crossed-crypto-scheme is proposed to be used in the communication of SCADA components.

Keywords: SCADA, Security Issues, Encryption, Crossed Crypto-scheme.

1 Introduction

Control systems is the combinations of components (electrical, mechanical, thermal, or hydraulic) that act together to maintain actual system performance close to a desired set of performance specifications. Open-loop control systems (e.g., automatic toasters and alarm clocks) are those in which the output has no effect on the input. Closed-loop control systems (e.g., thermostats, engine governors, automotive cruise-control systems, and automatic tuning control circuits) are those in which the output has an effect on the input in such a way as to maintain the desired output value. SCADA or Supervisory Control And Data Acquisition is a large scale control system for automated industrial processes like municipal water supplies, power generation, steel manufacturing, gas and oil pipelines etc. SCADA also has applications in large scale experimental facilities like those used in nuclear fusion.

SCADA systems monitor and control these operations by gathering data from sensors at the facility or remote station and then sending it to a central computer system that manages the operations using this information.

Conventional SCADA communications has been Point-to-Multipoint serial communications over lease line or private radio systems. With the advent of Internet

---

[*] Corresponding author.

T.-h. Kim et al. (Eds.): SIP/MulGraB 2010, CCIS 123, pp. 317–325, 2010.
© Springer-Verlag Berlin Heidelberg 2010

Protocol (IP), IP Technology has seen increasing use in SCADA communications. The connectivity of the Internet can give SCADA more scale which enables it to provide access to real-time data display, alarming, trending, and reporting from remote equipment.

On the Next parts of this paper, Control System is discussed, the conventional SCADA setup and the Internet SCADA. Advantages which can be attained using the Internet for SCADA are also covered. Security issues are being pointed out. We also suggest a security solution for a Web based Control System using symmetric key encryption.

2 Control Systems

A control system is a device or set of devices to manage, command, direct or regulate the behavior of other devices or systems. There are two common classes of control systems, with many variations and combinations: logic or sequential controls, and feedback or linear controls. There is also fuzzy logic, which attempts to combine some of the design simplicity of logic with the utility of linear control. Some devices or systems are inherently not controllable. An automatic sequential control system may trigger a series of mechanical actuators in the correct sequence to perform a task. For example various electric and pneumatic transducers may fold and glue a cardboard box, fill it with product and then seal it in an automatic packaging machine. In the case of linear feedback systems, a control loop, including sensors, control algorithms and actuators, is arranged in such a fashion as to try to regulate a variable at a set point or reference value. An example of this may increase the fuel supply to a furnace when a measured temperature drops. PID controllers are common and effective in cases such as this. Control systems that include some sensing of the results they are trying to achieve are making use of feedback and so can, to some extent, adapt to varying circumstances. Open-loop control systems do not make use of feedback, and run only in pre-arranged ways.

3 SCADA Systems

SCADA stands for supervisory control and data acquisition. It generally refers to industrial control systems: computer systems that monitor and control industrial, infrastructure, or facility-based processes, as described below:

* Industrial processes include those of manufacturing, production, power generation, fabrication, and refining, and may run in continuous, batch, repetitive, or discrete modes.
* Infrastructure processes may be public or private, and include water treatment and distribution, wastewater collection and treatment, oil and gas pipelines, electrical power transmission and distribution, Wind Farms, civil defense siren systems, and large communication systems.

* Facility processes occur both in public facilities and private ones, including buildings, airports, ships, and space stations. They monitor and control HVAC, access, and energy consumption.

Supervisory Control and Data Acquisition (SCADA) systems are now used in mining industries, modern manufacturing and industrial processes, public and private utilities, leisure and security industries. In these situations, telemetry is needed to connect systems and equipment separated by long distances. Some of this ranges to up to thousands of kilometers. In these cases, SCADA are sometimes connected through the Internet.

Telemetry is automatic transmission and measurement of data from remote sources by wire or radio or other means. It is also used to send commands, programs and receives monitoring information from these remote locations. SCADA is the combination of telemetry and data acquisition. Supervisory Control and Data Acquisition system is compose of collecting of the information, transferring it to the central site, carrying out any necessary analysis and control and then displaying that information on the operator screens. The required control actions are then passed back to the process. [1]. Typically SCADA systems include the following components: [2]

1. Operating equipment such as pumps, valves, conveyors and substation breakers that can be controlled by energizing actuators or relays.
2. Local processors that communicate with the site's instruments and operating equipment. This includes the Programmable Logic Controller (PLC), Remote Terminal Unit (RTU), Intelligent Electronic Device (IED) and Process Automation Controller (PAC). A single local processor may be responsible for dozens of inputs from instruments and outputs to operating equipment.
3. Instruments in the field or in a facility that sense conditions such as pH, temperature, pressure, power level and flow rate.
4. Short range communications between the local processors and the instruments and operating equipment. These relatively short cables or wireless connections carry analog and discrete signals using electrical characteristics such as voltage and current, or using other established industrial communications protocols.
5. Long range communications between the local processors and host computers. This communication typically covers miles using methods such as leased phone lines, satellite, microwave, frame relay and cellular packet data.
6. Host computers that act as the central point of monitoring and control. The host computer is where a human operator can supervise the process; receive alarms, review data and exercise control.

3.1 SCADA Hardware

PLCs or RTUs are also commonly used; they are capable of autonomously executing simple logic processes without a master computer controlling it.

A functional block programming language, IEC 61131-3, is frequently used to create programs which run on these PLCs and RTUs. This allows SCADA system engineers to perform both the design and implementation of a program to be executed on an RTU or PLC. From 1998, major PLC manufacturers have offered integrated HMI/SCADA systems, many use open and non-proprietary communications protocols. Many third-party HMI/SCADA packages, offering built-in compatibility with most major PLCs, have also entered the market, allowing mechanical engineers, electrical engineers and technicians to configure HMIs. [3]

3.2 SCADA Software

Supervisory Control and Data Acquisition software can be divided into proprietary type or open type. Proprietary software are developed and designed for the specific hardware and are usually sold together. The main problem with these systems is the overwhelming reliance on the supplier of the system. Open software systems are designed to communicate and control different types of hardware. It is popular because of the interoperability they bring to the system. [1] WonderWare and Citect are just two of the open software packages available in the market for SCADA systems. Some packages are now including asset management integrated within the SCADA system.

3.3 SCADA User Interface

SCADA system includes a user interface which is usually called Human Machine Interface (HMI). The HMI of a SCADA system is where data is processed and presented to be viewed and monitored by a human operator. This interface usually includes controls where the individual can interface with the SCADA system. HMI's are an easy way to standardize the facilitation of monitoring multiple RTU's or PLC's (programmable logic controllers).

Usually RTU's or PLC's will run a pre programmed process, but monitoring each of them individually can be difficult, usually because they are spread out over the system. Because RTU's and PLC's historically had no standardized method to display or present data to an operator, the SCADA system communicates with PLC's throughout the system network and processes information that is easily disseminated by the HMI. HMI's can also be linked to a database, which can use data gathered from PLC's or RTU's to provide graphs on trends, logistic info, schematics for a specific sensor or machine or even make troubleshooting guides accessible. In the last decade, practically all SCADA systems include an integrated HMI and PLC device making it extremely easy to run and monitor a SCADA system.

3.4 How SCADA Works

The measurement and control system of SCADA has one master terminal unit (MTU) which could be called the brain of the system and one or more remote terminal units (RTU). The RTUs gather the data locally and send them to the MTU which then issues suitable commands to be executed on site. A system of either standard or customized software is used to collate, interpret and manage the data.

SCADA as of now uses predominantly open-loop control systems, though some closed-loop characteristics are often built in. As this is an open-loop system, it means that SCADA system cannot use feedback to check what results its inputs have produced. In other words, there is no machine-learning.

4 Installation of SCADA

Supervisory Control and Data Acquisition (SCADA) is conventionally set upped in a private network not connected to the internet. This is done for the purpose of isolating the confidential information as well as the control to the system itself. Because of the distance, processing of reports and the emerging technologies, SCADA can now be connected to the internet. This can bring a lot of advantages and disadvantages which will be discussed in the sections.

4.1 Conventional Supervisory Control and Data Acquisition

The function of SCADA is collecting of the information, transferring it back to the central site, carrying out any necessary analysis and control and then displaying that information on a number of operator screens. Systems automatically control the actions and control the process of automation.

Conventionally, relay logic was used to control production and plant systems. With the discovery of the CPU and other electronic devices, manufacturers incorporated digital electronics into relay logic equipment. As need to monitor and control more devices in the plant grew, the PLCs were distributed and the systems became more intelligent and smaller in size. PLCs (Programmable logic controllers) and DCS (distributed control systems) are used as shown in Figure 1.

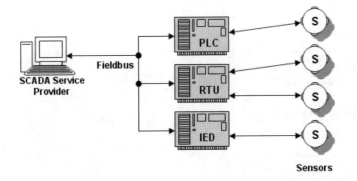

Fig. 1. Common SCADA Installation utilizing PLC/DCS, Sensors and Master Station connected using a fieldbus

4.2 Control System over the Internet

Companies and Operators are considering using the Internet for SCADA to provide access to real-time data display, alarming, trending, and reporting from remote equipment. Along with these advantages, there are three significant problems to overcome when implementing an Internet SCADA. Most SCADA operators find it difficult to connect their system to the Internet. A proposed solution to this problem is to connect the device to a PC and have the PC make the connection to the Internet via an Internet service provider using Secure Socket Layer.

Unfortunately, this solution may not meet the low-cost criterion and, depending on configuration, can lack reliability. An embedded solution can be used instead of a PC: a small, rugged, low-cost device that provides connectivity capabilities of a PC at a higher reliability and lower cost. It can be connected to the equipment via a serial port, communicates with the equipment in the required native protocol, and converts data to HTML or XML format. In cases where the equipment incorporates an electronic controller, it may be possible to simply add Web-enabled functionality into the existing microcontroller.

Conventional SCADA only have 4 components: the master station, plc/rtu, fieldbus and sensors. Internet SCADA replaces or extends the fieldbus to the internet. This means that the Master Station can be on a different network or location. In Figure 2, you can see the architecture of SCADA which is connected through the internet. Like a normal SCADA, it has RTUs/PLCs/IEDs. The SCADA Service Provider or the Master Station. This also includes the user-access to SCADA website. This is for the smaller SCADA operators that can avail the services provided by the SCADA service provider. It can either be a company that uses SCADA exclusively. Another component of the internet SCADA is the Customer Application which allows report generation or billing.

Along with the fieldbus, the internet is an extension. This is setup like a private network so that only the master station can have access to the remote assets. The master also has an extension that acts as a web server so that the SCADA users and customers can access the data through the SCADA provider website.

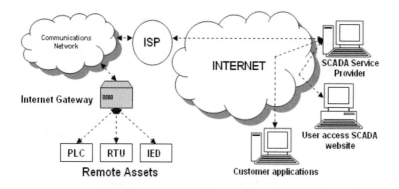

Fig. 2. Internet SCADA Architecture [4]

One may ask why we need to connect SCADA on the Internet even though there are a lot of issues surrounding it. The answer is because of many advantages it presents. [5]

5 Security Issues

Even before Control Systems was connected to the Internet, It is already surrounded by many security Issues and now the Internet has made them more vulnerable to attacks. Consequently, the security of SCADA-based control systems has come into question as they are increasingly seen as extremely vulnerable to cyberwarfare/cyberterrorism attacks.[6][7] Here are the common security issues in SCADA:

- The lack of concern about security and authentication in the design, deployment and operation of existing Control System networks.
- The belief that SCADA systems have the benefit of security through obscurity through the use of specialized protocols and proprietary interfaces.
- The belief that SCADA networks are secure because they are purportedly physically secured.
- The belief that SCADA networks are secure because they are supposedly disconnected from the Internet.
- IP Performance Overhead of Control Systems connected to the Internet.

6   The Crossed-Crypto Scheme

There are major types of encryptions in cryptography: the symmetric encryption and the asymmetric encryption. From the two major types of encryptions we can say that Asymmetric encryption provides more functionality than symmetric encryption, at the expense of speed and hardware cost.

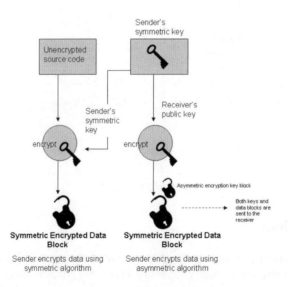

Fig. 3. Crossed crypto-scheme

On the other hand symmetric encryption provides cost-effective and efficient methods of securing data without compromising security and should be considered as the correct and most appropriate security solution for many applications. [8] In some instances, the best possible solution may be the complementary use of both symmetric and asymmetric encryption. Diagram of a crossed crypto-scheme is shown in Figure 3.

7   The Crossed-Crypto Scheme as Integrated to Control Systems

Theee scheme can be integrated in the communication of the SCADA master and SCADA assets. The algorithm presented here combines the best features of both the symmetric and asymmetric encryption techniques. The plain text data is to be transmitted in encrypted using the AES algorithm. Further details on AES can be taken from [9]. The AES key which is used to encrypt the data is encrypted using ECC. The cipher text of the message and the cipher text of the key are then sent to the SCADA assets. The message digest by this process would also be encrypted using ECC techniques. The cipher text of the message digest is decrypted using ECC technique to obtain the message digest sent by the SCADA Master. This value is compared with the computed message digest. If both of them are equal, the message is accepted otherwise it is rejected. Figure 4 depict this scenario.

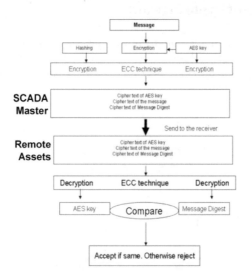

Fig. 4. Chain of operation

8   Conclusions

Interconnections of components forming system configurations which will provide a desired system response as time progresses. This is the concept of control systems.

A control system is set of electronic devices and equipment which are in place to ensure stability, accuracy and smooth transition of a process or any manufacturing activity. As a result of rapid development and advancement of technology, complicated control tasks are accomplished with a highly automated control system, mostly a PLC (Programmable Logic Controller) and if necessary a host computer.

SCADA stands for supervisory control and data acquisition is a good example of a control system. It generally refers to industrial control systems: computer systems that monitor and control industrial, infrastructure, or facility-based processes.

Control systems like SCADA can be connected to the internet.In this paper, we highlighted the security issues of setting up SCADA over the Internet. The utilization of a crossed-crypto scheme was discussed.

Acknowledgement. This work was supported by the Security Engineering Research Center, granted by the Korean Ministry of Knowledge Economy.

References

1. Bailey, D., Wright, E.: Practical SCADA for Industry (2003)
2. Hildick-Smith, A.: Security for Critical Infrastructure SCADA Systems (2005)
3. Wikipedia – SCADA,
   http://en.wikipedia.org/wiki/SCADA (accessed: January 2009)
4. Wallace, D.: Control Engineering. How to put SCADA on the Internet (2003),
   http://www.controleng.com/article/CA321065.html (accessed: January 2009)
5. Internet and Web-based SCADA,
   http://www.scadalink.com/technotesIP.htm (accessed: January 2009)
6. Maynor, D., Graham, R.: SCADA Security and Terrorism: We're Not Crying Wolf,
   http://www.blackhat.com/presentations/bh-federal-06/
   BH-Fed-06-Maynor-Graham-up.pdf (accessed: January 2009)
7. Lemos, R.: SCADA system makers pushed toward security (July 26, 2006),
   http://www.securityfocus.com/news/11402 (accessed: January 2009)
8. Balitanas, M., Robles, R.J., Kim, N., Kim, T.: Crossed Crypto-scheme in WPA PSK Mode. In: Proceedings of BLISS 2009, Edinburgh, GB. IEEE CS, Los Alamitos (August 2009) ISBN 978-0-7695-3754-5
9. Federal Information Processing Standards Publication 197, Announcing the ADVANCED ENCRYPTION STANDARD, AES (2001),
   http://csrc.nist.gov/publications/fips/fips197/fips-197.pdf
   (accessed: January 2009)

Author Index